Biotechniques for Air Pollution Control

PROCEEDINGS OF THE 3RD INTERNATIONAL CONGRESS ON BIOTECHNIQUES FOR AIR POLLUTION CONTROL, DELFT, THE NETHERLANDS, 28–30 SEPTEMBER 2009

Biotechniques for Air Pollution Control

Editors

Jan Bartacek, Christian Kennes & Piet N.L. Lens

CRC Press
Taylor & Francis Group
Boca Raton London New York

CRC Press is an imprint of the
Taylor & Francis Group, an **informa** business
A BALKEMA BOOK

Published by:
CRCPress/Balkema
P.O. Box 447, 2300 AK Leiden, The Netherlands
e-mail: Pub.NL@taylorandfrancis.com
www.crcpress.com – www.taylorandfrancis.com

ISBN: 978-0-415-58270-4 (hbk)
ISBN: 978-0-203-84792-3 (ebk)

Visit the Taylor & Francis Web site at
http://www.taylorandfrancis.com

and the CRC Press Web site at
http://www.crcpress.com

Contents

Biofiltration – keynotes

Emission control

Biofiltration – methane and ammonia

Bioenergy – resource recovery

Sulfur compounds

Poster presentations

Preface

In recent years, energy and feedstock materials for the chemical industry are in increasing demand. With constraints related to availability and use of oil, the energy and chemical industry is undergoing considerable changes. The need for the use of cheaper and widely available feedstocks, and the development of sustainable and environmentally friendly chemical processes is rapidly growing under both economical and public pressure.

Therefore, waste gas treatment has gradually been integrated into process design. Instead of discharging their waste gases into the atmosphere, industries increasingly attempt to become self-sufficient and recover compounds from their own wastestreams or use (upgraded) wastestreams of neighbouring industries as raw material.

Sustainable gas treatment concepts are under development, and can lead to the recovery of useful by-products like energy in the form of biogas, hydrogen or electricity, and chemicals in the form of e.g. fertilizers (ammonia, phosphates) or raw materials (elemental sulfur, sulfuric acid, a.o.). Obviously, adding value to waste gas by upgrading the recovered compounds will only be a reality if it is demonstrated that there is a fundamental basis and a tangible advantage in using these recovered compounds rather than buying raw materials and feedstock.

The 3rd International Symposium on Biotechniques for Air Pollution Control, held on 28–30 September 2009 in Delft (the Netherlands), aims to overview innovative biotechnology based processes for treatment of waste gasses.

The symposium is an opportunity for discussion about various innovative research aspects of environmental chemistry, environmental engineering and bioprocess technology amongst professionals as well as young researchers and Ph.D. students.

Jan Bartacek, Christian Kennes and Piet Lens
On behalf of the organizing committee

Acknowledgements

The organisers thankfully acknowledge the contribution of the following partners for supporting the 3rd International Congress on Biotechniques for Air Pollution Control:

Organisation

Organising committee

Jan Bartacek, *UNESCO-IHE, The Netherlands*
Christian Kennes, *University La Coruña, Spain*
Piet Lens, *UNESCO-IHE, The Netherlands*
Vera Schouten, *UNESCO-IHE, The Netherlands*

Scientific committee

Piet Lens, *UNESCO-IHE, The Netherlands* (Chair)
Christian Kennes, *University La Coruña, Spain* (Co-chair)

Ajit Annachhatre, *AIT, Thailand*
Yves Andrès, *Ecole des Mines de Nantes, France*
Jan Bartacek, *UNESCO-IHE, The Netherlands*
Marc Deshusses, *Duke University, USA*
Franz B. Frechen, *University Kassel, Germany*
Francisco Girio, *INETI, Portugal*
Michèle Heitz, *Université de Sherbrooke, Canada*
Laurence C.C. Koe, *NTU, Singapore*
Bart Kraakman, *Bioway, The Netherlands*
Janneke Krooneman, *Bioclear, The Netherlands*
Wim Maaskant, *BGP, The Netherlands*

Jan Paca, *ICT Prague, Czech Republic*
Korneel Rabaey, *University Queensland, Australia*
Sergio Revah, *University Autonoma Metropolitana, México*
Abdollah Rashidi, *PWUT, Iran*
Juan Pablo Silva Vinasco, *University Univalle, Colombia*
Richard Stuetz, *UNSW Sydney, Australia*
Peter van der Maas, *WLN, The Netherlands*
Herman Van Langenhove, *University Gent, Belgium*
María C. Veiga, *University La Coruña, Spain*

Waste gas treatment – keynotes

Chapter 1

BIOPROCESSES FOR WASTE GAS TREATMENT

Christian Kennes
Faculty of Sciences, Chemical Engineering Laboratory, University of La Coruña, La Coruña, Spain

ABSTRACT

Different technologies are available for waste gas treatment and air pollution control in general. Besides, conventional non-biological technologies, bioprocesses have appeared, somewhat more recently, as highly efficient and cost-effective treatment alternatives for the removal of low or moderate concentrations of pollutants from air. Bioprocesses are suitable for the removal of many different types of pollutants: organic, inorganic, hydrophobic, hydrophilic, as well as mixtures. Different types of bioreactors are available for the removal of air pollutants, *i.e.*, biofilters, one- and two- liquid phase trickling biofilters, bioscrubbers, membrane bioreactors, and one- and two- liquid phase suspended-growth bioreactors also called partitioning bioreactors. More recently, other bioreactor configurations have been considered as well but they are, so far, still in the development stage and are mainly being studied and optimized at laboratory-scale as, for example, the biodrums and rotating biodiscs. Multi-stage and hybrid systems have been studied as well and have been used in some full scale applications.

Keywords: Air pollution, biodegradation, bioreactors, biofilter, biotrickling filter, bioscrubber, membrane bioreactor, biodiscs, VOC, VIC, waste gases

INTRODUCTION

Air pollution has become a major concern worldwide since the twentieth century. It can sometimes be solved through bioprocesses. Bioprocesses used for the treatment of waste gases are mainly suitable for low to moderate pollutant concentrations (typically below 5–10 g/m^3) and over a wide range of air flow rates, typically up to a few hundreds cubic meters per hour (Kennes et al., 2009). The original application of such bioprocesses was mainly to solve odour problems. Recently, the number of applications has been broadened and new bioreactor configurations have been developed in order to better fit the needs of each specific air pollution problem and in order to try to improve performance while minimizing costs. Nowadays, bioreactors are also being used to treat industrial waste gases and, not only to remove odours, but also to biodegrade pollutants as CO, NOx, SOx, and different types of volatiles organic compounds (VOC) (Jin et al., 2005, 2009; Philip and Deshusses, 2003; Prado et al., 2008). Recent data and information on conventional and innovative bioreactor configurations will be overviewed. (More detailed information is available in: Kennes et al., 2009. "Bioprocesses for Air Pollution Control". J. Chem. Technol. Biotechnol. 84:1419–1436).

BIOREACTORS

CONVENTIONAL BIOFILTERS

The conventional biofilter is one of the oldest reactor configuration. It has been used for several decades, mainly for the removal of odours at composting facilities and wastewater treatment plants. The basic principal is quite simple. It consists in feeding polluted air through a filter bed colonized by a biocatalyst able to mineralize the pollutant(s). Clean air is thus released from the bioreactor to the atmosphere.

Soil, peat, or other similar natural materials are generally used as filter beds, although inert packing materials can be used as well. Natural materials already contain different microbial populations and inoculation is generally not required, unless highly recalcitrant pollutants must be removed. Such filter beds do also contain nutrients, meaning that those do often not need to be added to the system, unless they get depleted or their concentration becomes limiting. Bed replacement is often required after a few years. Such problem is basically avoided when using inert packing materials which remain chemically and physically stable over time. Pressure drop will then remain limited, unless excess biomass accumulates on the surface of the packing (Kennes and Veiga, 2002). Some inert materials that have been used in biofilters are perlite, lava rock, plastic rings, polyurethane foam, and activated carbon, among others (Kennes et al., 2009). Contrary to natural filter beds, inert ones do not contain microorganisms and need to be inoculated. They do generally not contain nutrients either. The intermittent feeding of an aqueous phase or a nutrient solution is then needed. Such aqueous phase also allows to maintain a near optimal moisture content. Fungi tolerate more extreme conditions in terms of water content than bacteria (Kennes and Veiga, 2004), but even when fungi are dominant in the biofilter, the moisture content still needs to be optimized (Estévez et al., 2005; Jin et al., 2007).

ONE- AND TWO- LIQUID PHASE TRICKLING BIOFILTERS

Conventional biofilters and trickling biofilters are quite similar. The major difference is that in the latter inert filter beds are always used and a water phase is continuously fed to the system. Trickling biofilters started becoming popular around the early nineteen eighties. Inert packing materials that have been used in such reactors are similar to those used in conventional biofilters. Biofilters can be used for highly hydrophobic pollutants as, for example, α-pinene (Jin et al., 2007). Conversely, trickling biofilters are best suited for somewhat less hydrophobic compounds, because of the continuous supply of a liquid phase.

Two-liquid phase trickling biofilters have been studied recently for waste gas treatment, although the first studies are quite recent (van Groenestijn and Lake, 1999) and only very few studies and publications are available so far (van Groenestijn and Lake, 1999; Arriaga et al., 2006; Bailón et al., 2009). In such systems, a binary liquid phase is used. It is composed of an aqueous fraction (water) and an organic fraction (silicone oil being the most popular one). Although only little research has been published on this, an organic to aqueous phase ratio of 5 to 10% is often enough to significantly increase the reactor's performance (Bailón et al., 2009). The addition of an inert organic phase, as silicone oil, allows to treat waste gases containing highly hydrophobic pollutants. The presence of such an organic phase is also interesting when dealing with highly fluctuating (shock) loads.

ONE- AND TWO- LIQUID PHASE SUSPENDED-GROWTH BIOREACTORS

Suspended-growth bioreactors as bioscrubbers and activated-sludge bioreactors are suitable for the treatment of air polluted with relatively hydrophilic compounds. In bioscrubbers, polluted air is first fed to a scrubber in which the pollutant is transferred from the gas phase to a liquid phase, usually water. The polluted water phase is then fed to a bioreactor, typically an activated sludge reactor, where the pollutants are biodegraded. Clean water is then available to be recirculated to the scrubber.

Another related bioprocess is based on the direct diffusion of the polluted gas phase to a suspended growth bioreactor, often an activated sludge unit (Burgess et al., 2001). If a bioreactor is already available for the treatment of wastewater *on-site*, that same system can simultaneously be used for air treatment. At the same time the polluted gas phase diffusing through the activated sludge-type reactor plays the role of an aeration system (Kennes and Veiga, 2001).

Suspended-growth bioreactors can be operated with two liquid phases, as in the case of trickling biofilters. A binary water-organic solution is then fed to the bioreactor instead of a single water phase. Here again, silicone oil is the most popular organic phase (Kennes et al., 2009), although other organic phases as, for example, hexadecane have been tested as well (Davidson and Daugulis, 2003). Ratios of 10% organic phase or less are often enough, although values as high as 33% (one third) have been used in some research studies (Davidson and Daugulis, 2003). Improvements in removal efficiency of as much as 300% or more have been observed when adding 10% organic phase to the liquid phase (Bailón et al., 2009).

OTHER BIOREACTORS

A few other bioreactor configurations have been developed. However, they have mainly been studied at laboratory-scale, *i.e.,* the membrane bioreactor (Kennes and Veiga, 2001; De Bo et al., 2002), the monolith bioreactor (Jin et al., 2006). A bioreactor for which full-scale applications exist but only in very limited number is the Rotating Biological Contactor (RBC), also called biodiscs (Kennes and Veiga, 2001). Hybrid and multi-stage systems have been used as well.

CONCLUSIONS

The conventional biofilter is still the most widely used reactor configuration at full-scale for air pollution control. Examples of full-scale applications of trickling biofilters and technologies based on air diffusion through suspended-growth bioreactors can also be found. Bioscrubbers are somewhat less used and are mainly suitable and efficient for relatively hydrophilic pollutants. The interest in broadening the application range of bioprocesses has led to the, often successful, study and development of innovative reactor configurations, such as the two-liquid phase reactors, membrane bioreactors, and hybrid-systems, among others.

ACKNOWLEDGEMENTS

Our research on innovative bioreactor configurations for waste gas treatment is presently being funded by the Spanish Ministry of Science and Innovation (project CTM2007-62700/TECNO) as well as European FEDER funds.

REFERENCES

1 Arriaga, S., Muñoz, R., Hernández, S., Guieysse, B. and Revah, S. 2006. Gaseous hexane biodegradation by Fusarium solani in two-liquid phase packed-bed and stirred-tank bioreactors. *Environ. Sci. Technol.* 40:2390–2395.

2 Bailón, L., Nikolausz, M., Kästner, M., Veiga, M.C. and Kennes, C. 2009. Removal of dichloromethane from waste gases in one- and two- liquid-phase stirred tank bioreactors and biotrickling filters. *Water Res.* 43:11–20.
3 Davidson, C.T. and Daugulis, A.J. 2003. The treatment of gaseous benzene by two-phase partitioning bioreactors: high performance alternative to the use of biofilters. *Appl. Microbiol. Biotechnol.* 62:297–301.
4 De Bo, I., Van Langenhove, H. and Heyman, J. 2002. Removal of dimetil sulfide from waste air in a membrane bioreactor. *Desalination.* 148:281–287.
5 Estévez, E., Veiga, M.C. and Kennes, C. 2005. Biofiltration of waste gases with the fungi Exophiala oligosperma and Paecilomyces variotii. *Appl. Microbiol. Biotechnol.* 67:563–568.
6 Groenestijn, van, J.W. and Lake, M.E. 1999. Elimination of alkanes from off-gases using biotrickling filters containing two liquid phases. *Environ. Prog.* 18:151–155.
7 Jin, Y., Veiga, M.C. and Kennes, C. 2005. Bioprocesses for the removal of nitrogen oxides from polluted air. *J. Chem. Technol. Biotechnol.* 80:483–494.
8 Jin, Y., Veiga, M.C. and Kennes, C. 2006. Development of a novel monolith bioreactor for the treatment of VOC-polluted air. *Environ. Technol.* 27:1271–1277.
9 Jin, Y., Guo, L., Veiga, M.C. and Kennes, C. 2007. Fungal biofiltration of α-pinene: Effects of temperature, relative humidity and transient loads. *Biotechnol. Bioeng.* 96:433–443.
10 Jin, Y., Guo, L., Veiga, M.C. and Kennes, C. 2009. Optimization of the treatment of carbon monoxide-polluted air in biofilters. *Chemosphere.* 74:332–337.
11 Kennes, C. and Veiga, M.C. 2001. *Bioreactors for Waste Gas Treatment.* Kluwer Academic Publishers, Dordrecht, The Netherlands, 312 pp. ISBN: 9-780-7923-7190-8.
12 Kennes, C. and Veiga, M.C. 2002. Inert filter media for the biofiltration of waste gases – Characteristics and biomass control. *Re/Views Environ. Sci. & Bio/Technol.* 1:201–214.
13 Kennes, C. and Veiga, M.C. 2004. Fungal biocatalysts in the biofiltration of VOC-polluted air. *J. Biotechnol.* 113:305–319.
14 Kennes, C., Rene, E.R. and Veiga, M.C. 2009. Bioprocesses for air pollution control. *J. Chem. Technol. Biotechnol.* 84:1419–1436.
15 Philip, L. and Deshusses, M.A. 2003. Sulfur dioxide treatment from flue gases using a biotrickling filter-bioreactor system. *Environ. Sci. Technol.* 37:1978–1982.
16 Prado, O.J., Veiga, M.C. and Kennes, C. 2008. Removal of formaldehyde, methanol, dimethylether and carbon monoxide from waste gases of synthetic resin-producing industries. *Chemosphere.* 70:1357–1365.

Chapter 2

BIOLOGICAL CONTROL OF INDUSTRIAL VOC EMISSIONS IN NORTH AMERICA – 20 YEARS OF EVOLUTION

James T. Boswell
Bio•Reaction Industries, LLC., Tualatin, OR, USA

The terms biofilter, biotrickling filter, bioscrubber, bio-oxidation, bio-treatment, and bio-reaction generally describe the equipment for and process of directing a contaminated airstream through a microbial 'bed' for collection and biodegradation of the various odorous compounds and/or VOCs/THCs. The earliest use of biofiltration techniques in North America was on wastewater odors beginning in the early 20th century. Until the early 1990s there were no bio-oxidation systems controlling industrial airstreams contaminated with VOCs/THCs in North America. Then with the passage of the Clean Air Act by the US Congress in 1990, there was great interest in a more cost effective way to control VOC/THC emissions. Unfortunately, because of the rather Draconian standards for VOC/THC control that had been set in the 1980s biofilter systems were hard pressed to be both efficient enough and capital cost competitive, even adding the fuel costs associated with the most used equipment, thermal oxidizers. Between 1990 and 1995 biofilter systems were placed on a variety of facilities ranging from wood products emissions, to fabric softening sheet manufacturing, to printing operations, and on vapors from saturated rags and still bottoms, all with relatively good results. The design and engineering for early applications on industrial emissions sources in the US were predominantly coming from Europe, specifically the Netherlands and Germany. Some of the companies involved in these early enterprises and still operating are: Bohn Biofilter; PPC Biofilter, and Bio•Reaction Industries. Other companies active in the US in the early 1990s are no longer in existence; like ClairTech, Monsanto Enviro-Chem Systems and Biofiltration Inc., while new ones have been developed like BioRem and others like Tri-Mer have added a bio-oxidation system to their line of scrubbers and precipitators. This initial implementation of bio-systems in the early 1990s was followed by a relative lull in installations during the late 1990s. Only a few additional units were installed, largely controlling fragrance emissions from the personal products industry and odors from the food and beverage industry. Several small units were also installed for control of household hazardous waste can puncturing operations for municipalities.

In the late 1990s the US Environmental Protection Agency (USEPA) proposed hazardous air pollutant (HAP) regulations for a number of industrial classifications, providing a potentially huge opportunity for renewed interest in bio-oxidation treatment systems. The US wood products panel board industry was one of those targeted for HAP control, specifically methanol and formaldehyde. These panel board emissions were generally well suited for control with a bio-oxidation system; however implementation of these regulations did not occur until 2006. Between 2000 and 2007, to my knowledge there were only a few biofilter installations, but several were noteworthy. Two systems were successfully placed on paint manufacturing operations and the largest unit that I am aware of was installed on a panel board facility (400,000 acfm, 679,000 m^3/hr). Finally, with the regulatory impact of the HAP regulations becoming necessary, activity heated up for installation of bio-oxidation systems. Beginning in 2007 and continuing through this year approximately 20 units have been installed representing more than 2,000,000 acfm (3,395,586 m^3/hr) of air emissions controlled by bio-oxidation.

During these approximately 20 years there have been numerous refinements to the systems. Some have incorporated units combining a biotrickling filter with a mulch-packed plastic sphere (patented) biofilter media; while others have utilized different types of synthetic biofilter media (inorganic and organic) and one company has developed a combined bioscrubber/biotrickling filter without a conventional biofilter bed. Units have been constructed out of stainless steel, concrete, fiberglass and reinforced polypropylene. Virtually all units are totally enclosed (in order to demonstrate regulatory compliance), but a few open-topped units remain on the market. Compounds that are being controlled range from hydrophilic ones like alcohols and aldehydes (methanol and ethanol; formaldehyde and acetaldehyde) and ketones (MEK and MIBK); to hydrophobic ones like the alkenes, terpene and limonene; and the aromatics like benzene and toluene. In North America today I estimate that there are between 75 and 100 bio-oxidation systems treating industrial air emissions, and virtually all of those have been installed within the past 15 years. We have a long way to go to reach the numbers currently operating in Europe, but I believe we are moving rapidly toward equality in volumes of air if not in numbers of units. Currently, the US industry desire to evaluate reducing green house gas emissions (GHG) is triggering a great interest in replacement of natural gas guzzling TOs with bio-oxidation systems. Replacement of a TO with a bio-oxidation system will eliminate between 80 and 85% of the GHG emissions being emitted by thermal control of VOCs/THCs. Additionally, of serious consideration is the operating cost savings when one shuts off all that natural gas that is being combusted in a TO. Therefore, the future of bio-oxidation applications in North American industry looks very bright, bright green that is.

Waste gas treatment – process fundamentals

Chapter 3

TWO LIQUID PHASE PARTITIONING BIOREACTOR SYSTEM FOR BIODEGRADATION OF PYRENE BY *MYCOBACTERIUM FREDERIKSBERGENSE*

Biswanath Mahanty, Kannan Pakshirajan & Veeranki Venkata Dasu
Department of Biotechnology, Indian Institute of Technology Guwahati, Guwahati, India

ABSTRACT

Biodegradation of pyrene by *Mycobacterium frederiksbergense* was evaluated in a two liquid phase partitioning bioreactor (TPPB) containing 20% v/v silicone oil in an aqueous phase medium. At an initial concentration of 400 mg l^{-1} in silicone oil, the mycobacterium could completely degrade pyrene within 6 d with a rate of 140 mg l^{-1} d^{-1}. When compared to the results of slurry phase and surfactant aided systems, the TPPB system was found superior not only in terms of pyrene removal efficiency and rate but also in terms of cost effectiveness of the process.

INTRODUCTION

Polycyclic aromatic hydrocarbons (PAHs) produced via natural and anthropogenic sources during the incomplete combustion of solid and liquid fuels in industrial activities are harmful to environment and human health.[1] However, due to low aqueous solubility/bioavailability, bioremediation of PAHs is largely limited. Two-phase partitioning bioreactors (TPPBs) have gained considerable attention, in which PAHs dissolved in an organic phase partitions into the aqueous phase based on equilibrium considerations and real-time demand of the microorganisms.[2,3]

In this present study, an efficient TPPB system was investigated for pyrene biodegradation by *M. frederiksbergense*. Its performance in terms of pyrene degradation rate and efficiency, and cost benefit over slurry phase and surfactant aided systems was also compared.

MATERIALS AND METHODS

CHEMICALS AND REAGENTS

Analytical grade pyrene and silicone oil used in the study were purchased from Sigma-aldrich Chemicals, India, and Loba Chemie, India, respectively. Non-ionic surfactant (Tween 80) and media components were purchased from Himedia laboratory, India.

MICROORGANISM AND CULTURE MEDIA

M. frederiksbergense NRRL B-24126 was obtained from USDA, USA. While nutrient Broth was used for its routine growth and maintenance, Bushnell Hass media was used in the pyrene biodegradation experiments.

SCREENING OF NON AQUEOUS PHASE LIQUID FOR USE IN TPPB SYSTEM

For selecting a suitable non aqueous phase liquid (NAPL) that is biocompatible with the microorganism in TPPB system, isopropyl myristate, n-hexadecane and silicone oil were considered. Five 125 ml flasks each containing 40 ml of BH medium, 5 ml of *M. frederiksbergense*, 5 ml of any of the solvents and 2 ml of soybean oil were taken; an additional flask containing all the ingredients except the solvents served as the positive control in this experiment. All the flasks were incubated at 28°C for 3 d, and at the end the biomass in the flasks were determined by the dry weight method.

PYRENE BIODEGRADATION EXPERIMENTS

TPPB system

All pyrene biodegradation experiments in the TPPB system were carried out in a 3 l autoclavable glass fermenter (Applikon®). Both the aqueous (1 L BH media) and organic phases (pyrene in silicone oil, 250 ml) were loaded together in the vessel, autoclaved in toto and allowed to cool down to room temperature. Thereafter 50 ml of overnight grown *M. frederiksbergense* culture was added as the inoculum, and the bioreactor operated at temperature 28°C, pH 7.0, aeration rate 1.5 vvm and agitation rate 600 rpm. During its operation, duplicate samples of 1.5 ml each were withdrawn from the fermenter for 15 d and remaining pyrene in the silicone oil samples was extracted with methanol to monitor its degradation in the system.

Slurry phase system

Pyrene biodegradation experiments in this system were carried out using the same fermenter mentioned in experiments under TPPB system, but by taking pyrene stock solution (25 g l^{-1} in acetone) in the empty vessel to give final pyrene content of 50 mg l^{-1}, which is based on the final aqueous phase volume. Other experimental conditions in this system remained the same as that of the TPPB system. Samples (1 ml each) from this slurry phase reactor were extracted twice with equal volumes of ethyl acetate before final analysis of pyrene.

Surfactant aided system

Based on our earlier study,[4] using Tween 80 as the surfactant , experiments in this system were carried out using the earlier mentioned 3 l fermenter with 1 l BH media containing pyrene (25 mg l^{-1}) that was solubilised by the surfactant. Again, the experimental conditions were maintained the same in this system as those of the two other systems.

ANALYTICAL METHODS

Pyrene analysis

Pyrene concentrations in all the samples, with or without pretreatment as mentioned before, were quantified using synchronous fluorescence spectroscopy using a FluoroMax-3® (HORIBA Jobin Yvon, USA) fluorescence spectrometer with the detection condition of $\Delta\lambda = 36.0$ nm, excitation peak maximum = 335 nm.

COST BENEFIT ANALYSIS OF THE THREE DIFFERENT SYSTEMS

Cost analysis of the three biodegradation systems investigated in this work was performed considering the costs involved in the preparation of inoculum and media, and power required for running fermenter and shake incubator in the experiments. These are presented in Tables 1 and 2, respectively.

Table 1 Preparation costs of inoculums and media used in the biodegradation experiments.

	For inoculums (50 ml)		*For media (1000 ml)*	
Chemical/power input	*Amount*	*Cost**	*Amount*	*Cost**
Nutrient broth	0.65 g	0.04	–	–
Power consumption in shaker (12 h)	41.4 unit	3.90	–	–
BH media	–	–	3.27 g	0.30
Trace element solution	–	–	2 ml	0.01
Total cost (in USD)	–	3.94	–	0.31

*All costs are represented in USD.

Table 2 Power requirements for running fermenter and shaker incubator in the experiments.

System	*Unit/Components (Power in kW)*	*Total power (kW)*	*Cost/h*†
Fermenter	Bioconsole (1.6), Controller (0.07), Agitator (0.04), Chiller unit (0.9), Air compressor (1.5)	4.11	0.38
Shaker incubator	Complete system (3.45)	3.45	0.07

†Cost in USD; 1 USD = 48.00 Indian rupees and Electricity charges are based on LT-commercial tariff of AERC.

RESULTS AND DISCUSSION

Selection of NAPL for use in the TPPB system

Solvent selection experiments were performed to choose a candidate solvent that is biocompatible but not bioavailable to *M. frederiksbergense*. Figure 1 presents the results in terms of relative metabolic activity (RMA) of the mycobacterium in presence of the different solvents, which clearly reveals that all except IPM (RMA = 1.7) were found non-bioavailable.

It is also apparent from Fig. 1 that relative metabolic activity in presence of silicone oil was comparable to that due to hexadecane. Hence, based on these results and economic considerations, silicone oil was selected as the suitable solvent in the study.

Silicone oil has traditionally been considered highly stable, hydrophobic and chemically resistant to oxidative attack, and has been employed successfully in several other TPPB studies[3,5,6], as well.

PYRENE BIODEGRADATION EXPERIMENTS

TPPB system

Pyrene biodegradation experiments employing the TPPB system was carried out at its initial concentration of 400 mg l^{-1}. Pyrene degradation profiles, shown in figure 2, reveal that pyrene was completely degraded in 6 d with a rate of 139 mg l^{-1} d^{-1} without any lag phase in this TPPB system. MacLeod and Daugulis reported complete degradation of pyrene (1 g) within 4 d at a rate of 138 mg l^{-1} d^{-1} by *Mycobacterium PYR-1*in a TPPB system containing bis(ethylhexyl) sebacate.[7] In another study, Vandermeer and Daugulis observed only 26% degradation of pyrene after 120 h of TPPB operation by a defined *Sphingomonas* consortium using a very high initial pyrene concentration of 1 g l^{-1} in dodecane.[8] These results clearly indicate the superiority of the NAPL silicone oil – *M. frederiksbergense* pair in biodegradation of pyrene over other TPPB systems reported in the literature.

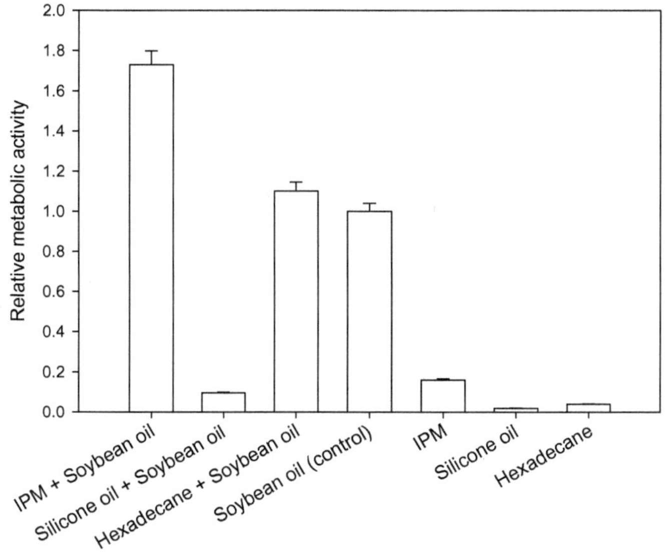

Figure 1 Results of solvent selection experiments.

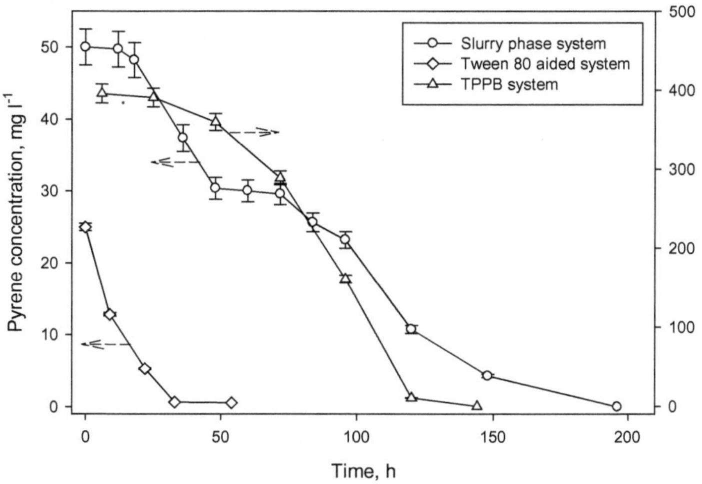

Figure 2 Pyrene biodegradation profiles obtained using slurry phase, surfactant aided and TPPB systems.

Slurry phase and Tween 80 aided systems

Results of pyrene biodegradation experiments in slurry phase system containing an initial concentration of 50 mg l^{-1} revealed complete degradation of the compound within 200 h (Fig. 2) with an overall degradation rate of 6 mg l^{-1} d^{-1}.

Table 3 Overall cost comparison of pyrene degradation employing the three different systems.

	Cost involved for (in USD)						
System	*Inoculums preparation*	*Basic media preparation*	*Pyrene delivery*	*Power*	*Total cost*	*Amount of pyrene removed (mg)*	*Cost per mg of pyrene removed (in USD)*
Slurry phase	3.94	0.46	–	77.51	81.91	75	1.10
Surfactant aided	3.94	0.31	0.04	13.95	18.24	25	0.73
TPPB	3.94	0.46	11.40	46.51	62.30	100	0.62

On the other hand, pyrene degradation profile with Tween 80 indicated near complete degradation within 50 h without any lag period with a degradation rate of 17.7 mg l^{-1} d^{-1}. It should be noted here that in both these two systems degradation of pyrene by the mycobacterium at initial concentrations above 50 mg l^{-1} were not investigated owing to either of the two limitations associated with such high concentration levels in these systems: no change in effective aqueous concentration of pyrene (slurry phase system) or huge requirement of surfactant (surfactant aided system). The chosen pyrene concentration of 25 mg l^{-1} in the present Tween 80 aided system was also based on our earlier study.[4]

COST BENEFIT ANALYSIS COMPARISON

As described earlier, cost benefit analysis was conducted considering the costs for preparing inoculum and media ingredients, and for power required to run fermenter and shaker incubator in these experiments. Table 3 comparatively illustrates the combined cost for the three systems, which clearly indicate that the TPPB system was the most economical among the three, and in which case only $0.62 was required for biodegrading a unit weight of pyrene.

REFERENCES

1 Mastral AM and Callen MS, A review on polycyclic aromatic hydrocarbon (PAH) emissions from energy generation. *Environ Sci Technol* 34:3051–3057 (2000).
2 Daugulis AJ, Two-phase partitioning bioreactors: a new technology platform for destroying xenobiotics. *Trends Biotechnol* 19:457–462 (2001).
3 Mahanty B, Pakshirajan K and Dasu VV, Biodegradation of pyrene by *Mycobacterium frederiksbergense* in a two-phase partitioning bioreactor system. *Bioresour Technol* 99:2694–2698 (2008).

4 Mahanty B, Sarma SJ and Pakshirajan K, Evaluation of different surfactants for use in pyrene biodegradation by *Mycobacterium frederiksbergense. Int J Chem Sci* 5:1505–1512 (2007).
5 Vanneck P, Beeckman M, De Saeyer N, D'Haene S and Verstraete W, Biodegradation of polycyclic aromatic hydrocarbons in a two-liquid-phase system. In *Bioremediation of recalcitrant organics*, ed by Hinchee RE, Anderson DB and Hoeppel RE. Battelle Press, Columbus, pp. 55–62 (1995).
6 Muñoz R, Guieysse B and Mattiasson B, Phenanthrene biodegradation by an algal-bacterial consortium in two-phase partitioning bioreactors. *Appl Microbiol Biotechnol* 61:261–267 (2003).
7 MacLeod CT and Daugulis AJ, Biodegradation of polycyclic aromatic hydrocarbons in a two-phase partitioning bioreactor in the presence of a bioavailable solvent. *Appl Microbiol Biotechnol* 62:291–296 (2003).
8 Vandermeer KD and Daugulis AJ, Enhanced degradation of a mixture of polycyclic aromatic hydrocarbons by a defined microbial consortium in a two-phase partitioning bioreactor. *Biodegradation* 18:211–221 (2007).

Biofiltration – organic pollutants

Chapter 4

α-PINENE REMOVAL FROM AIR IN ONE- AND TWO- LIQUID-PHASE THERMOPHILIC AND MESOPHILIC BIOTRICKLING FILTERS

María Montes, Eldon R. Rene, María C. Veiga & Christian Kennes
Faculty of Sciences, Chemical Engineering Laboratory, University of La Coruña, La Coruña, Spain

ABSTRACT

In the present study, two laboratory-scale biotrickling filters filled with the same type of packing material were operated at different temperatures (mesophilic and thermophilic conditions), either in presence or in the absence of an organic phase, in order to investigate their performance in treating α-pinene. The operating parameters considered in this study included the α-pinene loading rate and the flow rate (*i.e.*, EBRT) of the inlet waste stream. In absence of an organic phase, the thermophilic biotrickling filter removed α-pinene at loads up to 300 $g/m^3 \times h$ with high removal efficiencies, of more than 80%, reaching 33% higher EC than in the mesophilic reactor. With the addition of an organic phase, elimination capacities of α-pinene up to 2000 $g/m^3 \times h$ were obtained in the mesophilic biotrickling filter at high inlet concentrations and at a constant inlet air flow of 0.3 m^3/h. Similar results were obtained for the thermophilic biotrickling filter.

Keywords: alpha-pinene, biotrickling filter, organic phase, air pollution

INTRODUCTION

The treatment of polluted air in bioreactors allows the complete destruction of the contaminants under cost-effective conditions, contrary to what happens with some conventional physico-chemical processes.[1] Biotrickling filtration has proven to be an efficient technology for the elimination of many air pollutants, such as aliphatic and aromatic hydrocarbons.[2] Advantages of this kind of technology include effective removal, simple and low cost process and inexistence of negative side effect on the environment. In recent years, biotrickling filters have received increasing attention for the treatment of compounds difficult to degrade or compounds that generate acidic by-products, such as H_2S[2].

Many waste gases in industry have temperatures beyond the mesophilic range, but treatment of those gases has received only little attention.[1] Cooling of hot waste gases would be a solution, but this is expensive especially when the gas is saturated with water. The use of thermophilic microorganisms adapted to high temperatures is an interesting area of research, as it allows for treatment without prior cooling. α-Pinene was selected as a hydrophobic compound because of its high volatility, low water solubility, our recent past experience with this pollutant,[3,4] and because of its relevance to the application of biofiltration of air emissions in pulp and paper industries and the forest products industries.

In this study, several experimental runs were carried out in order to investigate the best operational conditions for reaching high removal efficiencies. The aim of this research was to investigate the biodegradation of α-pinene in a biological trickling filter with lava rock as packing material, at thermophilic (45°C) and mesophilic (23°C) temperatures, firstly without the addition of silicone oil and, secondly, with the addition of 5% silicone oil.

MATERIALS AND METHODS

Both biotrickling filters (BTF) were packed with 4 L lava rock and were inoculated with the leachate of a previously operated biofilter, containing predominantly *Ophiostoma stenoceras*, a fungus that has the ability to degrade α-pinene under a wide range of operating conditions.[3]

The experimental setup and operating conditions have been described elsewhere.[5] The nutritive solution described previously[3] was continuously recirculated through the reactors.

Gas-phase concentrations of α-pinene in the biotrickling filter were measured on a Hewlett-Packard 6890 gas chromatograph, using a 50 m TRACER column (TR-WAX, internal diameter 0.32 mm, film thickness 1.2 μm), equipped with a flame ionisation detector. Air samples were regularly taken from the inlet and outlet gas streams.

RESULTS AND DISCUSSION

Both in the biotrickling filter without oil and with oil, the removal of α-pinene was investigated in different phases corresponding to decreasing empty bed residence times (EBRTs) between 1 min and, as low as 14 sec. In the first step, without the addition of an organic phase, a gradual decrease in efficiency was observed as the pollutant's concentration was increased, showing a higher elimination efficiency under conditions of low flow rate (*i.e.* higher EBRT). The optimum inlet concentration and the corresponding removal efficiency appeared to be decreasing functions of the gas flow rate. The maximum elimination capacities reached were around 300 and 200 g/m^3 × h for the thermophilic and mesophilic bioreactor respectively at high residence times. These data could suggest that both the thermophilic and mesophilic biotricking filters would be highly efficient at relatively high inlet concentrations combined with moderate flow rates.

In waste gas treatment, the use of a second organic liquid phase is known to be able to significantly increase the bioreactors efficiency mainly in case of hardly water soluble pollutants and toxic compounds. As with the one-liquid phase systems, the inlet α-pinene concentration was gradually increased in order to evaluate both reactors' performance. Fig. 1 shows that the presence of a second, organic, liquid phase significantly improved the overall performance of the mesophilic reactor. When adding 5% silicone oil to the BTF, the maximum EC reached as much as 2000 g/m^3 × h to be compared to 200 g/m^3 × h in the one liquid phase system.

In the present study, in a biotrickling filter using lava rock as a packing material, excellent performance in the removal of α-pinene was thus achieved with the use of a biphasic liquid mixture composed of silicone oil and water, under both mesophilic and thermophilic conditions.

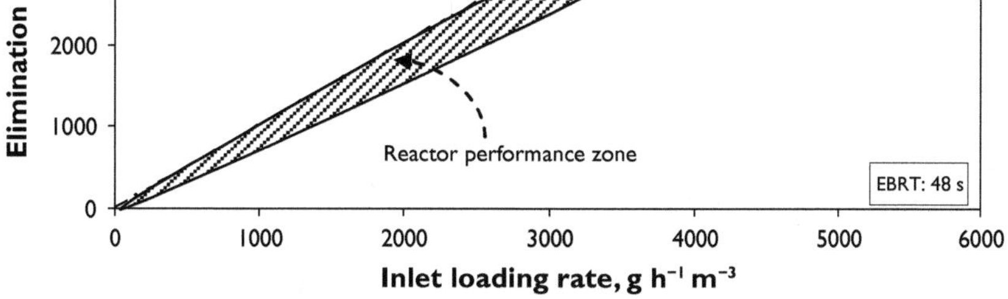

Figure 1 Effect of the addition of an organic phase in the mesophilic bioreactor at an EBRT of 48 s.

CONCLUSIONS

The results from this study showed that the treatment of α-pinene in gas-phase biotrickling filter using silicone oil as an organic phase can be performed with high efficiencies. The two-liquid-phase BTF performed always better than the one-liquid-phase bioreactor, under both mesophilic and thermophilic conditions. The maximum EC was 10 times higher when adding 5% silicone oil compared to the system without any addition of an organic phase.

ACKNOWLEDGEMENTS

The present research and the Ph.D. contract of M. Montes were financed through a project of the Spanish Ministry of Science and Innovation (CTM2007-62700/TECNO) and through European FEDER funds.

REFERENCES

1 Kennes C and Thalasso F, Waste gas biotreatment technology. *J Chem Technol Biotechnol* **72**: 303–319 (1998).
2 Kennes C and Veiga MC, *Bioreactors for Waste Gas Treatment.* Kluwer Academic Publishers: Dordrecht; Boston (2001).
3 Jin Y, Veiga MC and Kennes C, Performance optimization of the fungal biodegradation of α–pinene in gas–phase biofilter. *Process Biochemistry* **41**: 1722–1728 (2006).
4 Jin Y, Guo L, Veiga MC and Kennes C, Fungal biofiltration of α–pinene: effects of temperature, relative humidity and transient loads. *Biotechnol Bioeng* **96**: 433–443 (2007).
5 Bailón L, Nikolausz M, Kästner M, Veiga MC and Kennes C, Removal of dichloromethane from waste gases in one- and two- liquid-phase stirred tank bioreactors and biotrickling filters. *Water Res* **43**: 11–20 (2009).

Chapter 5

EFFECT OF POLYAMINE ON ADSORPTION AND DEGRADATION OF TOLUENE BY BIOCOMPOSITE BASED ON NATURAL FIBER

Vladimir A. Escobar Barrios & Olga B. Gutiérrez Acosta
División de Ciencias Ambientales, Instituto Potosino de Investigación Científica y Tecnológica A.C., San Luis Potosí, SLP, México

ABSTRACT

The effect of polyamine addition on biocomposite based on natural fiber and foamed polymeric matrix, which was used to support microorganisms for degradation of toluene, was investigated. The synthesized biocomposites was able to adsorb toluene in a very short period of time (1 day). The molecular weight of polyamine had a slight effect on toluene adsorption process, being the lower molecular weight more favorable for such process. The biocomposite without polyamine had a dramatic improvement for adsorption and biodegradation of toluene comparative with biocomposites with polyamines. The adsorption was very fast and biodegradation takes place in a slightly more effective way than adsorption. In the case of this biocomposite with microorganisms but without polyamine, CO_2 concentration was 390% higher than biocomposites with polyamine due to the higher degradation of toluene.

INTRODUCTION

Biofiltration is a quick gaining attention area since it is possible to degrade contaminants like toluene by microorganisms to less dangerous products such as water vapor and carbon dioxide. Toluene is widely used in diverse industries as a solvent, and usually it is emitted to air or water without any treatment, which promotes severe adverse effects on human health and environment. Therefore, treatment of toluene emissions is essential.[1,2]

The use of microorganisms for biodegradation of solvents has been studied and it is recognized that the supporting of such microorganisms is necessary in order to scale the results to commercial products. Several types of supports have been proposed, from natural to synthetic materials. In general, natural materials contain a source of nutrients for the microorganisms. However, this kind of materials can be degraded by microorganisms and became weak in terms of mechanical properties.[3,4] Regarding natural fibers, they can adsorb water several times their own weight. Thus, they facilitate the water retention of fiber-containing composites. In addition, usually, they are not easily degraded by microorganisms due to their composition and structure. In the case of mineral materials they are not degraded by microorganisms and have better mechanical resistance than natural materials but they do not serve as source of nutrients.[5,6] To facilitate the adsorption of toluene on composite surface, being more available to microorganisms present in such surface, this study proposed the synthesis of biocomposite based on natural fiber and synthetic polymeric matrix, adding two different polyamine types during the composite synthesis.

MATERIALS AND METHODS

MATERIALS

The prepolymer polyurethane was supplied by Bayer MaterialScience (Leverkusen, Germany), the natural fiber was provided by Tecnología Ambiental (San Luis Potosí, México). Polyamines with molecular weight of 25,000 g/mol and 800 g/mol were provided by BASF (Ludwigshafen, Germany), these polyamines were named A-1 and A-2, respectively. Toluene was analytical grade (J.T. Baker, N.J., USA). All these materials were used as they were received. The used inoculum was a microbial consortium that was previously isolated from a hydrocarbon contaminated site.

BIOCOMPOSITE SYNTHESIS

The biocompostes were synthesized with natural fiber (50% w/w) and polymeric matrix (50% w/w). The polyamine content was 18% by weight based on the biocomposite total weight.

BIOCOMPOSITE CHARACTERIZATION

Water retention capacity

The synthesized biocomposites, natural fiber and polymeric matrix were deposited in distillated water during 48 hours at room temperature, in order to saturate each sample. After that, the sample was analyzed by thermogravimetry, using Thermo Cahn TGA, with temperature ramp of 10° min^{-1} from room temperature up to 100°C and then it was kept isothermal at 100°C during 40 minutes to assure all the water lost.

Scanning Electron Microscopy (SEM)

The Scanning Electron Microscopy was carried out using an ESEM-QUANTA 200 (FEI Company). Samples were set in aluminum plate without any previous preparation, and analyzed using a LFD (Large Fill Detector) electron detector. Operation conditions were: 15 KV, 90 Pa as the chamber pressure with low vacuum mode and water atmosphere.

BATCH TESTS

Batch experiments were carried out using glass flasks (previously sterilized) with Teflon valves. In these flasks were deposited 0.5 grams of each synthesized biocomposite. Then, 250 microliters of mineral medium were added to three biocomposites and such samples were inoculated with 250 microliters of microorganisms consortium. The inoculated samples were named BCIA-1, BCIA-2 and BCI, corresponding to biocomposites with polyamine A-1, polyamine A-2 and without polyamine, respectively. Other three biocomposites samples, named BCA-1, BCA-2 and BC, were not added with mineral medium and were not inoculated. Afterward, all the samples were added with 2 microliters of toluene. The concentration of toluene was monitored along time and CO_2 concentration in the headspace of the flasks was measured from day 13 to the end of the experiment.

GAS CHROMATOGRAPHY

Toluene and CO_2 concentrations were determined by injecting 200 μL headspace samples into a 6890 series gas chromatograph (Agilent Technologies), using a FID Detector and capillary columns (DB-624). The temperature of the injector, oven and detector were maintained at 230, 60 and 230°C, respectively. For the case of CO_2 a thermal conductivity detector and a capillary column HP-PLOT Q were used. The injector, oven and detector temperatures were 180, 40 and 180°C, respectively.

RESULTS

WATER RETENTION CAPACITY

The obtained results confirm the hydrophobic character of polyurethane since it had the lower water retention capacity (29%), meanwhile for the natural fiber was 96%. The biocomposites had capacities from 37 to 66% in terms of water retention, being the highest for the BC composite and the lower for the BCA-1 composite.

BIODEGRADATION/ADSORPTION

The adsorption-biodegradation behaviors for the biocomposite with the polyamine with the higher molecular weight (BCIA-1) and the same biocomposite without microorganisms (BCA-1) are presented in Figure 1.

As it can be seen from Figure 1, there was an important adsorption of toluene (BCA1) at the beginning of the process and then had an asymptotic trend after day 4. It is important to point out that not significant reversible adsorption was detected, especially after day 12. In the case of the biocomposite inoculated with microorganisms (BCIA-1) it had a fast biodegradation of toluene and also it was achieved a steady behavior from day 3. After this time no more toluene degradation was noticed.

For the case of the biocomposites with polyamine A-2, the obtained adsorption-biodegradation behavior is shown in Figure 2.

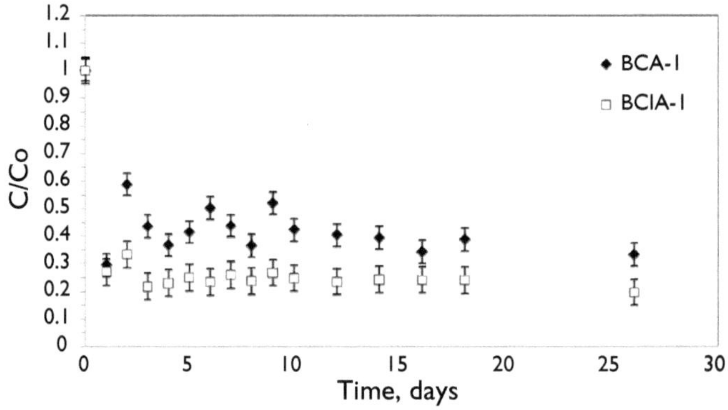

Figure 1 Toluene adsorption-biodegradation behavior, biocomposites BCA-1 and BCIA-1.

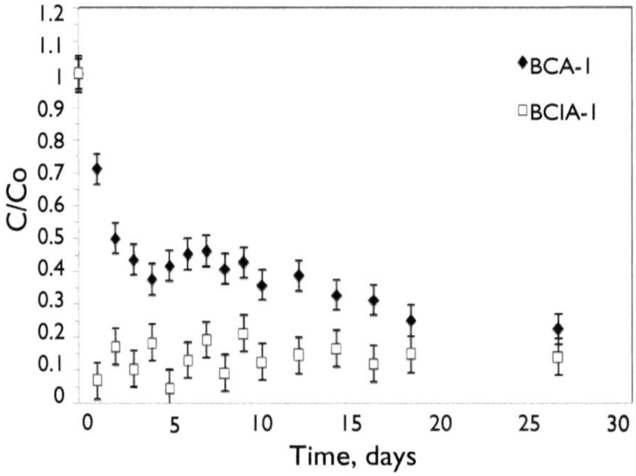

Figure 2 Toluene adsorption-biodegradation behavior, biocomposites BCA-2 and BCIA-2.

It is clear that there was an important adsorption but it was almost equal to that obtained with the biocomposite which had the polyamine with higher molecular weight, at the end of the experiment. However, regarding the biodegradation, the biocomposite BCIA-2, showed a better performance. The biodegradation was very effective from day 1 and from this day the behavior of biodegradation was quasi-steady with a C/Co ratio of 0.14 as mean value.

Regarding the adsorption- biodegradation of biocomposites without polyamine (BC and BCI) the Figure 3 shows the obtained behavior.

As it can be noted from Figure 3, there was a very fast adsorption of toluene by the composite, and it seems that it was more efficient than biodegradation during the first toluene injection. After 7 days, a second toluene injection was made since it was already depleted. In this case the biodegradation was more efficient than that of first injection, while the adsorption had almost the same behavior, it proceeded very fast and the toluene was exhausted in 6 days.

Finally, a third toluene injection was made and in this case the adsorbed toluene was lower than the adsorbed during injection 1 and 2, while the biodegradation effectiveness was lower than the previous injection (2) but better than injection 1. Nevertheless, the total biodegradation was achieved after 9 days.

CO_2 PRODUCTION

Biodegradation of contaminants such as toluene assumes that CO_2 and water are generated. The biocomposites that were not inoculated (BCA-1, BCA-2 and BC) have not a significant CO_2 production instead the obtained CO_2 production for the inoculated biocomposites at day 26, were important, especially for BCI, which was 390% higher than BCIA-1 and BCIA-2.

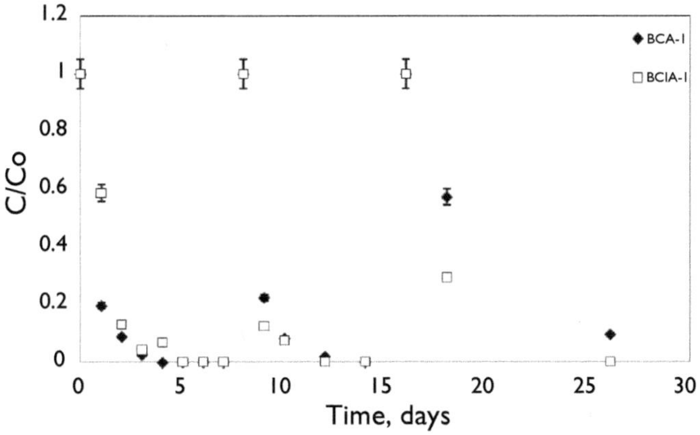

Figure 3 Toluene adsorption-biodegradation behavior, biocomposites BC and BCI.

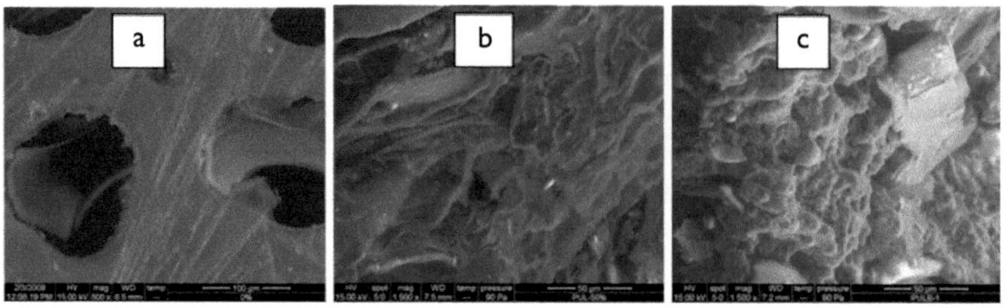

Figure 4 Electronic micrograph for (a) polymeric matrix; (b) non-inoculated biocomposite (BC) and (c) inoculated biocomposite (BCI).

MORPHOLOGY

The obtained micrographs, at the end of the experiment, for the samples without microorganisms neither polyamine (BC) and with microorganisms but without polyamines (BCI) are shown in Figure 4, since these samples are representative they showed the best performance in terms of adsorption and biodegradation.

As it can be seen from Figure 4, A and C, the morphology of the polymeric matrix was changed by the fiber, obtaining higher roughness.

Regarding the biocomposite with microorganisms (Figure 1-B), it is clear that there was an important growth of microorganisms on surface.

REFERENCES

1 Woertz JR, van Heiningen WNM, van Eekert MHA, Kraakman NJR, Kinney KA and van Groenestijn JW, Dynamic bioreactor operation: effects of packing material and mite predation on toluene removal from off-gas, *Appl Microbiol Biotechnol*, **58**: 690–694 (2002).

2 Alexander M, *Biodegradation and Bioremediation*, Academic Press, California, USA (1999).

3 Juteau P, Larocque R, Rho D and LeDuy A, Analysis of the relative abundance of different types of bacteria capable of toluene degradation in compost biofilter, *Appl Microbiol Biotechnol*, **52**: 863–868 (1999).

4 Song J and Kinney K, Microbial response and elimination capacity in biofilters subjected to high toluene loadings, *Appl Microbiol Biotechnol*, **68**: 554–559 (2005).

5 Jang JH, Hirai M and Shoda M, Styrene degradation by Pseudomonas sp. SR-5 in biofilters with organic and inorganic packing materials, *Appl Microbiol Biotechnol*, **65**: 349–355 (2004).

6 van Groenestijn J and Harkes M, Ceramic Materials in biofiltration, in *Proc USC-TRG Conf on Biofiltration*, ed by Reynolds FE, Los Angeles, pp. 317–324 (1995).

Chapter 6

BIOFILTRATION OF N-HEXANE IN THE PRESENCE OF DIFFERENT LEVELS OF BENZENE VAPORS

Ashraf Aly Hassan & George A. Sorial
Department of Civil and Environmental Engineering, University of Cincinnati, Cincinnati, OH

INTRODUCTION

Degrading hydrophobic volatile organic compounds in biofilters is very challenging. Research in our laboratories examined the degradation of hydrophobic contaminants by several means including changing the Empty Bed Resident Time (EBRT), utilization of surfactants and change of the effective microorganism consortium. As n-hexane is considered relatively soluble in benzene and alongside benzene is more soluble in water than n-hexane, it was anticipated that a mixture of both Volatile Organic Compounds (VOCs) feed to a Trickle Bed Air Biofilters (TBABs) would lead to an improved performance of n-hexane degradation.

The relative mixing ratios of n-hexane to benzene in the feed was studied using four parallel TBABs. The ratios of benzene: n-hexane studied based on volume are: 3:1, 2:1, 1:1 and 1:2. TBABs were operated at pH 4 to provide favorable conditions for fungi growth, while the other operating conditions were kept at the optimal performance as determined by earlier research in our laboratory including nutrient feed rate (2 L/day), air flowrate (1.4 L/min), bed depth (60 cm), EBRT (120 seconds), bed material (diatomaceous earth pellets) and room-temperature (20°C).

Keywords: Trickle Bed Air Biofilter, Volatile Organic Compound, n-Hexane, Benzene, mixture Hydrophobic Compound, Bacteria, Fungi

MATERIALS AND METHODS

VOLATILE ORGANIC COMPOUNDS

Two VOCs were used in this study. n-Hexane was obtained from (Fisher Scientific, Fair Lawn, NJ, pure 85%). Henry's Law Constant (HLC)

for hexane was reported at 20°C by different authors in the range of: 0.883 atm · m^3/mol,[1] 0.014 atm · m^3/mol[2] and 0.122 atm · m^3/mol.[3] Benzene was obtained from Fisher Scientific (Fair Lawn, NJ) with 99% purity. The Henry's law constant (HLC) for benzene at 25°C is 5.42×10^{-4} atm · m^3/mol^{-1}.[4] Four different TBABs were employed in this study. TBAB "A", "B", "C" and "D" received continuous feed of n-Hexane and benzene ratios of 1:3, 1:2, 1:1 and 2:1 of n-hexane: benzene based on volume, respectively.

TRICKLING BED AIR BIOFILTER

TBABs had similar operating conditions with the nutrient tank buffered at pH 4 using sodium acetate. The TBABs are constructed of seven cylindrical glass sections with an internal diameter of 7.6 cm and a total length of 130 cm. They are packed with palletized diatomaceous earth biological support media to a depth of about 60 cm (Celite® 6 mm R-635 Bio-Catalyst Carrier; Celite Corp., Lompoc, CA). The TBABs ran at a constant operating temperature of 20°C and operated in a flow switching from co-current to counter-current gas flow vs. liquid flow always downward. A schematic of the TBAB setup could be found in Figure 1.

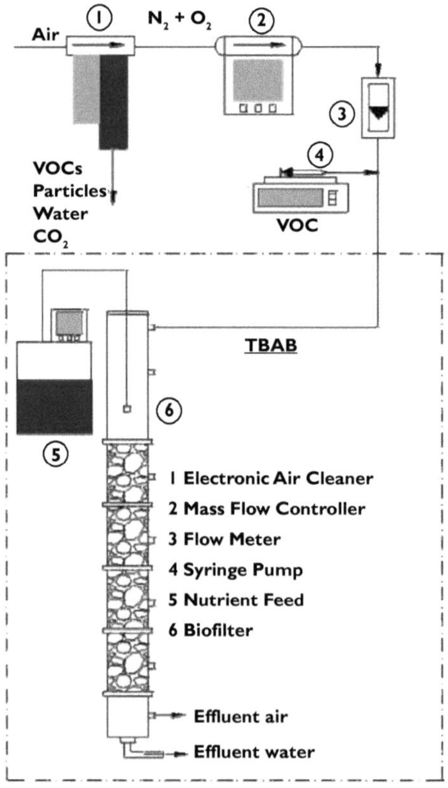

Figure 1 Schematic of the TBAB.

The air flow was set up at the rate of 1.36 L min^{-1} with a corresponding empty bed retention time (EBRT) of 120 seconds. Liquid n-hexane and benzene were injected via a syringe pump by the corresponding mixing volume, respectively, and vaporized into the air stream. Buffered nutrient solution was supplied at a rate of 2.0 L day^{-1}, the composition of the nutrient solution is similar to that reported by Sorial et al.[5]

Both TBABs were run using the stagnation technique for a period of 2 days per week with weekly flow switching of air flow as means of biomass growth control within the bed. This was proven to be the most effective technique for benzene removal.[6] The stagnant technique could be explained as no flow (VOC, nutrient, and air) passing through the TBAB.

ANALYTICAL METHODS

The concentrations of VOCs in the gas phase were measured by using a gas chromatograph (GC) (HP 5890, Series II, Hewlett-Packard, Palo Alto, CA) equipped with a flame ionization detector (FID). Effluent gas phase sample for CO_2 analysis were also taken by using gas-tight syringes through sampling ports in the TBAB. A GC (HP 5890, Series II, Hewlett-Packard, Palo Alto, CA) equipped with a thermal conductivity detector (TCD) was used for determining the CO_2 concentrations in the effluent gas phase. Liquid phase samples were analyzed for NO_3^--N, total carbon (TC), inorganic carbon (IC), and volatile suspended solid (VSS) concentration. NO_3^--N was determined according to Standard Methods 4500-NO_3^-[7] by using a Shimadzu UVmini 1240 UV-VIS spectrophotometer (Shimadzu Corp., Tokyo, Japan). TC and IC were determined by using a Shimadzu TOC 5050 analyzer (Shimadzu Corp., Tokyo, Japan) according to Standard Methods 5310.[7] The VSS concentrations in the effluent and backwashing water were determined according to Standard Methods 2540 G.[7]

RESULTS AND DISCUSSION

TBAB PERFORMANCE

All TBABs had similar runs at the lower influent concentrations. Fungi utilization greatly enhanced the performance of the TBABs. This was proven in an earlier study.[8] Generally, for all TBABs fluctuations in performance of n-hexane were noticed. The fluctuation in removal efficiency was reported previously without known reason.[9] The overall standard deviation in the removal efficiency of n-hexane ranged from 20% to 29%.

Figure 2 shows the elimination capacity vs. loading rate for all TBABs. It could be seen from this figure that both n-hexane and benzene performance deviated from the 100% performance line: for n-hexane at a loading rate of less than 10 g m^{-3} hr^{-1} and for benzene at loading rate of ca. 40 g m^{-3} hr^{-1}.

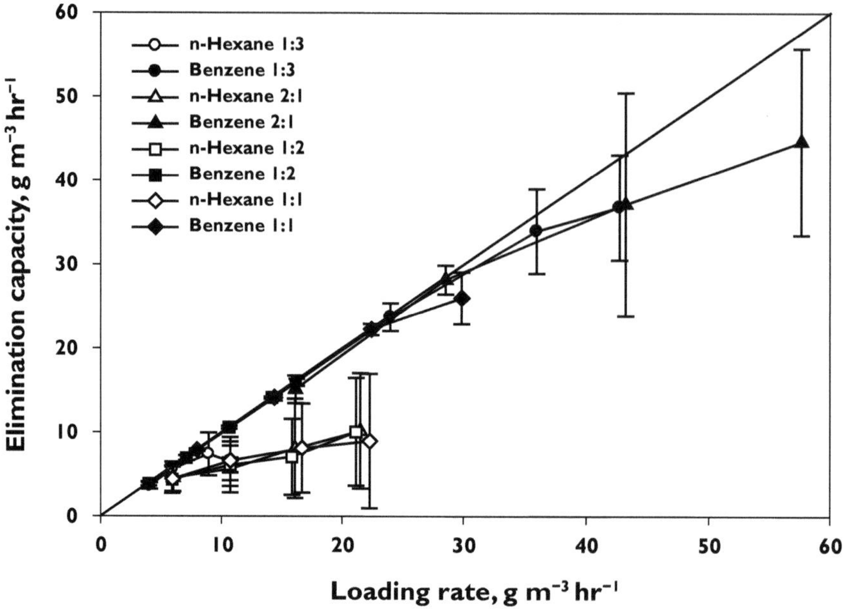

Figure 2 Elimination capacity vs. loading rates for different mixing ratios.

But the steady decrease in the elimination capacity at a higher loading rate meant that the performance was limited by the total loading of both compounds which exceeded the critical loading capacity of the TBAB. The different loading ratios did not have significant effect on the elimination capacity.

Liquid measurements of carbon performed on the bed showed that there were no VOCs available in the water phase. The carbon mass balance showed all the VOC concentrations were in the gas phase. This was true for both benzene and n-hexane. The later is insoluble in water. On the other hand the liquid phase flow of 2 L/day is a minor quantity as compared to the gas phase flow of 1.4 L/min (ca. 1000 times).

CONCLUSIONS

This study investigated the effect of pH on TBAB performance for removal of a mixture of n-hexane and benzene mixed in ratios of 1:3, 1:2, 1:1 and 2:1. Fungi utilization greatly enhanced the performance of n-hexane. The following points could be concluded out of the results of the study:

1 Flow switching with stagnation for a period of 2 days/week in the TBAB is a good tool to enhance the performance. This strategy could replace backwashing for low yield micro organisms as in the case of n-hexane.
2 Using different mixing ratios of n-hexane and benzene did not have significant impact on the biodegradation of both compounds.

ACKNOWLEDGMENT

The authors would like to thank the financial support from National Science Foundation under award # CBET 0852803.

REFERENCES

1 Montgomery, J.H.; Welkom, L.M., *Groundwater chemicals desk reference* 3rd Edition ed.; Boca Raton: CRC Lewis Publishers: 2000; p. 1345.

2 USEPA Superfund Chemical Data Matrix (SCDM). http://www.epa.gov/superfund/sites/npl/hrsres/tools/scdm.htm

3 EEIP *Prefered and alternative methods for estimating air emissions from wastewater collection and treatment*; Point Sources Committee, Emission Inventory Improvement Program: Morrisville, North Carolina, 1997; p. 98.

4 Shiu, W.-Y.; Ma, K.-C., Temperature dependence of physical—chemical properties of selected chemicals of environmental interest. I. mononuclear and polynuclear aromatic hydrocarbons. *Journal of Physical and Chemical Reference Data* **2000**; 29(1), 41–130.

5 Sorial, G.A.; Smith, F.L.; Suidan, M.T.; Biswas, P.; Brenner, R.C., Evaluation of trickle bed biofilter media for toluene removal. *Journal of the Air and Waste Management Association* **1995**; 45(10), 801–810.

6 Aly Hassan, A.; Sorial, G., Biological treatment of benzene in a controlled trickle air bed biofilter. *Chemosphere* **2008**, Submitted for publication August 11, 2008.

7 Eaton, A.D.; Clesceri, L.S.; Rice, E.W.; Greenberg, A.E.; Franson, M.A.H., *Standard Methods for Examination of Water & Wastewater*. 21st ed.; American Public Health Association, American Water Works Association, Water Environment Federation Washington, D.C., 2005; p. various pagings.

8 Aly Hassan, A.; Sorial, G. In *Trickle Bed Air Biofilter treating a mixture of n-hexane and benzene*, AWMA 103rd Annual Conference Detroit, MI, 2009; Detroit, MI, 2009.

9 Plaggemeier, T. Elimination der schwer wasserloslichen Modellabluftinhaltsstoffe n-Hexan und Toluol im Biorieselbettverfahren. Universitaet Stuttgart., Stuttgart, 2000.

Chapter 7

EFFECTS OF INLET CONCENTRATION AND SHUT DOWN ON THE BIOFILTRATION OF ISOAMYL ALCOHOL VAPORS

Juan Pablo Silva Vinasco
Faculty of Engineering, School of Environmental Resources, Universidad del Valle, Cali-Colombia

Astrid Jeritza Achipiz López & Maria Johanna Lopez Rojas
Universidad del Valle, Cali-Colombia

ABSTRACT

The effects of inlet concentrations and shut-down influence on the removal of an air- isoamyl alcohol (IA) mixture were studied in two identical bench scale compost biofilters. Variations on the efficiency and capacity of IA elimination with different inlet concentrations 0.72, 2.52, 4.33, 6.13 g m^{-3}, were investigated for 50 days. A maximum elimination capacity of 77.68 g m^{-3} h^{-1} for an EBRT of 42 s was obtained, while that the highest removal efficiencies found were 81.38% for an inlet concentration of 2.52 g m^{-3}. The influence of shut down (non use) on the stability of the biofilter and bacteria population has shown that for larger periods of shut down larger acclimation periods are required. In addition a decreasing of pseudomonas in the entire biofilter was found because the shut down.

INTRODUCTION

Isoamyl alcohol (IA) ($C_5H_{12}O$) is a VOC widely employed in industrial facilities, for instance, esters from isoamyl alcohol, such as isoamyl butyrate, and acetate are widely used and provide an economic valuable as chemicals for flavoring and fragrance manufacturing. However, when IA released to atmosphere, can lead to the formation of tropospheric ozone and smog, affecting both population health and environment. Thus, the industries have to put into practice control strategies that minimize the impacts of IA emissions. Biofiltration is a cost effective and reliable option in treating VOCs emitted from processes with large off gas volumes but low concentrations.[1] Biofiltration of alcohols has been the subject of many studies and has been applied in industry over a period of many years.[2,3] However, isoamyl alcohol biofiltration has been rarely studied and issues such as effects of inlet concentrations and shut-down of the system on efficiency removal of IA remain to be studied. Therefore, the objectives of this study were (a) Evaluation of both the variation of removal efficiency and stability of the biofilter to different inlet concentrations (b) Study the influence of shut-down on the stability of the biofilter.

METHODOLOGY

The packing material consisted of a mixture (60:40 v/v) of compost (0.99 mm and 2.36 mm size granules) and polystyrene inert particles (3 mm size). The inert material serves to increase the bed porosity and to ensure more homogeneous gas distribution across the filter bed. The filter material was not inoculated.

Figure 1 illustrates the schematic diagram of the biofilter. The biofilters were made of PVC tube with an internal diameter of 8.3 cm. The tube was packed to a height of 1 m. The filter material was supported by perforated plate. Sampling ports covered with rubber septa allowed collection of samples for analysis. Humidified air and isoamyl alcohol were mixed in a mixing chamber and the mixture was passed through the biofilters. Isoamyl alcohol concentration in the feed stream was varied by adjusting the flow rates of the air stream passing through the water and isoamyl solution. The biofilter was operated at different inlet feed concentrations 0.72, 2.52, 4.33, 6.13 g m^{-3}

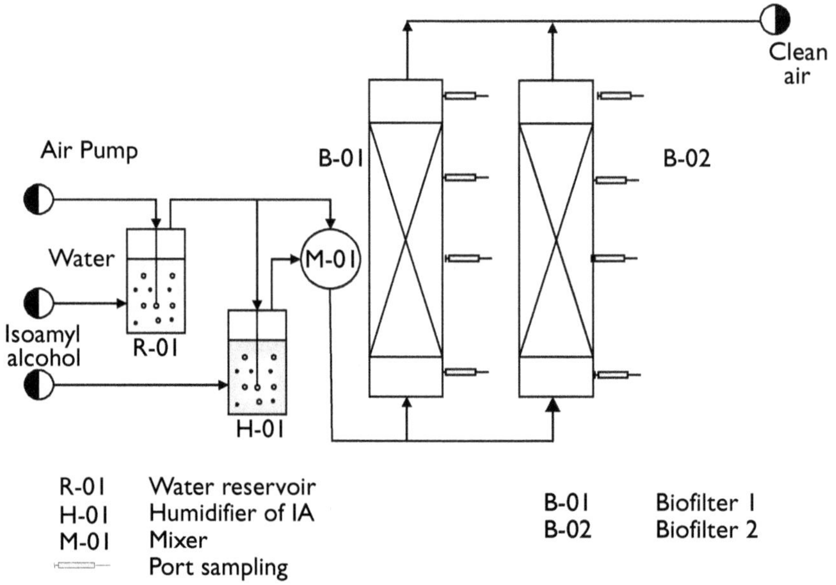

Figure 1 Schematic diagram of the experimental set-up.

maintaining a constant flow rate of 0.1623 $m^3 h^{-1}$ and an EBRT of 42 s. Gas samples were collected at regular intervals from the top of the biofilter as well as from the sampling ports using a gas tight syringe and analyzed for residual isoamyl alcohol.

The gas samples were analyzed by gas chromatography (Shimadzu 14B) using a Shimadzu packed column OV-17 and flame ionization detector. Nitrogen was used as the carrier gas at a flow rate of 20 ml min^{-1}. The temperature of the column was maintained at 150°C. Injector and detector temperatures were both 85 and 90°C respectively. The retention time of isoamyl alcohol was 3.5 minute under these conditions.

RESULTS

The physical and chemical properties of the packing material are shown in Table 1. The organic content of the compost in the biofilter bed provided all the nutrients necessary for the biomass growth; thus, there was not a previous inoculation of microorganisms into the biofilter. the microbial acclimation to isoamyl alcohol was achieved in two days by exposing the system to an inlet concentration (C_{in}) of 2.52 g m^{-3}, period in which a 82% removal efficiency constant was obtained.

The performance of the biofilter under steady state is shown in Fig. 2. It may be observed for inlet concentrations (C_e) less than 2.52 g m^{-3} that the system was quite stable in relation with the response of outlet concentration, contrasting with the behavior of the system when inlet concentrations applied were bigger than 4.33 g m^{-3}. Both microbial growth rate and biochemical reaction rate would be inhibited with increasing average inlet concentration.[4] These results are in accordance with other biofiltration studies that have reported the variability in the effluent concentrations like one of the major problems associated with the biofilter operation.

The steady state removal efficiency average of the biofilters at 30 m h^{-1} velocity (EBRT 42 s) are shown in Fig. 3. The removal efficiency decreased

Table 1 Physical and chemical properties of packing material.

Parameter	*Range*
pH	7.2
Density	1200 kg/m^3
Porosity	0.40
Water content	45%
Field capacity	51.32
Carbon	35%

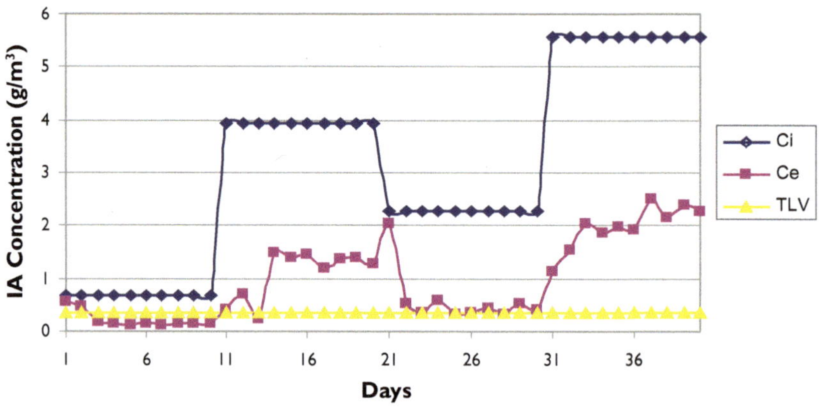

Figure 2 Response of the biofilter to inlet concentrations.

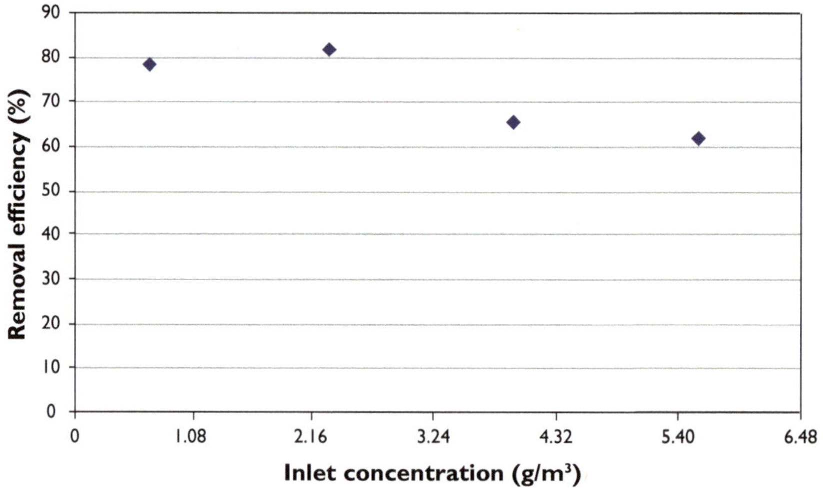

Figure 3 Removal efficiency at different inlet concentrations.

with increase of the inlet concentration isoamyl alcohol. At 2.52 g m^{-3} of inlet concentration 81.32% of the isoamyl alcohol was removed whereas for an inlet concentration of 5.50 g m^{-3}, the removal efficiency dropped 60%. Likewise, a maximum elimination capacity of 77.68 g m^{-3} h^{-1} for EBRT of 42 s was reached. Although, the removal efficiency was lower than reported in the literature for other alcohols[2,4] according with the results obtained the biofiltration could be considered effective for purify waste gases contaminated with IA fed at concentrations less 2.52 g m^{-3}, because the IA exit concentrations are lower compared to TLV of 0.36 g m^{-3} reported by OSHA (Fig. 2).

The influences of shut down and restart-up on the stability of the biofilter and bacteria population were investigated. First, the biofilter was shut down after 60 days of operation and restarted after 4 days (Fig. 4). As can be seen on 6 hours a 90% removal efficiency was reached likely by physicochemical effects in the bed, then on 30 hours this efficiency decreased till 40% and then levelled off 76% (2.9 days) because microbial activity.

Second, when the system reaches its stability again a shut- down of 8 days was applied (Fig. 5). The results showed that an acclimation period of 6 days was necessary to reach the steady state whereas only 2 days were enough for biofilter stability in the shut- down of 4 days.

In three different sections of the biofilters changes on pseudomonas population because shut- down periods were observed. As figure 6 shown, a decreasing on pseudomonas population from day 60 to day 84 (4 days shut down) at the bottom, medium and top sections was observed. This confirms that the biomass content in the biofilter was likely affected by the shutdown,

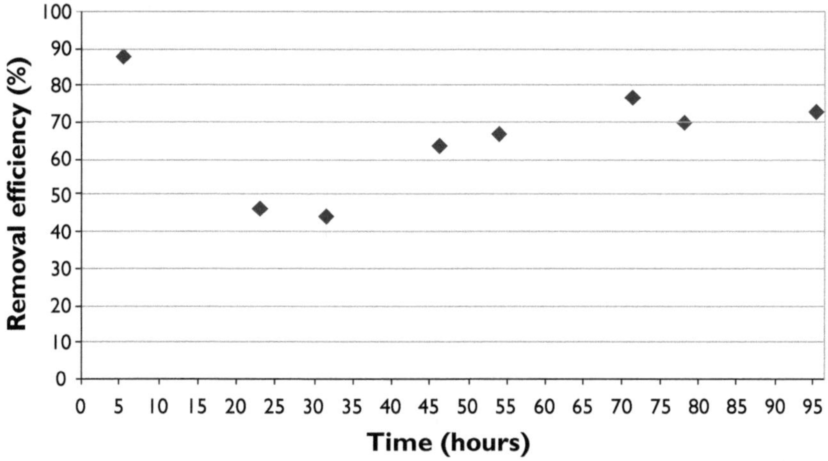

Figure 4 Shut down effect on removal efficiency (4 days).

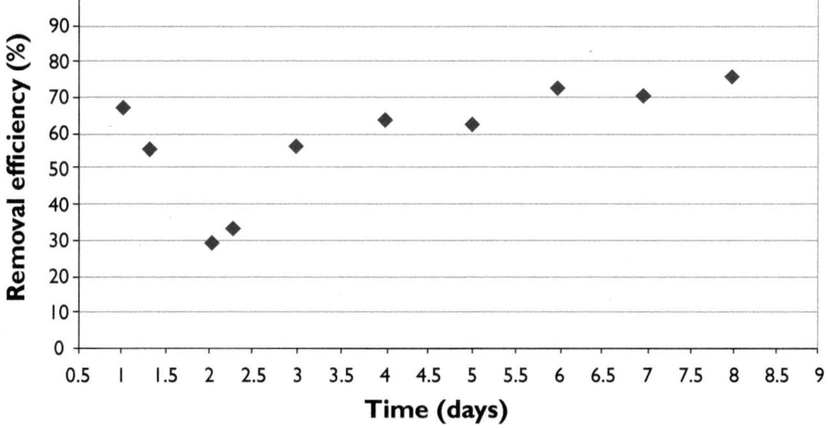

Figure 5 Shut- down effect on removal efficiency (8 days).

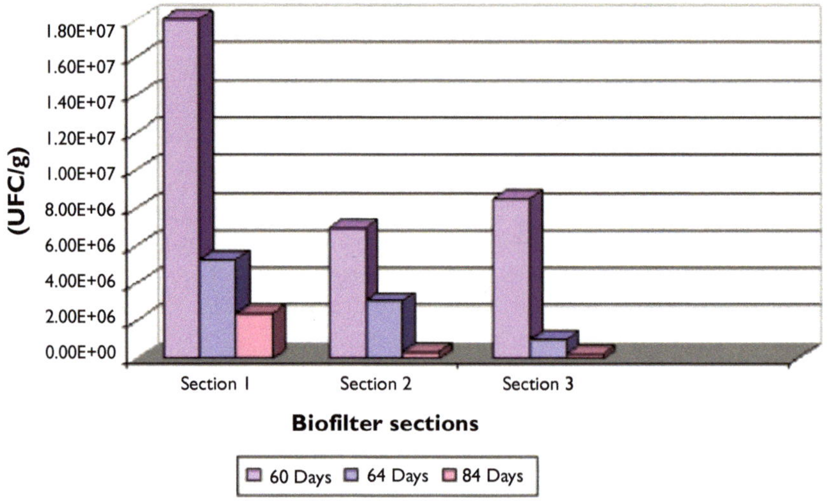

Figure 6 Pseudomonas variation during the experimentation.

although it was not critical for future re-acclimation (Figs. 4 and 5). A similar behavior was reported for toluene degradation in a biotrickling.[5]

CONCLUSIONS

This study provides a first attempt to apply a biofiltration approach for IA removal. Although the biggest efficiency reached WAS only 81.38% the biofiltration could be considered effective for purify waste gases contaminated with IA concentrations Lower than 2 g m^{-3}. Besides, the biofilter showed stability for future re-acclimation.

REFERENCES

1 Ottengraf SPP, Meester JJP, van der Oever AHC and Rozema HR, Biological elimination of volatile xenobiotic compounds in biofilters. *Bioprocess Eng* 1: 61–69 (1986).
2 Dastous P, Nikiema J, Soreanu G, Bibeau L and Heitz M., Elimination of Methanol and Ethanol by Biofiltration: An Experimental Study Water Air Soil Pollut: *Focus* (2007).
3 Arulneyam D and Swaminathan T, Biodegradation of ethanol vapour in a biofilter. *Bioprocess Engineering* 22: 63–67 (2000).
4 Chan W and Chang LY, Effects of Temperature and Inlet Concentration on Acetone Biofiltration in a Composite Bead Biofilter. *J Polym Environ* **DOI 10.1007/s10924-006-0046-z** (2006).
5 Cox HHJ and Deshusses MA, Effect of starvation on the performance and re-acclimation of biotrickling filters for air pollution control. *Environ Sci Technol* 36: 3069–3073 (2002).

Chapter 8

COMPARISON BETWEEN LABORATORY AND PILOT BIOTRICKLING FILTRATION OF AIR EMISSIONS FROM PAINTING AND WOOD FINISHING

Carmen Gabaldón, Vicente Martínez-Soria, Josep-Manuel Penya-roja, Feliu Sempere, Marta Izquierdo & Jordi Palau
Department of Chemical Engineering, Universitat de València, Burjassot, Spain

ABSTRACT

Painting and wood finishing is one industrial sector affected by the EU environmental regulations on volatile organic compound (VOC) air emissions. This study analyses the operation of a pilot biotrickling filter (BTF) installed in a furniture facility, and compares its performance with the removal efficiency obtained in the laboratory using a synthetic feed composed of n-butyl acetate, toluene and m-xylene to simulate industrial air emissions. Laboratory studies showed that fluctuating and oscillating VOC feeding adversely affected the performance of the bioreactor in comparison with stationary feeding. The operation of the pilot plant at 35 s of empty bed residence time (EBRT) remained within the legal limits. However, the installation of an activated carbon prefilter improved considerably the removal efficiency, with hourly average emissions kept below 40 mg C m^{-3} working at 24 s of EBRT. Biotrickling filtration has been shown as effective for controlling air emissions from a closed booth operated for automatically painting furniture pieces. Laboratory study can provide valuable data for the scale-up.

INTRODUCTION

Furniture manufacturing is one of the industries most affected by the environmental legislation on VOC air emissions into the atmosphere, since most of the wood coating paints and varnishes contain VOC solvents. In this sense, a closed booth operated to automatically paint pieces in Gamadecor SA (a furniture facility located in Villarreal, Spain) was selected as representative of the emissions of this industrial sector.

The main objective of this study was to provide data related to the removal of VOCs in air emissions from wood coating and painting applications by BTF. This work involves studies at both laboratory and pilot plant scales. A laboratory study was performed using a synthetic feed that simulated these kinds of industrial air emissions. Two feed regimes were tested: first, continuous and stationary loading, followed by discontinuous and oscillating loading. The pilot plant unit was connected to the industrial focus above mentioned. In this case, the main purpose was to show the robustness of the process and establish the minimum EBRT allowing the meeting of legal regulations. At both scales, an activated carbon prefilter was used to evaluate the influence of load equalisation on the overall removal efficiency.

MATERIALS AND METHODS

FOCUS CHARACTERISATION

The emissions of the closed booth were restricted to the operating hours of the factory (Monday to Friday, 7 a.m.–10 p.m.). The compounds of the polluted air stream were identified by gas chromatography (GC)-Mass Spectrometry. The major compounds detected were n-butyl acetate (20–60%), toluene (20–50%) and xylenes (5–20%). The VOC emission was very uneven, with high level emission episodes (>600–700 mg C m^{-3}) and average concentration of VOCs comprised between 140 and 200 mg C m^{-3}, surpassing Spanish regulatory limits (average VOC concentration: daily basis <75 mg C m^{-3}; hourly basis <110 mg C m^{-3}).[1]

LABORATORY BIOTRICKLING FILTER

The BTF was composed of three cylindrical modules of Plexiglas (120 cm of total bed length, 14.4 cm of internal diameter) and randomly filled with 1" of nominal diameter Flexiring™ polypropylene rings (Koch-Glitsch B.V.B.A., Belgium). Setup of the bioreactor was completed with a 14-L recirculation tank.[2] Contaminated air with a 2:1:1 weight mixture of n-butyl acetate, toluene and m-xylene respectively was introduced through the bottom and the recirculation nutrient solution was fed in a counter-current flow with respect to air flow. The recirculation solution was purged once a week, representing less than 2% of the weekly inlet load fed to the system. The total concentration of VOCs was measured using a Nira Mercury 901 total hydrocarbon analyser with a FID detector. The composition of the gas streams were determined by GC. In a first stage, the bioreactor was operated under continuous and stationary loading. Six operational conditions were tested: inlet load (IL) and EBRT were varied from 11 to 72 g C m^{-3} h^{-1} and 60 to 15 s, respectively (160 mg C m^{-3} to 350 mg C m^{-3} of VOC concentration). In a second stage, three set of discontinuous and oscillating loading conditions (16 hours per day from 6 a.m. to 10 p.m., seven days per week) were applied. Air flow rate was kept constant during night closures. In addition, the influence on the performance of a 0.825-L of activated carbon (AC) prefilter (Norit RB3) was evaluated.

PILOT UNIT

The pilot plant (VOCUS™) was provided by Pure Air Solutions b.v. (The Netherlands). The bioreactor vessel (packed volume of 0.750 m^3) was randomly filled with synthetic rings of 2" of nominal diameter. VOC polluted air from the factory was blown into the bottom of the column and recirculated water was sprinkled on top of the filter media. The trickled water was collected in a recirculation tank. A nutrient dosing system completed the equipment.[3] Inlet and outlet temperature, inlet and outlet VOC concentration, conductivity, pH and pressure drop between the gas inlet and outlet of the media bed were continuously monitored via modem communication. The pilot plant was operated during six months at EBRTs between 10 and 40 s. Air and water flow along with nutrient dose were maintained during nights and weekends, even though the booth was not running. Recycled water was drained and fresh water was added to replace it once a week. In the last 80 days, a 10-L filter of extruded AC (Norit RB3) was installed preceding the bioreactor.

RESULTS AND DISCUSSION

LABORATORY STUDY

The elimination capacity (EC) versus IL is plotted along with the removal efficiency (RE) versus inlet VOC concentration in Fig. 1. Even for the lower applied IL no complete removal was obtained, because of the low removal of toluene and m-xylene. A maximum EC of 32.5 g C m^{-3} h^{-1} with 44.8% RE was achieved working at 72.4 g C m^{-3} h^{-1} of IL and at 15 s of EBRT. RE increased with EBRT: at a VOC inlet concentration of around 170 mg C m^{-3}, RE increased from 53 to 78% in the range of 15 to 60 s of EBRT. At that concentration, the BTF was able to meet the regulatory limits (daily basis) even for the lower EBRT (IL of 39.6 g C m^{-3} h^{-1}). RE was adversely affected when inlet concentration was duplicated, showing that BTF was no competitive enough to meet the legal limits. In this case, RE increased slightly, from 45 to 55% when EBRT increased from 15 to 60 s.

From the VOC concentration profile data (not shown here), a complete removal of n-butyl acetate was observed in the first two-thirds of the column, whereas no degradation of the aromatic compounds was observed in the first third of the column. n-Butyl acetate is a hydrophilic and easily biodegradable compound; hence it could inhibit the growth of aromatic degrading bacteria in the top section where the n-butyl acetate is largely present. Aromatic

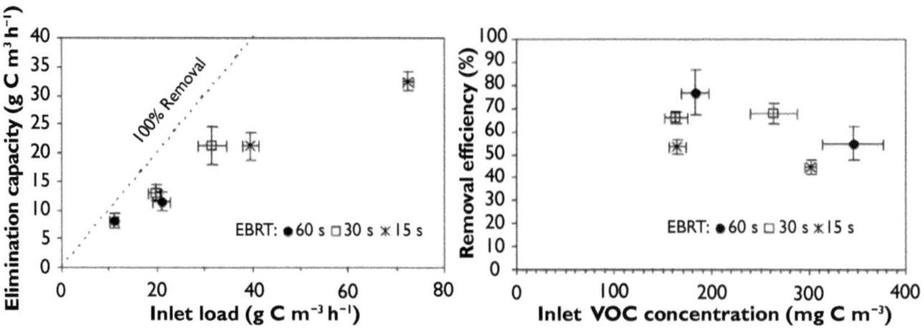

Figure 1 Performance results for the laboratory BTF under continuous loading.

Table 1 Performance from the laboratory BTF under intermittent loading.

	Stage I	*II*	*III*
Empty bed residence time (s)	30	15	15
Average inlet concentration (mg C m^{-3})*	220	105	157
Average elimination capacity (g C m^{-3} h^{-1})	13.8	11.4	15.6
Average outlet concentration (mg C m^{-3})	103	57	92
Total removal efficiency (%)	53.2	45.7	41.4
n-butyl acetate	>99	>99	92.7
toluene	22.9	9.9	8.2
m-xylene	20.3	9.2	6.2

*8 cycles of 2 h. Each cycle: 2 peaks of 30 min of duration with 2 or 3 times greater concentration than baseline.

compounds also present a slower degradation rate; so the BTF was not able to remove them in the part of the bed where n-butyl acetate was absent.

The fluctuation in the VOC concentration along with the night shutdowns adversely affected to the performance of the bioreactor, especially because of the decrease in the removal of the less biodegradable compounds (Tab. 1). As can be seen in Fig. 2, outlet emission fluctuated following same pattern than inlet stream. The deterioration of the performance implies that an average outlet concentration below 75 mg C m^{-3} could only be reached when inlet concentration during working hours was lower than 105 mg C m^{-3} (stage II).

The use of the AC prefilter with a volume 24 times lower than that of the bioreactor allowed the system to meet the legal regulations for all the experiments, even at the most severe conditions ($EBRT_{AC\text{-}prefilter} = 0.6$ s). As can be seen in Fig. 2, the AC prefilter temporarily accumulated during periods of high contaminant loading and then subsequently desorbed contaminants during intervals when concentration in the air was low, buffering the inlet emission into the BTF; and consequently smoothing the outlet emission during working

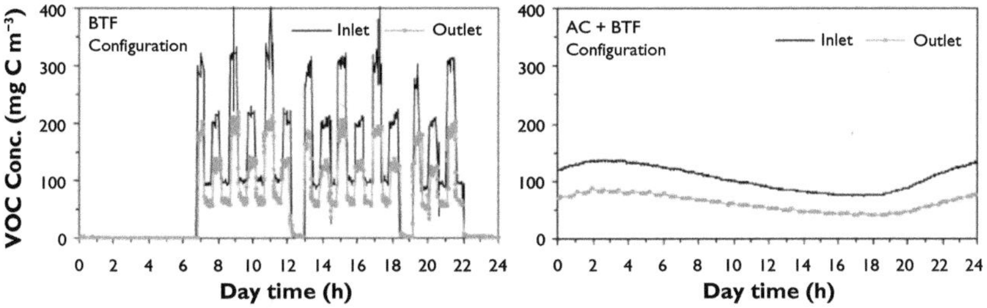

Figure 2 Monitoring data for the laboratory BTF under stage III of intermittent loading.

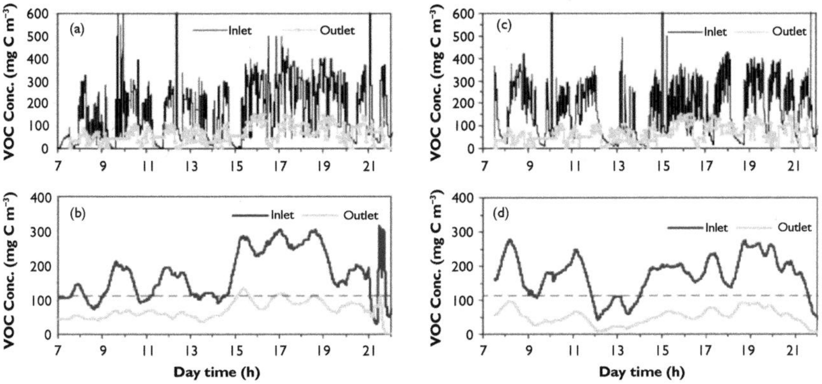

Figure 3 Pilot unit monitoring at two EBRTs: (a) instant and (b) hourly average values of a representative day at EBRT = 24 s; and (c) instant and (d) hourly average values at EBRT = 35 s. The discontinuous line corresponds with the legal limit.

hours. The use of a feeding composed only with toluene and m-xylene applied to the AC+BTF configuration (15 s of EBRT, avg. inlet conc. of 157 mg C m^{-3}) resulted in an average RE of 26%. The three-fold increase in the RE compared with the values obtained for aromatics in the experiments performed with a high concentration of n-butyl acetate indicates that the hydrophilic pollutant interferes with the degradation of these compounds.

PILOT UNIT OPERATION

The operation started at EBRT of 24 s. A transitory period of about 3–4 weeks with a progressive increase in the VOC removal was needed to reach a stable performance. Figures 3(a) and 3(b) show the results obtained in a typical day of the stable period. The performance of the system indicated that the regulatory limit (hourly basis) was not reached when inlet emission episodes with a high duration and concentration occurred. The air flow rate was then changed corresponding to EBRTs between 40 and 10 s. Figures 3(c) and 3(d) show the results obtained in a day operated at EBRT of 35 s As can be seen, in these conditions compliance with the legal limit (hourly basis) was reached for the entire daytime period.

The production of the factory was stable in terms of daily solvent emission, with similar inlet VOC concentrations for different days, with an average inlet concentration of around 180 mg C m^{-3}. The RE increased from 49 to 57% as EBRT increased from 10 to 24 s; only working at 35 s was it possible to meet the legal regulations (daily basis) with a RE of 70%. The RE-EBRT pairs matched with the trend obtained at the laboratory scale under continuous feeding (Fig. 1). Indeed, the performance of the pilot unit was better than that observed at laboratory scale when oscillating and intermittent VOC emissions were applied. The comparison between the laboratory results and those obtained herein indicates that laboratory studies could be a timesaving tool for obtaining valuable data to establish safe limits of the operation conditions allowing the legal requirements to be met.

In order to improve system performance, a passive AC prefilter was installed at the inlet stream, before the biotrickling reactor. The AC prefilter smoothed the VOC concentration spikes arising from sudden variations in the paint operation, and modulated the high and low pollutant loads. The VOC biotrickling emission met the regulation requirements working at EBRT of 24 s ($EBRT_{carbon\text{-}prefilter}$ ~ 0.4 s), with hourly outlet concentrations below 40 mg C m^{-3} for the entire inlet concentration range. In this case, the AC prefilter served as an effective buffer to stabilise the high variability both in concentration and in composition of the industrial emission.

ACKNOWLEDGEMENTS

Financial support by Ministerio de Educación y Ciencia (Spain, research project CTM2007-62511 with FEDER funds) is acknowledged. F. Sempere. M. Izquierdo and J. Palau thank Universitat de València and Generalitat Valenciana, Spain, for the award of a pre-doctoral fellowship.

REFERENCES

1 Real Decreto 117/2003. *Limitations of Emissions of Volatile Organic Compounds Due to the Use of Organic Solvents in Certain Activities and Installations.* Boletín Oficial del Estado, 7 de Febrero de 2003. Orden de 21 de Mayo de 2007, Diari Oficial de la Comunitat Valenciana 29.06.2007.

2 Sempere F, Gabaldón C, Martínez-Soria V, Penya-roja JM and Álvarez-Hornos FJ, Evaluation of a combined activated carbon prefilter and biotrickling filter system treating variable ethanol and ethyl acetate gaseous emissions. *Eng Life Sci,* DOI:10.1002/elsc.2009-00011 (2009).

3 Martínez-Soria VM, Gabaldón C, Penya-roja JM, Palau J, Álvarez-Hornos FJ and Sempere F, Soriano C. Performance of a pilot-scale biotrickling filter in controlling the volatile organic compound emissions in a furniture facility. *J Air & Waste Manage Assoc*, 59:998–1006 (2009).

Chapter 9

REMOVAL OF A MIXTURE OF VOCs USING IMMOBILIZED YEAST STRAIN *CANDIDA TROPICALIS* IN AN AIRLIFT BIOREACTOR

Hyeong-Kyu Namgung & Jeong Hee Kang
Department of Civil & Environmental Engineering, Sejong University, Seoul, Republic of Korea

Sun-Jin Hwang
Department of Environmental Engineering, Kyunghee University, Yongin, Republic of Korea

JiHyeon Song
Department of Civil & Environmental Engineering, Sejong University, Seoul, Republic of Korea

ABSTRACT

In this study, *Candida tropicalis*, a strain of yeast was used for the biodegradation of volatile organic compounds (VOCs). The pre-cultivated yeast strain was immobilized in novel polymer gel media that consisted of polyethylene glycol (PEG), sodium alginate, and powdered activated carbon. An airlift bioreactor containing the polymer gel media was operated, and bioreactor performance was tested to determine the feasibility of applying the yeast culture and the yeast-immobilized polymer media for the biodegradation of a mixture of toluene and methyl ethyl ketone (MEK). As bioreactor operation started, removal efficiencies of greater than 80% for both toluene and MEK were immediately achieved at various loading conditions. Maximum elimination capacities were 70.4 $g/m^3/hr$ and 56.4 $g/m^3/hr$ for toluene and MEK, respectively, when the bioreactor was subjected to the VOC mixture. Consequently, this novel bioreactor can be an alternative to conventional packed-bed biofilters.

INTRODUCTION

Many studies have shown that bioreactors such as packed-bed biofilters can successfully treat a wide range of volatile organic compounds (VOCs).[1,2] In most packed-bed biofilters, different mixed, bacterial-dominant populations may develop on the surface of packing materials depending on the composition of gas streams and physicochemical conditions.[3] However, these packed-bed biofilters are generally sensitive to micro-environmental conditions such as pH, moisture content and salinity.[4] In addition, operational problems such as excess biomass accumulation and biodegradation activity loss make these treatment methods less attractive.[3]

A strain of yeast, *Candida tropicalis*, was selected as an alternative microorganism to overcome these common problems encountered in packed-bed biofilter operation. The yeast strain is known to be relatively easy to control and cultivate, and it may be able to maintain biodegradation activity for VOCs even under unfavorable conditions. Nevertheless, a low growth rate of the yeast strain may result in a long acclimation period at the beginning of bioreactor operation.

Several studies have shown that entrapment of microorganisms within polymeric materials can achieve stable performance and high biodegradation rates.[5,6] The artificial immobilization provides higher concentrations of active microorganisms in bioreactors than packed-bed biofilters do. Biodegradation rates for various VOCs were much higher when using immobilized cells than those using free cells, and immobilized cells in a bioreactor tolerated higher concentrations of VOCs.[6]

Although technical advantages of using yeast cultures and cell immobilization methods have been demonstrated, the potential of a combined yeast-immobilization method for the biodegradation of VOCs has not been fully explored. In this study, the biodegradation of a mixture of VOCs was investigated in an airlift bioreactor containing the pure yeast culture *C. tropicalis* and immobilized cells. The polymer gel media, which were made of a mixture of alginate, polyethylene glycol (PEG), and powdered activated carbon (PAC), were used to entrap the yeast cells, and the media were employed as fluidized materials. The main objective of this study was to develop a novel treatment method by applying the yeast culture and the yeast-immobilized polymer media in an airlift bioreactor for the enhanced biodegradation of a mixture of VOCs.

MATERIALS AND METHODS

MICROBIAL CULTURE AND IMMOBILIZATION

The yeast strain, *C. tropicalis*, was obtained from the Korean Culture Center of Microorganisms (KCCM 50075). The strain was incubated and maintained on YM agar plates before transferring to a liquid nutrient medium. After several cycles of regeneration, the strain was transferred into a 120-mL serum bottle containing 50 mL of a nutrient solution. The nutrient solution consisted of a hydrocarbon minimal medium containing 1.36 g/L KH_2PO_4, 1.42 g/L Na_2HPO_4, 0.5 g/L $(NH_4)_2SO_4$, 3.03 g/L KNO_3, and a trace metal solution containing 0.25 mg/L $FeSO_4 \cdot 7H_2O$, 0.18 mg/L $MnCl_2 \cdot 4H_2O$, 0.02 mg/L $CuCl_2 \cdot 2H_2O$, 0.04 mg/L $ZnSO_4 \cdot 7H_2O$, 0.04 mg/L $CoCl_2 \cdot 6H_2O$, 0.02 mg/L $NiCl_2 \cdot 6H_2O$, 0.02 mg/L $Na_2MoO_4 \cdot 2H_2O$, and 0.02 mg/L H_3BO_4. Ten micro-liter of toluene was injected into the headspace of the serum bottle, and the culture was incubated in a shaker at 20°C. After several cycles of transfer and incubation, the enriched culture was used to inoculate the bioreactor as well as to immobilize in the polymer gel media.

The yeast-immobilized gel media were manufactured using a mixture of PEG, alginate and PAC, which referred to as "PACY" (PEG-Alginate-Carbon-Yeast) media. 10 g of sodium alginate and 180 g of polyethylene glycol were dissolved in 0.5 L of distilled water. The same volume of the yeast culture (OD_{600} of 2.0) was added in the polymer mixture and stirred thoroughly for 30 seconds. Then 300 μL of N,N,N,N-tetramethylenediamine (TMEDA) was added, and pH was adjusted to 6 using acetic acid. The final mixture was immediately poured into a mold and dried at room temperature for 30 minutes.

AIRLIFT BIOREACTOR CONFIGURATION AND OPERATION

The airlift loop bioreactor was operated with the yeast culture and the PACY media. The bioreactor was constructed as an internal loop column (width 7 × length 14 cm × height 30 cm) with a working volume of 2.0 L as illustrated in Fig. 1. An internal baffle was placed in the middle of the column with an opening of 3 cm at both the bottom and the top of the column to allow the media to be circulated along with the air flow.

An air stream generated by a compressor and controlled by a flow meter was contaminated with the mixture of VOCs by the slow injection of research-grade, pure chemical mixture using a syringe pump (KD Scientific, USA). The contaminated air stream was continuously introduced through a diffuser placed at one side of the bottom of the column. The air bubbles rose

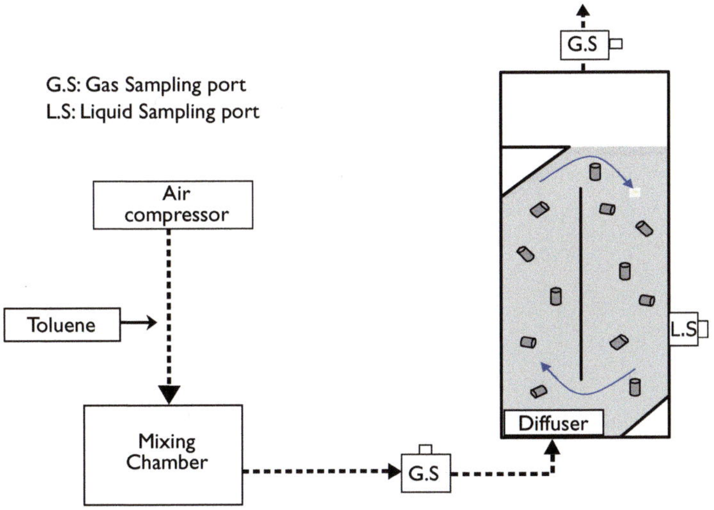

Figure 1 Schematic diagram of the airlift bioreactor used in this study.

through the column filled with the yeast culture and 50 g of the PACY media, and they completely mixed and fluidized in the airlift loop bioreactor. The gas flowrate was sequentially adjusted three times throughout the entire experimental period with empty bed gas residence times (EBRT) of 60, 30, and 15 seconds.

RESULTS AND DISCUSSION

AIRLIFT COLUMN PERFORMANCE

The airlift bioreactor containing the PACY media was operated for 12 days under four different VOC loading conditions sequentially. As indicated in Fig. 2, the operational period of the airlift bioreactor was divided into the following four phases based on the VOC inlet concentration and EBRT: 1) Run Acclimation (days 0–4) where the bioreactor was initially subjected to a toluene inlet concentration of 22.0 ppm and a MEK concentration of 23.3 ppm at an EBRT of 60 seconds, corresponding to toluene and MEK loading rates of 5.0 and 4.5 g/m^3/hr, respectively, 2) Run 1 (days 5–7) where the toluene and MEK concentration were doubled at the same EBRT as Run Acclimation, corresponding to toluene and MEK loading rates of 9.6 and 8.9 g/m^3/hr, respectively, 3) Run 2 (days 8–9) where the EBRT was shortened to 30 seconds by increasing the gas flowrate, corresponding to toluene and MEK loading rates of 19.1 and 17.8 g/m^3/hr, respectively, and

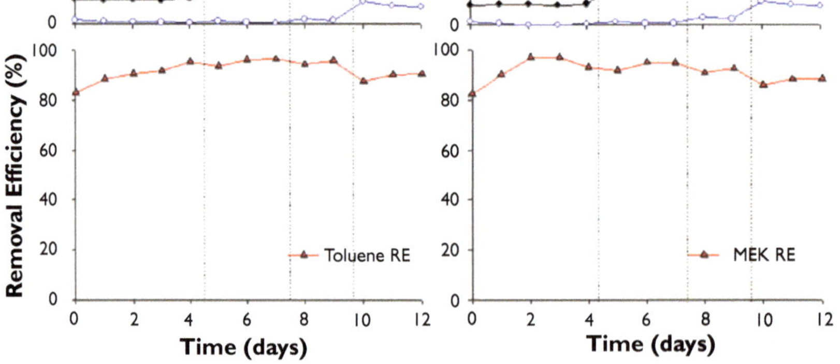

Figure 2 Removals of toluene and MEK in the airlift bioreactor.

4) Run 3 (days 10–12) where the EBRT was again shortened to 15 seconds, corresponding to toluene and MEK loading rates of 37.2 and 35.1 g/m³/hr, respectively.

Figure 2 illustrates inlet and outlet VOC concentrations and corresponding removal efficiencies for toluene and MEK in the airlift bioreactor containing the yeast culture and the PACY media. As bioreactor operation started, removal efficiencies of greater than 80% for both toluene and MEK were immediately achieved without a significant delay. And the removal efficiencies for both toluene and MEK increased to 95% during Run Acclimation.

The airlift bioreactor achieved stable performance with both toluene and MEK removal efficiencies of greater than 90% even at the shorter gas residence time during Runs 2 and 3, and pseudo-steady-states were observed almost immediately after the loading increases. These results indicate that the yeast cells inoculated in the liquid phase and immobilized in the PACY media were active for the VOC biodegradation. In addition, the stable performance at the short gas residence time implies that the airlift configuration resulted in an enhanced mass transfer of VOCs from the gas to liquid phases and an effective mixing of the PACY media and the recirculating liquid.

ELIMINATION CAPACITY

As another indicator of bioreactor performance, elimination capacity (EC) values were measured using the experimental data obtained at various inlet

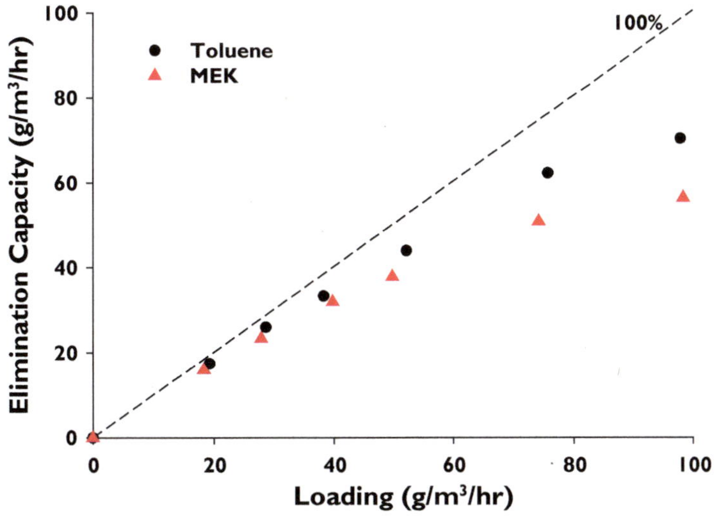

Figure 3 Elimination capacity curves for toluene and MEK in the airlift bioreactor.

loading rates. EC is defined as the amount of VOC removed biologically and physico-chemically per unit of effective reactor volume and time. The EC tests were conducted on day 13 after the end of Run 3 in the PACY airlift bioreactor. The EC curves for toluene and MEK showed maximum values of 70.4 and 56.4 $g/m^3/hr$, respectively, at the highest loading applied. The maximum EC for toluene, obtained in the PACY airlift bioreactor, was slightly higher than those reported for bacterial biofilters (typically 8–60 $g/m^3/hr$ for toluene).[7] In contrast, the maximum EC value for MEK obtained in this study was lower than that of toluene, but it was similar to values reported in the literature.[7] It is worth pointing out that, in the airlift bioreactor, the increase in EC was almost linear with increasing the VOC loading up to 100 $g/m^3/hr$.

The high EC value observed in the PACY airlift bioreactor was due to adsorption capacity and biodegradation activity of the yeast media. The novel polymer media can dampen the fluctuation of inlet loading as well as interact with active yeast cells in the liquid phase for the biodegradation of VOCs applied. Consequently, the PACY airlift bioreactor using the yeast culture, *C. tropicalis*, can be a successful method for the removal of the VOC mixtures.

REFERENCES

1 Devinny JS, Deshusses MA and Webster TS, *Biofiltration for air pollution control*. Lewis Publishers (1999).
2 Kennes C and Veiga MC, *Bioreactors for waste gas treatment*. Kluwer Academic Publishers (2001).
3 Song J and Kinney KA, Effect of vapor-phase bioreactor operation on biomass accumulation, distribution, and activity. *Biotechnology and Bioengineering* 68:508–516 (2000).
4 Kinney KA, Loehr RC and Corsi RL, Vapor-phase bioreactors: avoiding problems through better design and operation. *Environmental Progress* 18:222–230 (1999).
5 Chung YC, Huang C, Tseng C and Pan JR, Biotreatment of H_2S- and NH_3-containing waste gases by co-immobilized cells biofilter. *Chemosphere* 41: 329–336 (2000).
6 Shim H, Shin E and Yang ST, A continuous fibrous-bed bioreactor for BTEX biodegradation by a co-culture of Pseudomonas putida and Pseudomonas fluorescens. *Advanced Environmental Research* 7:203–216 (2002).
7 Deshusses MA and Johnson CT, Development and validation of a simple protocol to rapidly determine the performance of biofilters for VOC treatment. *Environmental Science and Technology* 34:461–467 (2000).

Chapter 10

EFFECTS OF ADDING OZONE ON A BIOFILTER TREATING GASEOUS CHLOROBENZENE:
Performance change and mechanism

Can Wang, Jin-Ying Xi, Hong-Ying Hu & Yuan Yao
Environmental Simulation and Pollution Control State Key Joint Laboratory, Department of Environmental Science and Engineering, Tsinghua University, Beijing, PR China

ABSTRACT

The recalcitrant volatile organic compounds with low biodegradability pose challenges for biofiltration technology. To investigate the effect of adding ozone on biofiltration process, a biofilter was operated continuously for more than 160 days under different inlet ozone concentrations. The experimental results showed that the chlorobenzene removal capacity of the biofilter with 10–50 mg $\cdot$ m^{-3} ozone was obviously higher than those with no ozone. Furthermore, a higher chlorobenzene removal capacity was achieved when the ozone concentration was increased. The mechanism of ozone on biofilter performance was also investigated. The average biofilm thickness and extra-cellular polymers content (EPS) were both reduced when the inlet ozone concentration was increased. The specific surface areas of the filter bed increased from 784 $m^2 \cdot m^{-3}$ (no ozone) to 820 $m^2 \cdot m^{-3}$ (10–30 mg m^{-3} ozone) and 880 $m^2 \cdot m^{-3}$ (30–50 mg $\cdot$ m^{-3} ozone). The microbial activity estimated by Biolog plate analysis was found to increase with higher ozone concentration. All the experimental results proved that ozone can promote the chlorobenzene removal capacity of the biofilter by optimizing the physical structure and microbial property of the filter bed.

INTRODUCTION

Volatile organic compounds (VOCs) are common air pollutants found in industrial air emissions, such as from chemical manufacturing plants and various hazardous sites. Most of those compounds are harmful to human health and cause severe environmental problems (Delhomenie et al., 2005). Among the various technologies for VOCs treatment, biofiltration is widely considered a cost-effective technology because of its low operating costs and convenient maintenance (Kennes and Veiga, 2004). However, some recalcitrant VOCs with low biodegradability pose challenges for biofilters. (Delhoménie and Heitz, 2003).

Some efforts have been proposed to address these challenges, which include adding a pretreatment before biofiltration. Congna et al. (2005) used granular activated carbon to control VOCs concentrations. This strategy was mainly proposed to reduce the inlet concentration of inhibitory compounds. The performance of the biofilter, actually, was not essentially improved. The interesting results came from Moussavi et al. (2007) and Wang et al. (2008). The ozone-producing ultraviolet lamps were used as pretreatments in prior to biofilters, which transferred some VOCs to more biodegradable products and thus improved the performance of the subsequent biofilters. However, a further study showed that the produced ozone of 10–60 mg m^{-3} could reduce bioaerosol emissions from the subsequent biofilter (Wang et al., 2009). These results indicated that the ozone could also significantly affect the microorganisms inside biofilters which attribute to the performance enhancement of biofilters.

Actually, ozone is a chemical substance which can play a considerable role on microorganisms due to its strong oxidation capacity. A few previous researches (Zhang et al., 2009; Dytczak et al., 2008) utilized ozone to disintegrate the excess activated sludge in waste water treatment process. Ozone destroys the zoogloea structures and converts the solid organic components of sludge to soluble substances, which can be further biologically degraded when the ozonated sludge is returned to the wastewater. Meanwhile, ozone is often regarded a measure to change the sludge properties. Dytczak et al. (2008) investigated the extra-cellular polymers (EPS) content and floccules shape of sludge after ozone treatment. The results showed that the sludge floccules structure became stronger, denser, and more ozone-resistant. However, these papers are mainly relevant to wastewater treatment process. The effects of ozone on the microorganisms in gas-solid phase bioreactors have rarely been reported. Moreover, ozone was a common disinfectant to deactivate the microorganism in wastewater (Tachikawa et al., 2009). Such deactivated effect is selectivity since some species of microorganisms have been reported to be resistant to ozone (Dytczak et al., 2008). Therefore, it is quite interesting to explore the changes of microbial characteristics in biofilters under ozone injection. Ozone can also be regarded a potential measure to regulate the microbial community in a mixed flora.

In this study, the biofilter performance without and with ozone injection was compared and the affecting mechanisms of ozone on the biofilter was further explored. The results of this study may be helpful to provide a new measure to improve the performance of biofilters and understand the affecting mechanism of ozone on biofilters.

MATERIALS AND METHODS

EXPERIMENTAL SET-UP AND OPERATION

The experimental set-up is shown in Figure 1. The synthetic waste gas stream was generated by evaporating liquid chlorobenzene into gas stream using a syringe pump (Shanghai Alcott Biotech Co, China). Ozone was produced from a generator and introduced to the biofilter mixed with the gas stream containing chlorobenzene.

The inlet concentrations of ozone were changed by varying the ratio of flow rates of the gas streams containing ozone and chlorobenzene. The detailed operating conditions were summarized in Table 1.

ANALYTICAL METHODS

Biofilm thickness measurement

The biofilm thickness of packing media samples from the biofilter were measured using a three-dimensional manual micromanipulator (Beijing Daheng

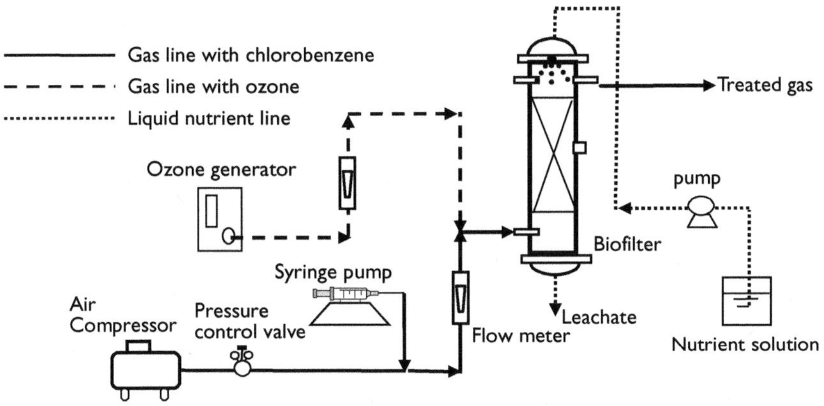

Figure 1 Experimental set-up for the biofiltration process with and without ozone.

Table 1 Summary of experimental conditions.

	Experimental conditions		
Parameter	*Phase 1*	*Phase 2*	*Phase 3*
Period of operation (Day)	0–50	51–120	121–168
Inlet ozone concentrations (mg m^{-3})	0	20±10	40±10
Inlet chlorobenzene concentration (mg m^{-3})	400–1000	400–1000	400–1000
EBRT of the biofilter (s)	122	122	122

Laser Equipment Co., Ltd., China). Five packing media were sampled at a height along the biofilter. Four individual measurements were done for each packing media. The average biofilm thickness from these 20 measurements at each height is calculated for a confidence interval of 95%.

EPS content analysis

EPS is the main components of microbial aggregates. The EPS of biofilm was extracted and analyzed by sonication and heating methods as described by Comte et al. (2006). Finally, the EPS content was expressed as the mass of total organic carbon (TOC) per gram of wet biofilm (mg-TOC g-1-biofilm). All samples on different days were taken from the bottom section of the filter bed (5–10 cm distance from the inlet). The average EPS content from duplicate samples is calculated.

Scanning Electron Microscopy (SEM)

Microbial samples were withdrawn from the biofilter, and then fixed and dehydrated as described by Chung et al. (2004). Micrographs were obtained from a Sirion200 scanning electron microscope (OXFORD Co., Ltd., British).

Analysis of the specific surface area of the filter bed

A pulse injection technique was established to estimate the specific surface area of filter beds. The detailed information about the pulse injection technique can be found in References (Xi et al., 2005). 1,2-dichloroethane (DCE) was used as the trace in this study. The time course of the tracer concentration in the outlet of the biofilter was define as the molecular retention time distribution (MRTD) curve. The average molecular retention time (v_1), the width of the MRTD curve (μ_2) and the specific surface area (a) were calculated based on the MRTD curves.

Biolog plates analysis

The Biolog ECO-microplates were used in this study to analyze microbial metabolic activities. The detailed sampling and measuring process was same as described in References (Garland and Mills, 1991). The microbial suspension extracted from the packing media was used to inoculate Biolog plates. The plates were read during the inoculation with the sample community. The average well color development (AWCD) was calculated according to eq. (1) (Garland and Mills, 1991):

$$\text{AWCD} = \frac{1}{3}\sum_{i=1}^{31}(R_{it} - R_{0t}) \qquad (1)$$

where R_{it} and R_{0t} are the absorbencies of the sole carbon source i and the water blank at time t.

RESULTS AND DISCUSSION

EFFECTS OF OZONE ON BIOFILTER PERFORMANCE

The variation of the chlorobenzene removal efficiency was shown in Figure 2. During the first 50 days (phase 1), the biofilter was operated without ozone injection. The chlorobenzene removal efficiencies fluctuated around 40%. From day 51, the ozone with 10–30 mg m^{-3} was injected into the biofilter and was completely eliminated (data not shown). The removal efficiencies increased to around 70% (phase 2). From day 121, further increase of ozone concentration to 30–50 mg m^{-3} made the removal efficiencies reach to 90% (phase 3). The outlet ozone was detected below 5 mg m^{-3}.

These results indicated that ozone remarkably improved the removal performance of chlorobenzene by the biofilter. However, little is known to illustrate why the performance of the biofilter was enhanced. Therefore, it is interesting to investigate the detailed mechanisms in this case.

EFFECTS OF OZONE ON BIOFILM PROPERTIES

In our previous experiment, ozone can hardly directly remove chlorobenzene in a sterilized biofilter with no active microorganisms (Data not shown). Thus, gaseous chlorobenzene is mainly removed by biofilm inside the biofilter. Therefore, the effects of ozone on the biofilm properties were firstly investigated.

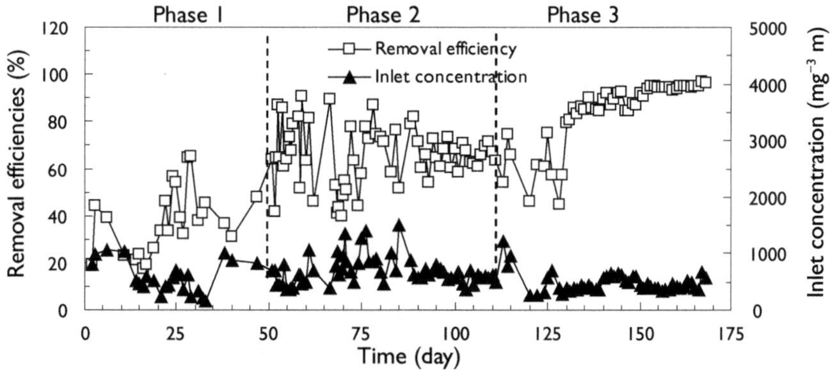

Figure 2 Chlorobenzene removal efficiencies of the biofilter during phase 1 (day 0–50: without ozone injection), phase 2 (day 51–120: with ozone injection 10–30 mg m^{-3}), phase 3 (day 121–168: with ozone injection 30–60 mg m^{-3}): ▲ inlet chlorobenzene concentration; □ removal efficiency.

BIOFILM THICKNESS

Biofilm thickness is an important parameter to describe the biofilm properties. Figure 3 shows the biofilm thickness results along the height of the biofilter during various operating phases. The biofilm thickness in phase 1 was significantly higher than those in phases 2 and 3. That was caused by ozone injection during these phases. Ozone is often regarded as a disinfectant and strong oxidant to inhibit microorganisms in both water and gas phases (Tachikawa et al., 2009). In this way, the biofilm was probably oxidized and the thickness was reduced by the ozone.

EPS CONTENT

The EPS contents of biofilm during operation were analyzed in Figure 4. During phase 1, the EPS content increased from 3.5 to 4.9 mg-TOC g-1-biofilm,

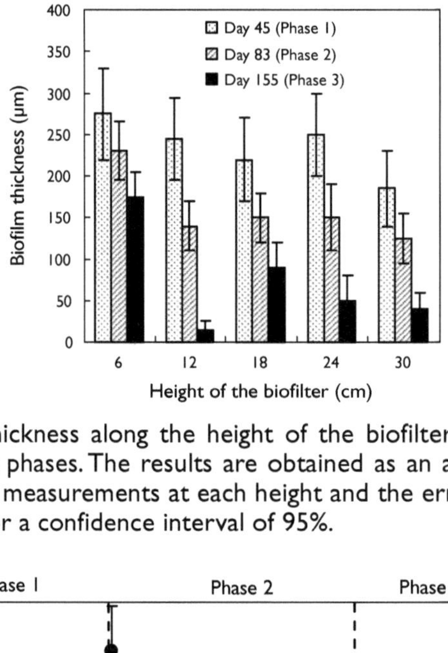

Figure 3 Biofilm thickness along the height of the biofilter during various operating phases. The results are obtained as an average from 20 individual measurements at each height and the error bars are calculated for a confidence interval of 95%.

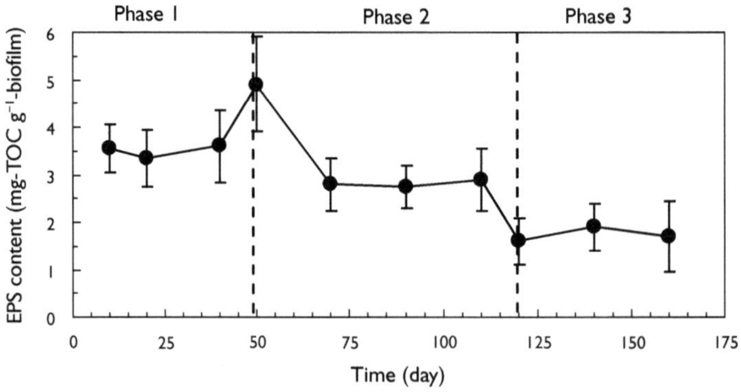

Figure 4 EPS content of biofilm during various operating phases. The results are obtained as an average from duplicate samples. Error bars represent standard deviation based on duplicate samples.

probably due to utilization of chlorobenzene as substrate by microorganisms. Starting on day 51, ozone was injected into the biofilter. The EPS content was reduced to around 3 mg-TOC g-1-biofilm. In phase 3, further increase of ozone concentration resulted in a continuous decrease of EPS content to 1.5–2 mg-TOC g-1-biofilm.

The results suggested that the action of ozone on biofilm was oxidizing the surface part EPS. Dytczak et al. (2006) reported that the EPS content can form a dense gel that resists the transfer of oxygen and nutrients in the biofilm. Therefore, the reduction of both biofilm thickness and EPS content by ozone can promote the pollutant transfer in the biofilm, which might attribute to the performance improvement of the biofilter.

CHANGES OF THE SPECIFIC SURFACE AREA OF THE FILTER BED

The average molecular retention time (v_1), the width of the MRTD curve (μ_2) and the specific surface area (a) were calculated based on the MRTD curves and the result was shown in Table 2. The increases in specific surface area were estimated to be caused by the increase in void fraction due to the reduction of biofilm thickness by ozone oxidation. Moreover, a higher specific surface area improved the mass transfer and biodegradation rate in the filter bed. This should also be one of reasons to explain why the performance of the biofilter was improved by ozone.

CHANGES OF MICROBIAL ACTIVITY IN THE BIOFILTER

To quantify the microbial metabolic activities in the biofilter, the developments of AWCD under different operating conditions were investigated (Figure 5). The slopes of AWCD curves from 20 to 40 hours during the cultivation period represented the average metabolic activities of microbial samples. The slopes of AWCD curves were calculated as 0.66, 0.88 and 0.95 $cm^{-1}\ h^{-1}$ for the samples on days 45, 83 and 155, respectively. These results suggested that ozone stimulate the microbial metabolic activities in the biofilm.

Table 2 Theoretic parameters calculated from the MRTD curves.

Parameters	*Day 45*	*Day 83*	*Day 155*
v_1 (min)	6.8	9.0	9.4
μ_2 (min^2)	53.4	40.7	41.8
a ($m^2\ m^{-3}$)	784	820	880

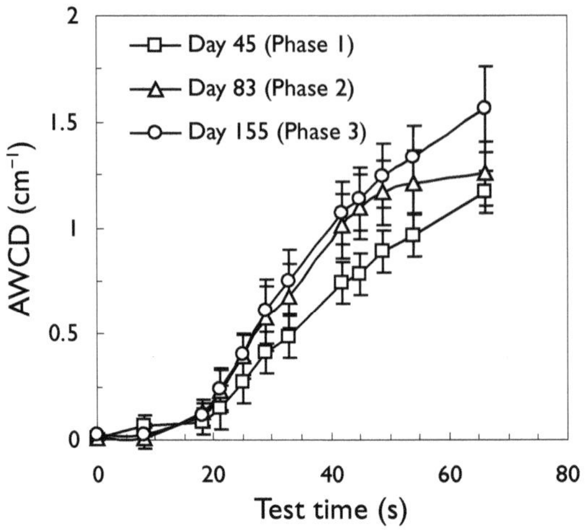

Figure 5 Development of AWCD during incubation period of microbial samples under various operating phases. The results are obtained as an average from duplicate samples. Error bars represent standard deviation based on duplicate samples.

CONCLUSIONS

1 Ozone ranging from 10 to 50 mg m^{-3} had a remarkable promotive effect on the performance of a biofilter treating gaseous chlorobenzene.
2 Ozone can oxidize the biofilm in the biofilter, which lead to biofilm thickness reduction, EPS content decrease and specific surface area increase. All these change could promote the oxygen, nutrients and pollutant transfer in the filter bed and therefore could enhance the chlorobenzene removal capacity.
3 The microbial community with ozone showed higher metabolic activities, which could also promote chlorobenzene removal capacity of the biofilter.

REFERENCES

1 Chung, Y.C., Lin, Y.Y. and Tseng, C.P. Operational characteristics of effective removal of H2S and NH3 waste gases by activated carbon biofilter. *J. Air Waste. Manage. Assoc.* 2004, 54(4), 450–458.
2 Comte, S., Guibaud, G. and Baudu, M. Relations between extraction protocols for activated sludge extracellular polymeric substances (EPS) and EPS complexation properties Part I. Comparison of the efficiency of eight EPS extraction methods. *Enzyme. Microb. Technol.* 2006, 38(1–2), 237–245.
3 Congana, L. and William, M.M. Activated carbon load equalization of discontinuously generated acetone and toluene mixtures treated by biofiltration. *Enviorn. Sci. Technol.* 2005, 39(7), 2349–2356.
4 Delhomenie, M.C. and Heitz, M. Biofiltration of air: a review. *Crit. Rev. Biotechnol.* 2005, 25(1–2), 53–72.

5 Delhoménie, M.C. and Heitz, M. Elimination of chlorobenzene vapors from air in a compost-based biofilter. *J. Chem. Technol. Biotechnol.* 2003, 78(5), 588–595.

6 Dytczak, M.A., Londry, K., Siegrist, H. and Oleszkiewicz, J.A. Extracellular polymers in partly ozonated return activated sludge: impact on flocculation and dewaterability. *Water Sci. Technol.* 2006, 54(9), 155–164.

7 Dytczak, M.A. and Oleszkiewicz, J.A. Performance change during long-term ozonation aimed at augmenting denitrification and decreasing waste activated sludge. *Chemosphere.* 2008, 73(9) 1529–1532.

8 Garland, J.L. and Mills, A.L. Classification and characterization of heterotrophic microbial communities on the basis of patterns of community-level sole-carbon-source utilization. *Appl. Environ. Microb.* 1991, 57(4), 2351–2359.

9 Kennes, C. and Veiga, M.C. Fungal biocatalysts in the biofiltration of VOC-polluted air. *J. Biotechnol.* 2004, 113(1–3), 305–319.

10 Moussavi, G. and Mohseni, M. Using UV pretreatment to enhance biofiltration of mixtures of aromatic VOCs. *J. Hazard. Mater.* 2007, 144(1–2), 59–66.

11 Tachikawa, M., Yamanaka, K. and Nakamuro, K. Studies on the disinfection and removal of biofilms by ozone water using an artificial microbial biofilm system. *Ozone-SCI. ENG.* 2009, 31(1), 3–9.

12 Wang, C., Xi, J.Y. and Hu, H.Y. A novel integrated UV-biofilter system to treat high concentration of gaseous chlorobenzene. *Chinese Sci. Bull.* 2008, 53(17), 2712–2716.

13 Wang, C., Xi, J.Y. and Hu, H.Y. Reduction of toxic products and bioaerosol emission of a combined UV-biofilter process for chlorobenzene treatment. *J. Air Waste. Manage. Assoc.* (In press).

14 Xi, J.Y., Hu, H.Y. and Qian, Y. Effects of adding inert spheres into the filter bed on the performance of biofilters for gaseous toluene removal. *Biochem. Eng. J.* 2005, 23(2), 123–130.

15 Zhang, G.M., Yang, J. and Zhang, J. Sludge ozonation: disintegration, supernatant changes and mechanisms. *Bioresource. Technol.* 2009, 100(3), 1505–1509.

Biofiltration – keynotes

Chapter 11

BIOFILTRATION – FROM PAST TO FUTURE – A SCIENTIFIC VIEW

Franz-Bernd Frechen
Department of Sanitary and Environmental Engineering (DESEE), Kassel University, Kassel, Germany

Wolfram Franke
Yara Research Centre, Herøya Industripark, Porsgunn, Norway

ABSTRACT

The use of biofilters for air treatment is a common method since decades. Science has the duty to examine existing technologies and find new concepts. This has to be done with regard to the roots and basics of biofiltration and to actual and future demands. Hence, research and development firstly have to conduct performance analysis under laboratory and large-scale conditions with special regard to the use of surrogate parameters suitable or at least promising concerning process control as well to real load parameters. In addition, science has to explain the phenomena observerd, resulting in a better process understanding and, if possible, modelling of processes. Derived from that, predictions for possible future applications are needed as a guideline for the commercial biofilter market development. This paper shows different approaches to fulfil these demands. Results from large-scale applications as well as laboratory scale applications will be shown.

INTRODUCTION

Biofiltration is a well-known method of treating polluted air since decades. The first biofilter in Germany went intooperation some 40 years ago, inspired by a so called soil filter operated in Switzerland. Basic rules for the method are fixed in US Patent 4086167 (1978) [1]. During the years technology improved and the actual state of the art is described for example by Burges et al. (2001) [2]. Biofiltration is commonly used if waste gases are polluted with low concentrations. R&D has the task to adopt the method for higher polluted gases. Therefore new fields of application have to be described, including demands for loads and treatment aims.

One possibility is to replace ordinary biofilter material with "advanced media" in large-scale application. Using comparative investigation methods, the benefits can be verified. After successful tests modelling and deriving further optimisation steps have to be done.

Another possibility is the use of laboratory scale systems. Doing so, the surrounding conditions can be controlled and the number of influencing parameters can be decreased. Therefore it is often necessary to use substitutes for the loads, as laboratory and real site are apart.

At DESEE, several approaches have been undertaken so far to improve biofiltration technology during the past five years. Large-scale tests have been performed at WWTP Frankfurt Niederrad. The results were presented in Frechen et al. (2007) [3], and here just a short summary will be given.

In addition, measurements were carried out at real-world, but small biofilters, which are used in manhole of sewerage systems in order to investigate the performance of these devices. First results were presented in Frechen and Franke (2009) [4].

Last but not least lab scale tests using substitute and real load parameters have been performed. First results were presented in Franke (2009) [5], and in this paper some more information will be given. Full results will be presented soon within a Ph.D. thesis at DESEE.

Scientific work includes the use of sophisticated measurement devices. In our work, we included advanced measurement methods like multi sensor measurement ("E-Nose") as well as sensory measurements via olfactometry.

LARGE-SCALE TESTS

LARGE FILTRATION SYSTEMS

This type of investigation gives the possibility to improve technology in that place where it will be used in future time. Hence all surrounding parameters – known and unknown – are set up as needed and so a big surprise will not take place later on. So the major task is to find the right configuration of testing methods answering the following questions:

- Performing tests sequentially or in parallel?
- Performing random or on line measurement?
- What happens if the test fails?
- Financial situation secured?

The results given in Frechen et al. [3] were derived from a R&D project. The investigated site consisted of a biofilter with two filter segments operated in parallel, which made a comparative study possible. A direct comparison between two filter materials loaded in nearly the same way under the same surrounding conditions was done. Baseline as well as variation due to modification of material was measured. Online hydrogen sulphide (H_2S) measurement and random sampling for measuring odour concentration according to EN 13725:2003 [6] were performed. The data revealed that the advanced material performed better than the standard. As the period of time was nearly 1 year, the estimation for further use on long term was possible.

SMALL FILTRATION SYSTEMS

Test with manhole filters were performed. Manhole filters are used to filter air from sewer networks. Most interesting question on this application is: Do manhole filters just block the air stream or do they really treat air?

A special manhole cover hood was constructed. Using different air flows and measuring the resulting pressures gave the dynamic pressure-flow-relationship for manhole covers with and without filters installed.

In addition, olfactometric measurements were undertaken with isolated filters to find out how these devices, if installed properly and tight, are able to biologically treat the foul sewer air.

From the tests it could be derived that manhole filters are able to treat air depending on the surrounding condition. However, their effect often is just the closure of the manholes.

LABORATORY SCALE TEST

Laboratory tests have the benefit to allow a good control of most surrounding parameters. As laboratory and real application are usually not in the same place, mostly artificial polluted air (air with synthetic load) has to be used.

Artificial pollutant substitutes have to be found for load measurement as well as for efficiency measurement. As biofiltration is commonly used for odour degradation, a substitute for odour concentration is needed. A comparison of several odour related parameters is given in Franke et al. (2008) [7].

In addition, in lab scale tests the microbial situation can be investigated easier than in large-scale systems. For such tests, very selective setups using inert carrier materials and precisely defined nutrient solutions as well as defined loads can be used. With such a laboratory setup Kraakman et al. (1997) [8] were able to investigate fungi-filters for degradating Styrene. Mitzkat (2001) [9] separated different micro organism populations. The result was that a biofilter works with externally fed organisms and –necessary for long term use-an additional internal metabolism system living on the external fed organisms. BARGEL (2004) [10] investigated a monoculture. The result here was that this is not helpful; and this results underlined Mitzkat's results.

Using a laboratory scale setup with real waste gas loads was implemented at DESEE. A 4-line biofilter system was built in our experimental hall at WWTP Kassel. This biofilter system was predominantly with waste gas from stripping odour from the raw wastewater. In addition, Ammonia (NH_3) as well as organic gases (measured via total organic carbon (TOC) and volatile organic compounds (VOC)) were added for extra investigations. The setup is shown in Fig. 1.

The four lines were:

- Reference,
- Pretreatment with ultra violet light (UV) on 254 nm (ozone free),
- Pretreatment with an organic solubility enhancer (cyclodextrine) and
- Organic impregnated material

From the investigation several results were derived:

- The ozone free UV pre treatment delivered just a slight benefit. For TOC and NH_3 a slight improvement was recognized. Odour quality got worse, odour concentration increased. This pre treatment is not recommendable for sewer air. However this method is cheap and easy to use, so for specific cases it might be an option.

Figure 1 Setup with a 4-line biofilter in lab scale; loads are produced in a wastewater fed stripping tank.

- The pre treatment with cyclo dextrine had a slight benefit for NH_3 and TOC degradation. As odour quality and odour concentration was not affected, no general benefit was observed. From an economical point of view, this technology is too expensive for continuous operation, but acceptable to decrease peak loads.
- The treatment in an organic material with impregnation of activated carbon and ferrous salt had slight effect on the tested low loads; maybe higher loads would have led to a benefit. Here the economical benefit strongly depends on downsizing of filters, what is possible due to very high specific surface as well as additional physical adsorption and chemical absorption of hazardous gas compounds.

This lab scale setup will be used for further investigations at DESEE. Especially the UV pre-treatment with ozone production at 185 nm and for industrial application will be focussed. Additionally new filter materials will be checked.

CONCLUSION

As biofiltration basically is a well-known method since a couple of decades, basic knowledge exists. Further development can be achieved using improved filter materials as well as new ways of raw air conditioning and pretreatment to enable this technology to replace chemical scrubbers and activated carbon filters due to economical and ecological reasons.

Therefore laboratory and large-scale tests are necessary. Concerning analytical possibilities a wide range of measurement methods has to be used, including olfacotmetry, on-line multisensory arrays and tracer gas long term monitoring. Nevertheless, as in most cases efficiency of odour degradation is decisive, olfacotmetry stillplays a key role, and measuring tracer gases do not deliver the full picture. Continuing this work will hopefully deliver easy to handle surrogate parameters for biofiltration monitoring and future process control options.

REFERENCES

1 United States Patent 4086167 (1978): *Biofilter.* Tapola, Esko (Imatra, SF), Sten, Matti (Imatra, SF) und Koistinen, Onni (Imatra, SF). http://www.freepatentsonline.com/4086167.html (11.03.2009)
2 Burges, J.E., Parsons, S.A., Stuetz, R.M. (2001): Developments in odour control and waste gas treatment biotechnology: a review. *Biotechnology Advances* 19 (2001), pp. 35–63. Elsevier.
3 Frechen, F.-B., Franke, W., Scholl, B. (2007): Odour and H_2S degradation in a full scale bio-filter with a mineral based organic coated filter media. Proceedings of the "*2nd Congress on Biotechniques for Air Pollution Control*", A Coruna/Spain, 3–5 October 2007. ISBN: 978-84-9749-258-4.

4 Frechen, F.-B., Franke, W. (2009): Geruchsemissionen aus Entwässerungsnetzen, DWA-WasserWirtschafts-Kurs N/1 – Entwässerungskonzepte. S. 349–388. DWA, Hennef 2009. ISBN 978-3-941089-55-6.
5 Franke, W. (2009): Abluftbehandlung zur Geruchsminderung in einstufigen Biofiltern – aktuelle Verfahrenstechnik, Seminar Wasser, Abfall, Umwelt, Universität Kassel, 29. Januar 2009.
6 DIN EN 13725:2003–07: Luftbeschaffenheit – Bestimmung der Geruchsstoffkonzentration mit dynamischer Olfaktometrie/Air quality – Determination of odour concentration by dynamic olfactometry; German version EN 13725:2003. Beuth Verlag. Berlin. 2003.
7 Franke, W., Frechen, F.-B., Giebel, S., H_2S, VOC, TOC, electronic noses and odour concentration: Use and comparison of different parameters for emission measurement on Air Treatment Systems. *3rd IWA International Conference on Odours and VOCs*, Barcelona, Spain, 8–10 October 2008.
8 Kraakmann, N.J.R., Groenestijn, J.W., Koers, B., Heslinga, D.C. (1997): Styrene removal using a new type of bio-reactor with fungi. Biological Waste gas cleaning = Biologische Abluftbehandlung. *Proceedings of an international symposium*, Maastricht, The Nederlands, 28–29 April 1997. ISBN 3-18-990026-4.
9 Mitzkat, L. (2001): Die Optimierung des mikrobiellen Abbaus von Limonen in Biofiltern. Inaugural-Dissertation zur Erlangung des Grades eines Doctor medicinae veterinariae (Dr. med. vet.) durch die Veterinärmedizinische Fakultät der Universität Leipzig. Institut für Tierhygiene und öffentliches Veterinärwesen. Leipzig, 2001.
10 Bargel, P. (2004): Bio-filter – biologische, kinetische und thermodynamische Charakterisierung. Im Department Chemie und Chemietechnik der Universität Paderborn vorgelegte Dissertation zur Erlangung des akademischen Grades eines Doktors der Naturwissenschaften. Paderborn, 2004.

Chapter 12

BIOFILTRATION – FULL SCALE APPLICATIONS – CHALLENGES AND SOLUTIONS

Terry J. Schulz
The Odour Unit Pty Ltd., Australian Technology Park, Eveleigh, NSW, Australia

ABSTRACT

This Keynote presentation will examine design and operational aspects associated with full-scale biofilter applications. It discusses challenges faced by designers in converting laboratory or pilot-scale biofilter research data into designs for full-sized biofilters treating medium to large airflows. By its nature the presentation will be broadly cast, with all pertinent design and operational issues discussed. The list of these issues below summarises key design and operational parameters that need to be followed, if sustainable odour destruction performance is to be achieved. The presentation will focus more on physical design and operational issues, in the understanding that the scale-up of lab or pilot plant designs to full scale is typically limited by physical rather than biological factors.

Biofilter structure

Fully enclosed cells
Free-form in-ground and above-ground

Loading rates

Volumetric airflow loading rates ($m^3/m^3/hr$)
Surface airflow loading rates ($m^3/m^2/hr$)
Residence times (empty bed residence times)
Organic loading rates ($mg/m^2/hr$)
Odour loading rates ($ou/m^2/s$)

Removal efficiency rates (mg removed/m^3/s)
Current best practice loadings

Air distribution

Slotted pipe and manifold systems (biofilter)
Plenum floor systems (Biofilter)
Air inlet ducting designs
Odour collection system designs (point-source capture)

Biofilter medium selection

Role of the medium
Biological activity
Pressure loss
Moisture retention
Robustness
Longevity
Ease of loading/unloading
Cost

Moisture control

Psychrometric charts (moisture holding capacity of air)
Foul air pre-humidification
Sub-plenum cooling/RH increases
Humidifier designs
Surface irrigation
Consequences of dry medium

Pressure loss and energy considerations

Low pressure distribution duct designs
Breakdown of pressure losses through distribution system and medium
Advances in low pressure design
Energy/GHG emission reduction examples

Thermal limitations

Medium life vs operating temperature issues

Emission control

Chapter 13

COMPOSITION OF AIR EMISSIONS FROM TOBACCO FACTORIES AND DEVELOPMENT OF THE BIOCATALYST FOR ODOR CONTROL

N.A. Zagustina, A.K. Kulikova, M.E. Romanov, A.O. Ruzhitsky, V.G. Zhukov & V.O. Popov
A.N.Bach Institute of Biochemistry of the Russian Academy of Sciences, Leninskiy, Moscow, Russia

N.I. Krikunova, T.A. Misharina & M.B. Terenina
N.M.Emmanuel Institute of Biochemical Physics of the Russian Academy of Sciences, Kosygina, Moscow, Russia

N.A. Zagustina, A.A. Veprizky & V.G. Zhukov
Innovational Biotechnologies Ltd., Leninskiy, Moscow, Russia

INTRODUCTION

Tobacco factories are venting to the atmosphere smelly and potentially harmful emissions which are irritating for the majority of the local population. In the course of processing of tobacco raw materials (steam treatment, fluffing, cutting, drying, etc.) volatile organic compounds (VOCs) emitted from fermented tobacco leaves, including nicotine, are transferred to the aspirating ducts and air conveyors. Nicotine, the main alkaloid extracted from tobacco leaves, is poisonous, readily absorbed by a human organism and may lead to serious disfunctions. Thus, control of the air emissions of tobacco factories is required.

Biofiltration is a technique of choice for odor control of industrial emissions, and in some cases, e.g. for the large and diluted volumes of the laden air, is the only economically feasible technology[1]. Deodouration is a rather specific biofiltration application as it requires abatement of the numerous VOCs which, as a rule, are not properly identified and characterized, vary in time and could be present only in trace amounts. Usually organoleptic tests are used to evaluate deodouration efficiency, however quantitative and qualitative estimation of individual VOCs is required to develop a really efficient process.

Previously we have described a number of successful applications of biotrickling filters for odor and VOC abatement in industrial and municipal air emissions[2,3]. Trickling bed filters show a number of advantages over conventional biofilters, in particular they enable full process control, enhanced efficiency per unit volume, and have a small footprint. However they incur higher capital costs and are subject to fouling with excess biomass[4,5]. The goal of the present study is the extension of the biotrickling filter technology into the area of tobacco processing and development of a model, laboratory scale trickling biofilter for deodouration and purification of the emissions of a tobacco factory, as well as characterization of the volatile components of the emission.

EXPERIMENTAL

IDENTIFICATION OF VOLATILE SUBSTANCES OF FERMENTED TOBACCO LEAVES

To characterize chemical compounds that determine specific "tobacco smell" and are to be removed by biofiltration, the volatiles of the fermented tobacco leaves were isolated by co-distillation with water and diethyl ether[6]. Blends of tobacco varieties routinely used for manufacturing of tobacco products were employed in the present study.

GC assay was performed with a capillary gas chromatograph HP 5730A ("Hewlett Packard", USA) equipped with a flame ionization detector and a quartz capillary column SPB-1 (50 m × 0.25 mm, phase layer 0.3 microns) in a temperature gradient mode. The retention times of the components of the test samples and of N-alkanes were used to obtain values of the individual retention indexes. The peak areas of the individual substances were normalized to the internal standard and expressed in μg per 100 g of tobacco.

Sniffing analysis for evaluation of the odor of the individual chromatographic zones of the eluate was performed under similar conditions. With a flow splitter attached to the end of the chromatographic column a half of the eluate was directed to the detector, while the other through the heated capillary was forwarded for evaluation of the smell of the eluted components. Odor characterization of specific eluted zones was performed by three trained tasters.

LAB-SCALE TRICKLING BIOFILTER FOR DEODOURATION

For the development of the process of biological deodouration laboratory scale trickling biofilters similar to that described previously[2] were used. Glass biofilter columns (200 × 30 mm) were packed with a fibrous polyamide carrier and bacterial cultures were immobilized thereon. The biocatalyst was continuously moistened with a circulating nutrient solution containing mineral salts, necessary for bacterial activity.

The active microbiological consortium was made of the degrading strains obtained from the enrichment cultures that contained in the growth medium the extractive compounds of the tobacco leaves, and further augmented with the strains of *Pseudomonas, Rhodococcus* and *Nocardia* from the in-house culture collection maintained at the A.N.Bach Institute of Biochemistry, Russian Academy of Sciences.

IDENTIFICATION OF THE VOLATILES AT THE INLET AND OUTLET OF THE BIOFILTER

A real industrial emission was simulated by passing the air through the crushed and fermented tobacco leaves constantly heated to 60°C. The obtained VOC laden air with a tobacco smell was analyzed on a GC/MS instrument Shimadzu QP2010 ("Shimadzu", Japan). The capillary quartz column MDN-1 (30 m × 0.32 mm) filled with methyl silicone ("Supelco", USA) was used. Individual compounds were identified using NIST mass spectra library.

Conversion rates for the individual compounds were calculated as ratios of the respective peak areas at the inlet and outlet of a laboratory biofilter. Nicotine (the most toxic component of the emission) was used as a standard compound to evaluate the efficiency of the process of VOC bioconversion by the biofilter. The degree of removal of smelly compounds was also estimated by analysis of the organic substances adsorbed on a cartridge placed at the biofilter outlet and packed with Chromosorb 106 and further eluted by diethyl ether. The ether fractions were stripped up to 50 µl and analyzed by GC/MS.

RESULTS AND DISCUSSION

COMPOSITION OF VOLATILE SUBSTANCES OF FERMENTED TOBACCO LEAVES

The chemical compounds present in tobacco smoke are well identified and characterized, while the information concerning volatile components of the raw material, fermented tobacco leaves, is rather scarce.

Overall more than 80 individual compounds belonging to different classes were thus identified in the tobacco emissions: C4–C20 organic acids, branched and normal hydrocarbons, aliphatic and aromatic aldehydes and ketones, alcohols. Sniffing analysis allowed to identify individual volatile components that defined the smell of the leaf tobacco and thus the odor of the emissions of tobacco factory. The principal compounds responsible for the tobacco smell were shown to be solanone and isomeric megastimatrienones (tabanones). Aromatic aldehydes and ketones, geranyl acetone, tetramethyl hexadecenol, franezene, ledole, damascenone and ionone were important contributors to smell formation. Some other compounds, namely cyclopentanes, cyclohexanes and pyranes substituted with alcohol, keto and aldehyde groups participated in smell formation as well.

SET-UP AND EFFICIENCY OF THE LAB-SCALE BIOFILTERS FOR ABATEMENT OF ORGANIC SUBSTANCES FROM THE TOBACCO EMISSIONS

To develop a biological process for deodoration of tobacco smell an artificial air emission simulated as described under Experimental section was created and fed to the series of laboratory scale trickling biofilters. A composition of VOCs in the real emission of a tobacco factory (in the local exhaust systems and pneumatic transport systems) was investigated. The GC/MS analysis of the artificial emission showed that it was rather similar to the VOC composition of the real tobacco factory exhaust.

Laboratory biofilters with immobilized bacterial cultures were operated continuously for a period of 12 months at 20–25°C and specific productivity (defined as volumes of the air passing through a volume of the bed media per hour) of 1000 h^{-1} (3.6 s contact time). The biofilters demonstrated a stable rate of conversion of nicotine exceeding 90%. Use of the preselected bacterial consortium accounted for a short acclimation period of the reactors that did not exceed a week.

The results obtained (Table 1) demonstrate that nicotine was depleted 8.5-fold, while 3,7,11,15-tetramethyl-2-hexadecen-1-ol – the compound mainly responsible for the odor, was diminished 11-fold. Some substanses completely disappear from the outlet, while some new volatiles absent in

Table 1 GC/MS of the major volatile components of leaf tobacco (simulated emission) at the inlet and outlet of the biofilter.

Name	Retention time	*Inlet* Area. %	*Outlet* Area. %	$Area_{inlet}/Area_{outlet}$
C_nH_2n+2	8.33	2.32	14.00	0.80
5-Hexen-2-one	8.54	8.37	8.49	4.74
C_nH_2n+2	10.06	2.27	14.57	0.75
C_nH_2n+2	11.48	1.08	7.44	0.70
Nicotine	13.05	57.19	32.43	8.48
Butanoic acid, anhydride	15.16	0.74	2.54	1.40
C_nH_2n+2	15.26	0.21	–	–
3,7,11,15-Tetramethyl-2-hexadecen-1-ol	16.67	24.98	11.02	10.90
C_nH_2n+2	18.88	2.85	–	–
	Extra volatile components traced at the biofilter outlet			
3-Hexanone-2,2-dimetyl	11.68	–	1.41	–
C_nH_2n+2	12.67	–	2.75	–
Fatty acid, ME	17.06	–	1.52	–
Phtalate	17.16	–	1.58	–
ND	18.07	–	2.25	–

the initial emission are detected. These are considered to be non-hazardous transformation products devoid of odor.

Concentration of the odorous substances in the simulated air emission is relatively small, close to the detection level even of the highly sensitive GC/MS equipment used in the present work. In order to obtain a more accurate estimation of the degradation efficiencies for smelly substances the simulated emission (biofilter inlet) and the filtered air (biofilter outlet) were concentrated on Chromosorb 106, adsorbed compounds than eluted by diethyl ether and resulting eluates analyzed by GC/MS as described above. The analysis of thus obtained eluates verified that more than 90% of nicotine as well as other odorous substances is removed by biofiltration.

STUDY OF THE COMPOSITION OF THE BIOFILTER TRICKLING LIQUID

After 5 months of bioreactor operation without full replacement of the trickling solution, the liquid phase was analyzed. The major components detected in the liquid phase were dibutyl phthalate, pentanoic acid, 2,2,4-trimethyl-3-carboxyisopropyl, isobutyl ester, ethyl isopropyl ether. Other compounds were present in trace amounts. The GC/MS analysis of the biofilter trickling liquid phase showed that neither nicotine no other toxic substances – possible intermediate products of the destruction of the VOCs present in the simulated emission, were detected. Since the trickling liquid phase in the industrial biofiltarion unit will be periodically drained, the results obtained ensure that the technology under development will not result in the contamination of the environment by secondary degradation products.

Pilot plant trials scaled down to the 1:50 ratio from the projected capacity of the industrial unit have been initiated at the tobacco factory.

CONCLUSIONS

Thus, in the present study we investigated the composition of the volatile compounds of leaf tobacco and determined the substances primarily responsible for tobacco odor. We showed that biotrickling filter technology provides a viable solution for deodouration of the emissions of tobacco processing factories and verified this by successful operation of laboratory scale bioreactors for a period of one year. The results obtained laid down a basis for the pilot plant trials with the real emission of a tobacco factory.

REFERENCES

1 Popov VO and Zhukov VG, Odor removal in industrial facilities, in *Biotechnology for Odor and Air Pollution Control*, ed by Shareefdeen. Z, Singh A, Springer-Verlag, Heidelberg-New York, Germany, pp. 305–326 (2005).

2 Zhukov VG, Rogozhin IS, Ushakova NA, Zagustina NA, Popov VO and Bezborodov AM, Development of microbiological technology for deodoration of waste air under laboratory and industrial conditions with the use of pilot plant. *Appl. Biochem. Microbiol 34*: 370–376 (1998).
3 Popov VO, Khomenkov VG, Zhukov VG, Cavanagh M and Cross P, Design, construction and long-term performance of novel type of industrial filters for VOC and odour control. *International Congress Biotechniques for Air Pollution Control*, La Coruña (Spain), October 5–7 Congress proceedings, pp. 257–262 (2005).
4 McNevin D and Barford J, Biofiltration as an odour abatement strategy. *Biochemical Engineering J 5*: 231–242 (2000).
5 Nikiema J, Dastous PA and Heitz M, Elimination of volatile organic compounds by biofiltration: *Rev Environ Helth 22*: 273–294 (2007).
6 Misharina TA and Golovnya RV, Efficiency of extraction of volatile substances from foodstuff by different methods. *Zhurn Analit Khimii 47*: 650–659 (1992).

Chapter 14

CHARACTERIZATION OF AIRBORNE PARTICULATE MATTER IN GREATER CAIRO AREA (EGYPT) USING NEUTRON ACTIVATION ANALYSIS TECHNIQUE

Abou Bakr A. Ramadan
National Center for Nuclear Safety and Radiation Control, Cairo, Egypt

ABSTRACT

Aerosol samples for PM10 were collected by high volume samplers from September 2006 until July 2008 at three aerosol stations located in greater Cairo area in the north-south direction. Instrumental Neutron Activation Analysis and Atomic Absorption Spectrophotometer techniques were used to measure the concentration of about 35 elements, while a Liquid Chromatography was used to measure ions (SO_4, NO_3 and NH_4). PM10 mass concentrations are generally high through the greater Cairo area. PM10 values obtained for all sampling events at the three monitoring sites are generally high and exceeded the Governmental Egypt's (GOE) Low Number 4 of 1994 limit of 70 $\mu g/m^3$ (24-hour average). PM_{10} concentrations maximize in spring associated with sand storms during Khamseen conditions and in autumn with longer duration due to trapping of pollutant emissions within a shallow planetary boundary layer. A number of sources including traffic and also biomass/waste burning contribute to the formation of the pollution episodes. The analysis of PM10 samples has shown wide variation in toxic pollutants. The results reveal that the concentrations of several pollutant elements in dust samples from these three sites are highly variable. Some elements showed a dependence on the day of the week, while others did not. The highest mean concentrations of organic carbon (18.6 $\mu g/m^3$) and elemental carbon (11.3 $\mu g/m^3$) in PM10 were found at Shoubra Kheima among the three sites. The major fractions of sulfate at the three monitoring sites are non-sea-salts sulfate. The monitoring results suggest that control of particulate matter in greater Cairo area may be a difficult problem.

INTRODUCTION

Aerosol samples for PM10 were collected by high volume samplers from September 2005 until July 2007 at three aerosol stations located in greater Cairo area in the north-south direction. Instrumental Neutron Activation Analysis and Atomic Absorption Spectrophotometer techniques were used to measure the concentration of about 35 elements, while a Liquid Chromatography was used to measure ions (SO_4, NO_3 and NH_4). PM10 mass concentrations are generally high through the greater Cairo area. PM10 values obtained for all sampling events at the three monitoring sites are generally high and exceeded the Governmental Egypt's (GOE) Low Number 4 of 1994 limit of 70 μg/m^3 (24-hour average). PM_{10} concentrations maximize in spring associated with sand storms during Khamseen conditions and in autumn with longer duration due to trapping of pollutant emissions within a shallow planetary boundary layer. A number of sources including traffic and also biomass/waste burning contribute to the formation of the pollution episodes. The analysis of PM10 samples has shown wide variation in toxic pollutants. The results reveal that the concentrations of several pollutant elements in dust samples from these three sites are highly variable. Some elements showed a dependence on the day of the week, while others did not. The highest mean concentrations of organic carbon (18.6 μg/m^3) and elemental carbon (11.3 μg/m^3) in PM10 were found at Shoubra Kheima among the three sites. The major fractions of sulfate at the three monitoring sites are non-sea-salts sulfate. The monitoring results suggest that control of particulate matter in greater Cairo area may be a difficult problem.

In urban areas in many developing countries, PM concentrations can be much higher than in developed countries on a regular basis. PM_{10} concentrations in the range from a few hundred μg/m^3 to a few mg/m^3 are capable of exerting adverse effects on human health. Greater Cairo is considered the most important city in Egypt from demographical point of view. It encompasses 27% of the Egyptian population, 64% of the industry and 45% of motor vehicles. Greater Cairo area lies in a sub-tropical region with a dry climate, warm summers and mild winters. Dust storms also occur during April and May when the Khamasin winds blow over the Egyptian western desert with a wind speed of the order of 10 ms^{-1}. Relative humidity fluctuates between 59% in June and 71% in December, with visibility of about 5 km. This restricted visibility is the result of the presence of solid particles in the atmosphere. The strong emissions of trace gases and aerosol particles by vehicles traveling on the city's narrow roads, industry and resuspended soil dust, together with secondary aerosol, coupled with the unfavourable natural conditions of dispersion, are responsible for the high concentrations of pollutants observed in the greater Cairo metropolitan area. Greater Cairo areas suffer from high ambient concentrations of atmospheric pollutants.[1,2]

In this article we aimed to study the chemical pollutants in greater Cairo atmosphere compared to the international levels and to support environmental decision-making in greater Cairo to control air pollution with particulate matter (PM).

SAMPLING AND SAMPLING ANALYSIS

Total particulate matters were collected from September 2005 until July 2007 at three sites in greater Cairo area as shown in Fig. 1. The three stations (Shoubra El Khama, Mokatam Hill, and Helwan) cover a north-south distance of about 30 km within the greater Cairo area.

Epithermal neutron activation analysis was used a basic method for elemental determination in air filters as well as in blank filter material. Short-lived radionuclides were determined using small parts of filters (2.5×2.5 cm^2) packed in sealed polyethylene packs. Long-lived radionuclides were determined using activation with the epithermal neutrons. Data processing and element concentration determinations were performed. Organic and elemental carbons were analyzed using the thermal manganese dioxide oxidation (TMO) method. The quality control was provided by using certified reference materials Urban Dust (IAEA, Vienna) for NAA.

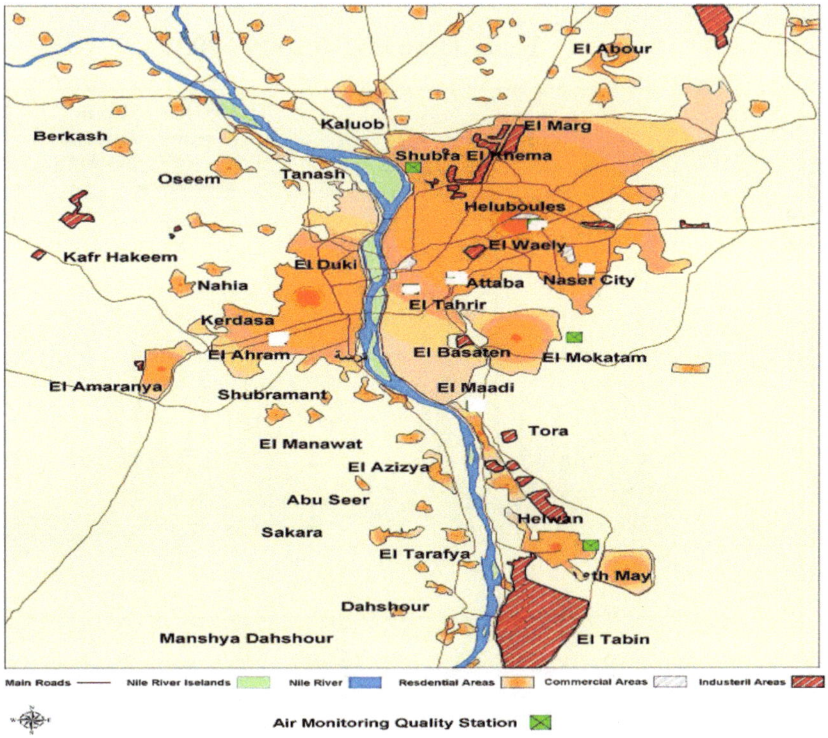

Figure 1 Locations of the sampling sites.

ANALYSIS OF PARTICULATE MATTER IN THE GREATER CAIRO AREA

Figure 2 shows the daily averaged measured PM10 (particle size less than 10 μm) concentrations (μg/m^3) for Shoubra El Khama, Mokatam Hill, and Helwan during the period from September 1, 2005, through July 12, 2007. The three stations are located within the greater Cairo area and are approximately 15 km from each other. Because the prevailing wind direction in Cairo is northeast, we selected three measurement sites on a line oriented approximately north-south (Fig. 1).

The hourly measurements at the three stations are clearly correlated with each other. The figure reveals two distinct peaks during the period of the analysis. One peak appears during the spring (March–May) and the second appears during the fall (October–December). The spring maximum is related to the Khamasin conditions over Egypt and the associated sand storms, while the fall maximum is related to more local pollution conditions. Noticeable also is the relatively longer duration of the fall pollution episode (although with smaller values) than the spring pollution episode (with higher values); this reveals the nature of the sand storms, with sharp snapshots that last for a few days. Khamasin conditions are regional phenomena that extend beyond the urban scale.

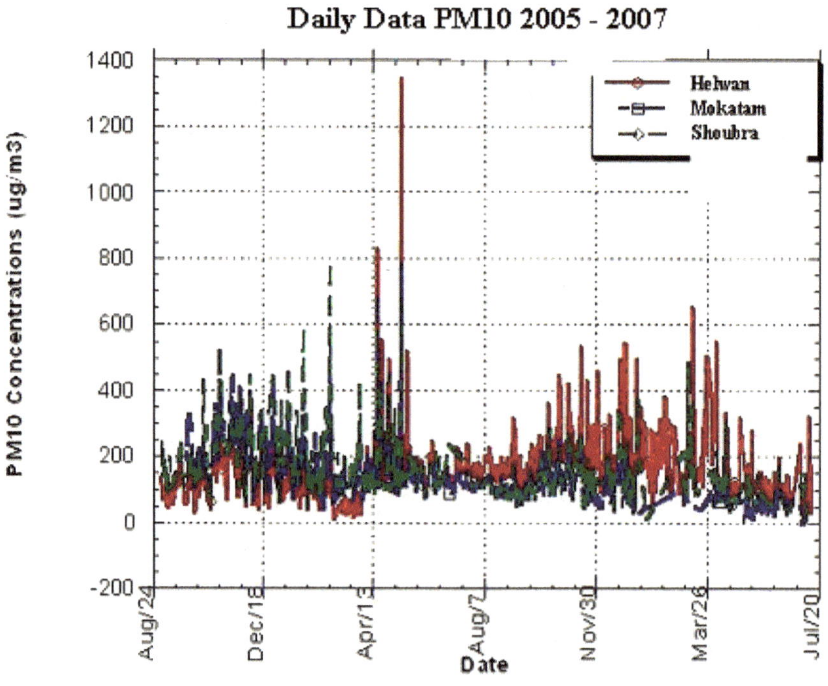

Figure 2 Measured hourly PM10 concentrations at Shoubr El Khama, Mokatam Hill and Helwan (September 1, 2005–July 12, 2007).

CHEMICAL AND MINERALOGICAL COMPOSITION

The general trend in the elemental concentrations determined for the three aerosol measurement sites is the pronounced increase when moving from the northern site to the southern site (Fig 3). This could be explained by the north-south wind direction in winter. The significant increases in the elemental pollutant concentrations in aerosols along this north-south line in the greater Cairo area was determined to be from 10 times greater (for rare earth elements) to 12 times greater (for some heavy metals and major marine elements). Some elements showed a dependence on the day of the week, while others did not. Scandium (Sc), lanthanum (La), and potassium (K) are among the elements that did not show this dependence, which evidently reflects their crustal origin. Other elements did show this dependence, with the last two days of the week (Thursday–Friday) experiencing the lowest values of pollutants, and the beginning of the week (Saturday–Sunday) showing the highest values (Fig 4). The weekly cycles of air pollution noted in our work for the Egypt, with its weekends on Thursday and Friday, are in good agreement with the behavior of air pollution in the western world, with its weekends on Saturday–Sunday.[3,4]

According to elemental and mineralogical analysis of the dust collected from all traps in the greater Cairo, the dust particles were composed mostly of quartz and calcite, with smaller amounts of dolomite and clay minerals. The major constituents were Si and Ca; minor constituents were Al, Fe, Mg, K and Na. The characteristic features of the analysis can be summarized as follows: Silicon and calcium are regarded as crustal element. They are important components of PM_{10} measured in Cairo. The two elements hold about 23% of PM_{10} mass. The presence of coarse silicon can be ascribed to erosion from desert land surrounding monitoring station. Ongoing building activities and degradation of building elements can be considered as additional sources

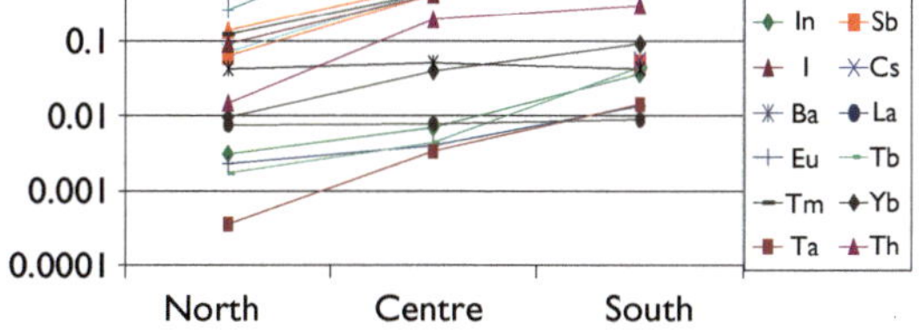

Figure 3 Greater Cairo area aerosol elemental content.

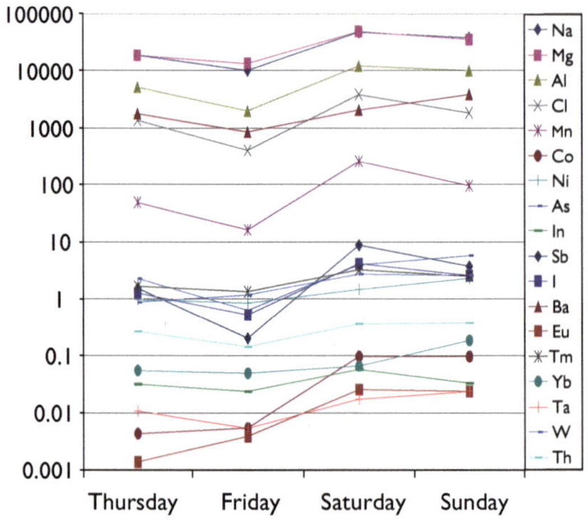

Figure 4 Irregularity of aerosol elemental content in the South of the Greater Cairo area.

for increasing silicon level. Calcium concentrations are much similar in all locations except for increased coarse fraction in Helwan site which are to be ascribed to partly degradation of buildings, partly to emanating from industries such as cement plants.

High concentrations of soluble potassium (K^+), a tracer for vegetative burning, were also observed. The average K^+/K ratio was roughly 0.6, indicating significant vegetative burning in the greater Cairo area. Low chlorine and sodium concentrations can be understood as a strong sea-derived contribution. Small concentration elements of aluminum, zinc, manganese and iron indicating the importance of soil derived dust of traffic origin. Crustal components (Si, Ca, Fe and Al) were significant at all sites. Helwan and Shoubra El Khama had the highest PM_{10} crustal levels due to resuspended road dust at Shoubra El Khama and the nearby cement plants at Helwan.[5,6] Finally, the composition of dust deposited in greater Cairo city during Khamasin storm was similar to that deposited during normal weather conditions, indicating that the composition of dust deposited during the storm over a wide area of the desert around greater Cairo was uniform.

ORGANIC AND ELEMENTAL CARBONS

Organic carbon (OC) and elemental carbon (EC) were major components of PM10 at all sites. Potential sources include mobile emissions and vegetative burning. The highest average PM10 OC and EC levels were observed at Shoubra El Khama. The high emission of OC from Shoubra El Khama (18.6 μg/m³) might be due to the traffic flow in Shoubra mixed with the emission from industrial area. Relatively low concentrations of OC and EC were determined in Helwan due to low traffic.[5]

REFERENCES

1 Nasralla MM, Air Pollution in Greater Cairo. Comparing the health risk in Cairo, Egypt (Vol. 3) Annex G, submitted to USAID/Egypt project 398-0365 (1994).
2 Elminir H, Dependence of urban air pollutants on meteorology, *Science of the Total Environment*, Vol. 359 (1–3) p. 231(2005).
3 Cerveny RS and Balling RC, Weekly cycles of air pollutants, precipitation and tropical cyclones in the coastal NW Atlantic region. *Nature*, 394/6, 561–563 (1998).
4 Frontasyeva M, Ramadan A and Galinskaya T, *Weekly cycles of elements pollutants in air of the greater Cairo area (Egypt) studied by Neutron Activation Analysis.* International aerosol conference, Moscow, Russia, 19–23 of June (2003).
5 Khoder MI, Atmospheric conversion of sulfur dioxide to particulate sulfate and nitrogen dioxide to particulate nitrate and gaseous nitric acid in an urban area *Chemosphere*, Vol. 49, p. 675 (2002).
6 Borai EH and Soliman AA, 'Monitoring and statistical evaluation of heavy metals in airborne particulates in Cairo, Egypt', *J. Chromatog.* A 920, p. 261 (2001).

Biofiltration – methane and ammonia

Chapter 15

BIOFILTRATION OF METHANE FROM THE PIGGERY INDUSTRY – INFLUENCE OF THE METHANE AND NITROGEN CONCENTRATIONS

Matthieu Girard, Josiane Nikiema & Michèle Heitz
Faculty of Engineering, Department of Chemical Engineering and Biotechnological Engineering, Université de Sherbrooke, Québec, Canada

Gerardo Buelna
Centre de Recherche Industrielle du Québec, Québec, Canada

INTRODUCTION

In Canada, the piggery industry is an essential part of the agricultural sector. In 2007 alone, this industry produced over 31 million hogs and comprised more than 64 000 jobs, with exports worth over 3 billion dollars.[1,2] However, the main waste product of this industry, swine slurry, causes severe environmental problems. The anaerobic storage conditions of this waste product lead to the emission of methane (CH_4), an important greenhouse gas (GHG). In fact, in Canada in 2005, swine slurry management was responsible for the release of 1.6 million metric tons of carbon dioxide (CO_2) equivalent of CH_4.[3] In terms of climate change, CH_4 has a global warming potential 21 times that of CO_2.[4] Typical concentrations of CH_4 from pig houses vary between 7 and 150 ppmv (5 and 100 mg/m^3) and depend essentially on the ventilation flow rate. Methane concentrations from covered slurry storages with no aeration can reach 65% v/v (425 g/m^3), but storage covers are rarely airtight and concentrations usually vary from 150 to 30600 ppmv (0.1 to 20 g/m^3).[5]

In order to limit the piggery industry's impact on climate change, GHG emissions must be reduced or treated. It is possible to collect the gas and burn it using a flare, but concentrations are rarely high enough for direct combustion which requires a minimal concentration of 20% v/v6. An interesting approach uses biofiltration. Several studies have been carried out on the biofiltration of CH_4 originating from sanitary landfills; Nikiema et al. (2007)[6] conducted an extensive review on this topic. However, very few papers on the biofiltration of CH_4 from the piggery industry are available. Two studies have been published on the treatment of CH_4 from slurry storages using biofilters packed with organic materials 5, 7. For CH_4 concentrations up to 35000 ppmv, theses studies obtained maximum elimination capacities (EC) of 9 and 20 g/m^3/h with inlet loads (IL) up to 30 g/m^3/h. Although CH_4 biofiltration is usually carried out using organic packing materials, interesting results have been obtained with inorganic materials.[6] When using this type of filter bed, nutrients such as nitrogen are not present naturally and must be supplied by an exterior source (with a nutrient solution for example).

The objectives of this study were to determine the influence of both the methane inlet concentration and the nitrogen concentration in the nutrient solution on the performance of a biofilter packed with an inorganic material.

MATERIAL AND METHODS

The biofilter used in these experiments was made up of 3 identical sections of Plexiglas tubing with an internal diameter of 15 cm, as shown in Figure 1. Each section was filled with 33 cm an inorganic gravel material for a total filter bed height of 1 m and a volume of approximately 18 L.

A synthetic gas composed of humidified air and pure CH_4 (supplied in gas cylinders by Praxair Inc., Quebec, Canada) was injected at the base of the biofilter. The CH_4 concentration was maintained at values representative of the piggery industry, i.e., between 250 and 2100 ppmv (165 and 1375 mg/m^3). The total gas flow rate was kept at 0.25 m^3/h (4.2 L/min) throughout the

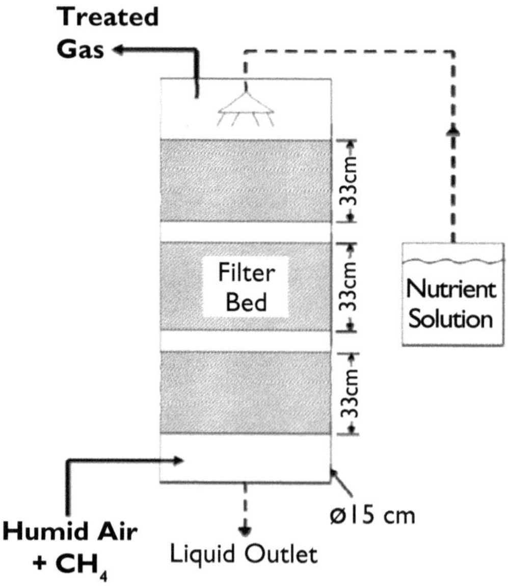

Figure 1 Lab-scale biofiltration system.

entire study, which corresponds to an empty bed residence time (EBRT) of about 4.2 minutes.

A synthetic nutrient solution was sprayed at the top of the biofilter at a rate of approximately 1.5 L/day to ensure proper filter bed moisture and to provide the nutrients necessary for microbial growth. The nutrient solution is similar to that presented by Fox et al. (1989)[8] and contains mainly nitrate, phosphorous, potassium and various trace elements. Nitrate was used as the nitrogen source since it seems to be the preferred type of fixed nitrogen for CH_4 biofiltration.[6]

RESULTS AND DISCUSSION

Figure 2 presents the EC as a function of the inlet CH_4 concentration for a nitrate concentration in the nutrient solution of 0.5 gN/L.

These results show that by varying the inlet CH_4 concentration from 250 to 2100 ppmv, the EC steadily increased from 1.0 to 9.4 g/m³/h. However, the CH_4 conversion actually remained relatively stable, between 41 and 48%. The maximum EC obtained corresponds to the lower value (9 g/m³/h) in the range reported by the other studies on the biofiltration of CH_4 from the piggery industry.[5,7] This could be due to the relatively low EBRT used in this study (4.2 minutes) as compared to the other studies at values between 7 and 80 minutes.

By measuring the distribution of the main anions in the nutrient solution over the height of the biofilter, it was possible to calculate the nutrient uptake.

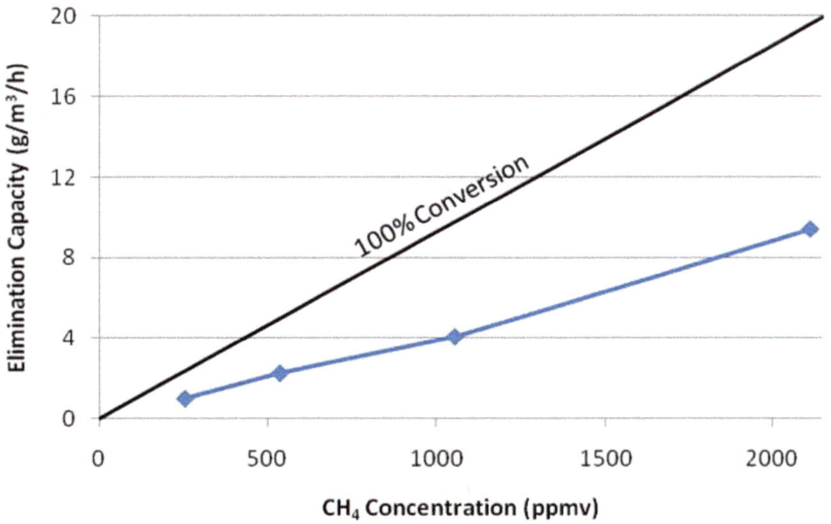

Figure 2 Elimination capacity as a function of the methane inlet concentration.

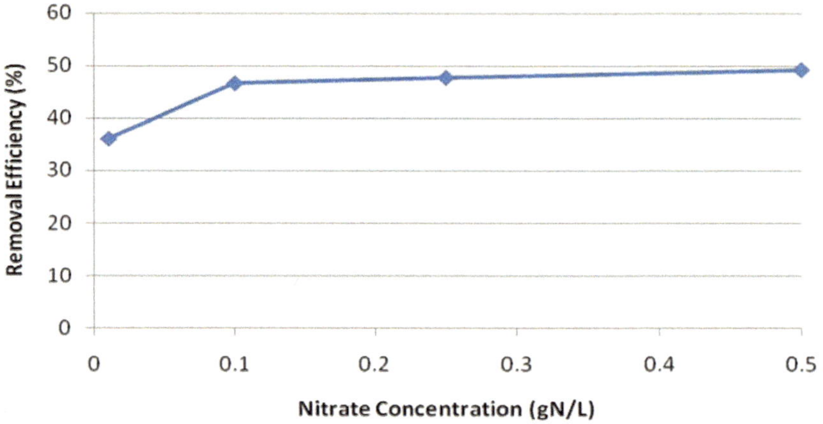

Figure 3 Removal efficiency as a function of the nitrate concentration for a CH_4 IL of 14 g/m³/h.

For a CH_4 concentration between 500 and 1000 ppmv, the nitrogen uptake was around 20% while the phosphorous uptake varied between 14 and 24% and the sulphate uptake was about 14%.

Regarding the influence of the nitrogen concentration, Figure 3 presents the CH_4 removal efficiency as a function of the nitrate concentration in the nutrient solution for an inlet CH_4 concentration of 1500 ppmv and an IL of 14 g/m³/h. When the nitrate concentration was reduced from 0.5 to 0.1 gN/L, the CH_4 removal efficiency decreased only slightly, from 49 to 47%. However, once the nitrate concentration was reduced to 0.01 gN/L, the CH_4 removal efficiency fell to 36%. A previous study on CH_4 biofiltration using an inorganic packing material found that the optimum nitrate concentration decreases with the CH_4 IL (0.75 gN/L for ILs between 55 and 95 g/m³/h and 0.50 gN/L for ILs between 20 and 55 g/m³/h).[9] According to this paper, the optimal nitrate concentration should be around 0.50 gN/L for an IL of 14 g/m³/h. However, since there is little difference in the CH_4 removal efficiency, the nitrate concentration should be set at 0.1 gN/L to lower operating costs.

CONCLUSION

This study showed the potential of biofiltration for the treatment of methane originating from the piggery industry using an inorganic packing material. Biofilter performance was influenced both by the CH_4 inlet concentration and the nitrogen concentration in the nutrient solution. The EC was found to increase with the CH_4 inlet concentration and a maximum EC of 9.4 g/m³/h was obtained for a CH_4 concentration of 2100 ppmv. Regarding the influence of the nitrate concentration in the nutrient solution, the removal efficiency decreased slightly for concentrations between 0.5 and 0.1 gN/L, but fell significantly when the nitrate concentration reached 0.01 gN/L. Results also showed that a nitrate concentration of 0.1 gN/L is sufficient for proper biofilter operation at a CH_4 IL of 14 g/m^3/h.

ACKNOWLEDGEMENTS

The authors wish to thank the National Sciences and Engineering Research Council of Canada (NSERC) for the strategic research grant that funded this study and for Matthieu Girard's scholarship (NSERC Postgraduate Scholarship – Doctorate).

REFERENCES

1 Canadian Pork Council, Statistics and Market Reports. Available from http://www.cpc-ccp.com/industry/statistics.html (2009) [Accessed in August 2009].
2 Agriculture and Agri-Food Canada, Trade in Pork and Swine Between Canada and the United States. Available from http://www.agr.gc.ca/itpd-dpci/amr/4875-eng.pdf (2007) [Accessed in August 2009].
3 Jaques A, Rapport d'inventaire national 1990–2005: Sources et puits de gaz à effet de serre au Canada, Environnement Canada, ISBN: 978-0-662-07181-5 (2007).
4 CITEPA – Centre Interprofessionnel Technique d'Études de la Pollution Atmosphérique, Pouvoir de réchauffement global - PRG. Available from: http://www.citepa.org/emissions/nationale/Ges/ges_prg.htm (2009) [Accessed in August 2009].
5 Melse RW, and Van der Werf AW, Biofiltration for mitigation of methane emission from animal husbandry. *Environmental Science and Technology*, **39**: 5460–5468 (2005).
6 Nikiema J, Brzezinski R, and Heitz M, Elimination of methane generated from landfills by biofiltration: a review. *Reviews in Environmental Science Biotechnology* **6**(4): 261–284 (2007).
7 Canadian Pork Council, Demonstration project on a commercial farm of a technology that capture and oxidize methane from manure storage facilities. Canadian Pork Council Greenhouse Gas Mitigation Program – Final Project Report. Available from http://www.cpc-ccp.com/industry/pdf/ghgmp-reports/FPPQ_Lennoxville_GHGMP%20Final%20Report%20CPC.pdf (2006) [Accessed in August 2009].
8 Fox BG, Froland WA, Dege JE, and Lipscomb JD, Methane Monooxygenase from Methylosinus trichosporium OB3b – Purification and Properties of a Three-Component System with High Specific Activity from a Type II Methanotroph. *The Journal of Biological Chemistry*, **264**(17): 10023–10033 (1989).
9 Nikiema J, *Attenuation of greenhouse gas emissions by means of methane biofiltration: Optimization of the operating parameters. Ph.D.* thesis. Department of Chemical Engineering, Université de Sherbrooke (2008), 189 pages.

Chapter 16

EFFECT OF SILICONE OIL FRACTION AND STIRRING RATE ON METHANE DEGRADATION IN A STIRRED TANK REACTOR

José Rocha-Rios
Departamento de Ingeniería de Procesos e Hidráulica (IPH), Universidad Autónoma Metropolitana-Iztapalapa, México D.F., México

Raúl Muñoz
Departamento de Ingeniería Química y Tecnología del Medio Ambiente, Universidad de Valladolid, Valladolid, España

Sergio Revah
Departamento de Procesos y Tecnología, Universidad Autónoma Metropolitana-Cuajimalpa, México D.F., México

ABSTRACT

Methane is an important greenhouse gas and its biological oxidation constitutes a cost-effective alternative for the treatment of low concentration emissions. Nevertheless, due to its low aqueous solubility (similar to oxygen), the limiting step during the biological degradation of methane is often the mass transfer from the gas phase. Two-phase partition bioreactors are especially devoted to increase the biodegradation of poorly soluble compounds. In this work, the effect of the stirring rate (200, 500 and 800 rpm) and the silicone oil fraction (0, 5 and 10% v/v) on the abiotic volumetric oxygen transfer coefficient and the methane elimination capacity by a methanotrophic consortium in a stirred tank was evaluated. The results showed that, under abiotic conditions, the overall volumetric mass transfer coefficient (k_La) increased when increasing both factors considered, which resulted in an enhanced *EC* of the system during the degradation experiments.

INTRODUCTION

Methane is an important greenhouse gas exhibiting a global warming potential 21–25 times higher than CO_2 and an atmospheric concentration increasing twice as fast as that of CO_2.[1] The use of CH_4 in biogas for energy production is not economically feasible when its concentrations are lower than 30%, and alternative abatement methods must be considered when treating low flow rates or concentration below 20% as is the case in old and/or small landfills.[2] In this particular case, biological oxidation would be a cost-effective alternative for CH_4 abatement. However, the biodegradation of this compound is limited by its low solubility in the aqueous phase (where the methanotrophic microorganisms are present), which constitutes the bottleneck of the process. In this context, the addition of an organic phase with a higher affinity for CH_4 than water can be used to improve the removal efficiency of the bioprocess. This organic phase is commonly called "mass transfer vector",[3] and the resulting systems are referred as Two-Phase Partitioning Bioreactors (TPPBs).[4] Hence, the addition of 10% (v/v) of silicone oil in both a stirred tank and a biotrickling filter reactor increased methane elimination capacity (*EC*) by approximately 41% and 106%, respectively.[5] However, that work was devised as a proof of concept study and the elucidation of the mechanisms responsible for the *EC* enhancement was out of the scope of the study. The elucidation of the effect of key operational variables such as the stirring rate (*N*) and the vector fraction (ϕ) in the transfer of gaseous substrate is required to better understand the mechanisms of enhancement in TPPBs. In this context, the volumetric mass transfer coefficient (k_La) is crucial to accomplish the above mentioned objective since it considers the different factors involved in mass transfer in a bioreactor: geometrical and operational characteristics of the vessel, media composition (i.e., ionic strength), biomass concentration, stirring rate, organic fraction, etc.[6]

The aim of this study was the quantification of the effect of the stirring rate and the silicone oil fraction both on the abiotic k_La for oxygen and on the enhancement in *EC* during CH_4 biodegradation by a metanotrophic bacterial consortium in a stirred tank reactor.

MATERIALS AND METHODS

MICROORGANISMS AND CULTURE CONDITIONS

The methanotrophic consortium used in this experiment was obtained from an activated sludge sample from the wastewater treatment plant of UAM-Iztapalapa (México City). Culture maintenance and inoculum preparation were carried out as previously described.[5]

CHEMICALS

Methane gas at 99.9% was purchased from Praxair (Mexico). Silicone oil (polydimethylsiloxane, 200 cst) was obtained from Sigma-Aldrich (USA).

STIRRED TANK REACTOR (STR)

The experimental set-up consisted of a 3.5 L fermentor (Bioflo III New Brunswick, USA) with a working volume of 2 L. The bioreactor was equipped with an agitation system composed of two Rushton turbines separated 0.08 m

($D_i = 0.057$ m and $w_i = 0.011$ m) and operated at 3 stirring rates (N = 200, 500 and 800 rpm) and 3 silicone oil fractions ($\phi = 0$, 5 and 10% v/v).

EXPERIMENTAL ABIOTIC $k_La_{O_2}$ DETERMINATION

Abiotic k_La for oxygen were determined using the static method according to Moo-Young et al.[7] Each experiment was carried out in duplicate and the standard error reported.

CONTINUOUS METHANE BIODEGRADATION

Methane at 5.5 g m^{-3} (in air) was introduced in the system at 0.42 L min^{-1} via a perforated ring (four orifices with diameter of 10^{-3} m) located at the bottom of the bioreactor. This gas flow rate corresponded to 0.21 vvm or an empty bed residence time (EBRT) of 4.8 min and resulted in average methane loading rates of 65 g m^{-3} h^{-1}. Mineral medium was continuously added at a dilution rate of 0.05 d^{-1} (0.1 L day^{-1}) in order to maintain a constant bacterial concentration of 3 g L^{-1}. Temperature (T) and pH were controlled at 30°C and 7.0, respectively. The dissolved oxygen concentration (C_{O_2}), pH, T and N were monitored on-line using the software Fermentor Control Program® version 4.3 (FoxyLogic, Denmark).

ANALYTICAL PROCEDURES

Methane concentration in the gas phase was measured using a gas chromatograph (Agilent Technologies 6890 N, USA) equipped with an AT-WAX 25 m × 0.25 mm × 0.2 m column (Alltech, USA) and a FID detector. Carbon dioxide concentration was measured using the same chromatograph with a Porapack Q 80/100 6' × 1/8" column and a TCD detector according to Rocha-Rios et al.[5] Biomass concentration in the reactor was estimated from protein concentration measurements using the Lowry's method (BioRad kit, USA) and considering a 50% protein content. All determinations (gas chromatography and biomass) were done in triplicate. The C_{O_2} was monitored using an Inpro6820 probe (Mettler Toledo, Mexico). Since the $k_La_{O_2}$ values were in all cases lower than 0.03 s^{-1}, it was assumed that the response of the oxygen electrode to the variations in oxygen concentration was fast enough and did not affect the accuracy of the determination.[8]

RESULTS AND DISCUSSION

EFFECT OF STIRRING RATE AND OIL FRACTION ON EXPERIMENTAL ABIOTIC $k_La_{O_2}$

The experimental abiotic $k_La_{O_2}$ determined at 200, 500 and 800 rpm for: 1) tap water, 2) 3 g L^{-1} of inactive cells (sterilized) in tap water, 3) 5% of silicone oil in tap water, and 4) 10% of silicone oil in tap water. The $k_L a_{O_2}$ values increased with the addition of inactive cells and silicone oil compared to the values obtained in tap water; nevertheless, unlike in the experiments carried out with tap water, a reduction in $k_L a_{O_2}$ was observed when increasing stirring rate from 500 to 800 rpm in the added systems (figure 1).

This reduction in $k_L a_{O_2}$ was less significant when inactive cells were added but notably marked at increasing silicon oil fractions. In turn, the results in figure 2 indicate that $k_L a_{O_2}$ increased with increasing the oil fraction for a constant stirring rate, this effect becoming more important at high stirring rate (above 200 rpm). A detailed review of the literature available revealed that other authors have also reported similar reductions in $k_L a_{O_2}$ at increasing stirring rate in presence of cells[6,8] and the organic vector fraction.[9–11] These authors hypothesized that this reduction was due to the blockage of oxygen transfer as a result of an agitation-promoted adsorption of the cells or the organic solvent to the air bubbles. The low diffusivity of oxygen in the viscous

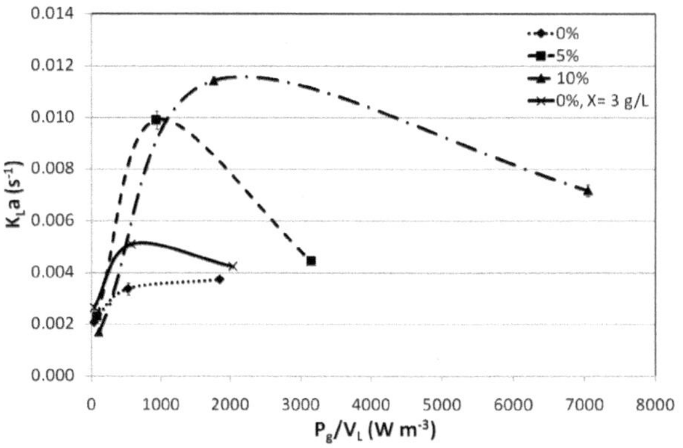

Figure 1 Effect of stirring rate on $k_L a_{O_2}$.

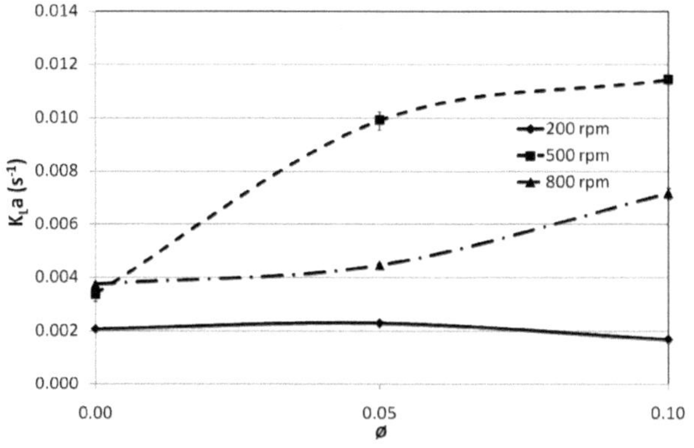

Figure 2 Effect of silicone oil fraction on $k_L a_{O_2}$.

organic vector severely hindered oxygen mass transport from the gas to the aqueous phase, where the oxygen probe is located. This blockage in the oxygen transfer increases at increasing stirring rates as a result of the decreasing organic drop diameters and of the higher frequency of collision between the vector and the air bubbles.

BIODEGRADATION METHANE EXPERIMENTS

The biodegradation of methane was carried out at a constant biomass concentration of approximately 3 g L^{-1} and an average methane load of 65 g m^{-3} h^{-1}. A steady state was achieved for each operational condition tested (Table 1).

The stirring rate was the dominant variable determining methane biodegradation performance while the effect of oil fraction was only significant above 200 rpm. Nevertheless, unlike the abiotic $k_L a_{O_2}$ determination, neither EC nor RCO_2 decreased when the stirring rate was increased from 500 to 800 rpm (at 5 and 10% of silicone oil). This could be explained by the adhesion of active cells to the gas-oil and gas-cells interphases due of the hydrophobic nature of the cell wall.[11] This adhesion to both gas and oil interphases might increase methane bioavailability as a result of the higher methane concentrations present in the organic phase (tentatively 10 times higher)[5] and the potential direct uptake of methane and oxygen from the gas bubbles and oil drops. In this context, the specific methane elimination capacities (per gram of biomass) increased at 800 rpm by 48.5% with $\phi = 5\%$ and by 47% with $\phi = 10\%$. A mass balance for oxygen transfer at 800 rpm with silicone oil addition (5 and 10%) suggest that oxygen transferred to maximal gradient (calculated with the experimental abiotic $k_L a_{O_2}$), is not sufficient to satisfy the oxygen requirements needed to mineralize the CH_4 corresponding to the ECs recorded, which indicates that alternative pathways may be established including direct cell contact with the oil. In addition, no direct correlation between the abiotic for oxygen and EC was established, probably due to the complex interfacial effects governing substrate mass transport in TPPBs (1 continuous and 3 dispersed phases).[11] A similar conclusion was proposed by other authors.[9,10] Finally, it is important to highlight that a reduction in

Table 1 Elimination capacity (*EC*), CO_2 production rate (*RCO_2*) and percent removal efficiency (*%RE*) at different stirring rates and silicone oil fractions.

	EC (*g m^{-3} h^{-1}*)			*RCO_2* (*g m^{-3} h^{-1}*)			*%RE*		
N (rpm)	*0%*	*5%*	*10%*	*0%*	*5%*	*10%*	*0%*	*5%*	*10%*
200	6 ± 1	5 ± 1	6 ± 2	6 ± 1	9 ± 1	11 ± 1	8 ± 2	8 ± 2	9 ± 2
500	20 ± 3	24 ± 2	30 ± 3	24 ± 6	34 ± 3	43 ± 4	28 ± 2	38 ± 3	44 ± 1
800	33 ± 1	41 ± 3	48 ± 4	39 ± 2	54 ± 4	75 ± 5	59 ± 3	65 ± 2	70 ± 2

k_La at higher stirring rates does not necessarily implies a reduction in overall methane consumption as other factors, such as oil drop diameter changes may vary affecting the direct cell uptake.

CONCLUSIONS

The abiotic $k_La_{O_2}$ increased with the addition of silicone oil or inactive cells compared to systems supplied exclusively with water. Nevertheless, an increase in the stirring rates from 500 to 800 rpm reduced the abiotic volumetric oxygen transfer coefficient with silicone oil or inactive cells addition in the reactor. More research is however needed to understand the mechanisms underlying this unexpected experimental finding. The increase in both stirring rate and oil fraction did however enhance CH_4 elimination capacity and CO_2 production rate in the system. In this regard, the increase in specific *EC* and specific RCO_2 at increasing stirring rates from 500 to 800 rpm suggest an enhanced bioavailability of oxygen and methane for the metanotrophic cells due to a facilitated uptake of these compounds directly from the oil drops or air bubbles.

ACKNOWLEDGEMENTS

The authors appreciate the financial supports from the Mexican Council of Science and Technology (José Rocha-Rios grant #295496) and from the Spanish Ministry of Education and Science (RYC-2007-01667 and CONSOLIDER-NOVEDAR CSD 2007-00055).

REFERENCES

1 U.S. Environmental Protection Agency (USEPA), Global mitigation of non-CO_2 greenhouse gases (2006).
2 http://www.epa.gov/climatechange/economics/downloads/GlobalMitigationFullReport.pdf
3 Nikiema J, Brzezinski R and Heitz M. Elimination of methane generated from landfills by biofiltration: a review. *Rev Environ Sci Biotechnol* 6:261–284 (2007).
4 Quijano G, Revah S, Gutierrez-Rojas M, Flores-Cotera L and Thalasso F. Oxygen transfer in three-phase airlift and stirred tank reactors using silicone oil as transfer vector. *Process Biochem* 44:619–624 (2009).
5 Daugulis A, Two-phase partitioning bioreactors: a new technology platform for destroying xenobiotics. *Trends Biotechnol* 19:457–462 (2001).
6 Rocha-Rios J, Bordel S, Hernández S and Revah S. Methane degradation in two-phase partition bioreactors. *Chem Eng J* 152:289–292 (2009).

7 Galaction A, Cascaval D, Oniscu C and Turnea M. Enhancement of oxygen mass transfer in stirred bioreactors using oxygen-vectors. 1. Simulated fermentation broths. *Bioprocess Biosyst Eng* 26:231–238 (2004).
8 Moo-Young M and Blanch H. Design of biochemical reactors. Mass transfer criteria for simple and complex systems. *Adv Biochem Eng* 19:1–69 (1981).
9 Galaction A, Cascaval D, Turnea M and Folescu E. Enhancement of oxygen mass transfer in stirred bioreactors using oxygen-vectors. 2. *Propionibacterium shermanii* broths. *Bioprocess Biosyst Eng* 27:263–271 (2005).
10 Nielsen D, Daugulis A and McLellan P. A restructured framework for modelling oxygen transfer in two-phase partitioning bioreactors. *Biotechnol Bioeng* 91:773–777 (2005).
11 Nielsen D, Daugulis A and McLellan P. A novel method of simulating oxygen mass transfer in two-phase partitioning bioreactors. *Biotechnol Bioeng* 83:735–742 (2003).
12 McLeod C and Daugulis A. Interfacial effects in a two-phase partitioning bioreactor: degradation of polycyclic aromatic hydrocarbons (PAHs) by an hydrophobic *Mycobacterium*. *Process Biochem* 40:1799–1805 (2005).

Chapter 17

FUNCTIONAL AND STRUCTURAL RESPONSE OF AMMONIA AND VOC CONVERTING BIOFILM TO VARIATIONS IN AIR LOADING AND WATER MANAGEMENT

Susanne Juhler & Lars D.M. Ottosen
Department of Biological Sciences, Microbiology, University of Aarhus, Aarhus C, Denmark

Lise Bonne Guldberg
SKOV A/S, Roslev, Denmark

Lars Peter Nielsen, Andreas Schramm & Niels Peter Revsbech
Department of Biological Sciences, Microbiology, University of Aarhus, Aarhus C, Denmark

ABSTRACT

In situ activity, distribution, and stability of nitrifying and heterotrophic processes were investigated under various air load and water management strategies in a full-scale biotrickling filter designed for the degradation of NH_3 and odorous organic compounds in the ventilation air from a pig facility. The filter biofilm was examined through a broad in situ approach combining microscale profiling of O_2 and pH by electrochemical sensors, fluorescence in situ hybridization (FISH), calculation of NH_3 and NO_2^- conversion rates based on air and water analysis, and batch incubation measurements of NH_3 oxidation capacities.

High oxygen consumption by other processes than ammonia oxidation were observed in a surface biofilm layer of 50–100 μm extending throughout the filter. Ammonia-oxidizing bacteria (AOB) were constrained to an underlying biofilm layer, resulting in a steep pH decline from the air-biofilm interface to the AOB zone. While NH_3 uptake from the air phase depended on the interface pH, AOB activity was controlled by the lower within-biofilm pH, which through protonation of NO_2^- and NH_3 subjected AOB to HNO_2 inhibition and NH_3 limitation.

The magnitude and zonation of non-AOB oxygen consumption in the filter indicated high concentrations in the air of an unidentified and moderately soluble substrate different from ammonia and known VOC's in pig farm ventilation air. A nearly constant 1:1 production ratio of NO_3^- and NO_2^- throughout 10 weeks of variations in air load, water drainage, and biofilm biomass and a low abundance of known nitrite-oxidizing bacteria (<3% of the AOB cell number) suggested the action of an alternative pathway of nitrate and nitrite production.

Chapter 18

REGULATION OF AMMONIA OXIDATION IN BIOTRICKLING AIR FILTERS

Lars D.M. Ottosen & Susanne Juhler
Department of Biological Sciences, Microbiology, University of Aarhus, Aarhus C, Denmark

Lise Bonne Guldberg
SKOV A/S, Roslev, Denmark

Lars Peter Nielsen, Andreas Schramm & Niels Peter Revsbech
Department of Biological Sciences, Microbiology, University of Aarhus, Aarhus C, Denmark

ABSTRACT

Ammonia removal in bio trickling air filters takes place through dissolution in a biofilm with ammonia oxidizing bacteria (AOB) which partially oxidize ammonia to nitrite. Products of protonation (NH_4^+) and oxidation (NO_2^-) wash out via filter water drainage. AOB activity is limited by availability of their substrate NH_3 (not NH_4^+) and concentration of the oxidation product HNO_2 (not NO_2^-). Mechanisms controlling AOB activity in a bio trickling filter remains only partially understood. Recent clarification on separated inhibition kinetics for activity and growth suggest that when growth is fully inhibited by HNO_2, activity remains only 40–50% inhibited. This must have specific implications for AOB growth and activity patterns in bio filters with high ammonia load. The recent findings on AOB inhibition kinetics in general and specific studies of filter microbiology indicate that:

1 The AOB biomass should stabilize in relation to the average load of NH_3. If NH_3 load decreases, inhibition will increase and biomass will decrease. If NH_3 load increase, inhibition will decrease and biomass increase.
2 AOB activity will respond immediately to changes in NH_3 load, at constant biomass.
3 Outlet air NH_3 concentration is approximately in equilibrium with filter water and is therefore governed by its $[NH_4^+]$ and pH, which again is controlled by filter water conductivity and AOB activity respectively.

Effects of experimentally varied NH_3 loads on AOB growth and activity were investigated. Ammonia removal, filter water pH and conductivity, were measured on line during manipulations of ammonia load in a full scale biotrickling filter constructed for removal of NH_3 and odor from ventilation air at a pig farm. On line recordings were supplemented by point measurements of odor removal. Effects of changes in NH_3 load were studied over short periods (hours) and long periods (weeks).

The proposed auto regulation of biomass and activity was in principle confirmed by observations on filter response to experimental changes in NH_3 and outlet NH_3 concentrations, but gain and loss of biomass happens surprisingly slow over weeks due to low growth rates and even lower decay rates.

Implications of the function of the AOB community in bio trickling filter is that biomass adapt over weeks to the average load of NH_3, and will adjust activity to most of the short term (hours) variation typically seen in ventilation air rates from livestock production houses. The presence of high concentrations of organic gasses (e.q. in the same range as NH_3) may result in less predictable response to NH_3 load variations but the general mechanisms remains.

Bioenergy – resource recovery

Chapter 19

MICROALGAE AS FEEDSTOCK FOR BIODIESEL PRODUCTION:

Carbon dioxide sequestration, lipid production and biofuel quality

Érika Cristina Francisco

School of Chemical Engineering, State University of Campinas, Campinas-SP, Brazil

Eduardo Jacob-Lopes

School of Agricultural Engineering, Federal University of Pelotas, Pelotas-RS, Brazil

Débora Balthazar Neves & Telma Teixeira Franco

School of Chemical Engineering, State University of Campinas, Campinas-SP, Brazil

ABSTRACT

The objective of this paper was to evaluate the carbon dioxide sequestration capacity, biomass production, lipid content, lipid productivity and biodiesel quality of six microalgal strains cultivated photosynthetically in a bubble column photobioreactor. Lipid productivity was the criteria for selection species; for the best specie, carbon dioxide sequestration rates, biomass productivity, lipid content and lipid productivity of 17.8 mg/L · min, 20.1 mg/L · h, 27.0% and 5.3 mg/L · h were obtained. Qualitative analysis of the fatty acid methyl esters demonstrates the predominance on saturated (43.5%) and monounsaturated (41.9%) fatty acids. The critical parameters for quality properties of biodiesel evaluated (ester content, 99.85%; cetane number, 56.73; iodine value, 65.00 $gI_2/100$ g; degree of unsaturation, 74.07% and cold filter plugging point, 4.54°C) comply with the US Standard (ASTM 6751), European Standard (EN 14214), Brazilian National Petroleum Agency (ANP 255) and Australian Standard for biodiesel. The novelty of this paper is the estimation of the fuel properties of the microalgal biodiesel and the comparative study of conventional sources used as feedstock for biodiesel manufacture, facilitating worldwide advances in this research area.

Keywords: photobioreactor, biomass, microalgae, carbon dioxide sequestration, single cell oil (SCO), biodiesel

Chapter 20

TOLUENE GAS PHASE BIOFILTRATION BY *PAECILOMYCES LILACINUS* FOR BIOMASS PRODUCTION AND RECOVERY OF A HYDROPHOBIN PROTEIN

Gabriel Vigueras & Keiko Shirai
Departamento de Biotecnología, Universidad Autónoma Metropolitana-Iztapalapa, México D.F., México

Marcia Morales & Sergio Revah
Departamento de Procesos y Tecnología, Universidad Autónoma Metropolitana-Cuajimalpa, México D.F., México

ABSTRACT

Paecilomyces lilacinus consumed toluene as the sole carbon source in a gas-phase biofilter obtaining an average elimination capacity of 50 g m^{-3} h^{-1} and a removal efficiency of 53%. From the mycelia produced in the biofilter, a hydrophobin (PLHYD) of 10.6 ± 0.3 kDa was purified by electrobubbling, and identified by peptide mass fingerprinting. The yield was 1.1 mg PLHYD g $biomass^{-1}$. PLHYD production was also evaluated when grown in liquid (SmF) and solid-state fermentations (SSF) with n-hexadecane (HXD). The yield in SSF-HXD, 0.8 mg PLHYD g $biomass^{-1}$, was lower than the obtained in the toluene biofilters but in SmF-HXD the PLHYD was not detected. The respiratory activity in SmF-HXD (*V*max 1.16 g CO_2 m^{-3} h^{-1}) was two times higher than the SSF-HXD (*V*max 0.57 g CO_2 m^{-3} h^{-1}). The PLHYD showed surface activity to modify the hydrophobicity of Teflon from 130.1 (±2)° to 57.0 (±5)° and tolerated hot SDS washing.

INTRODUCTION

Biofilters are open systems and growth of filamentous fungi is consistently found. Consequently the number of fungal strains isolated and tested successfully in biofilters has increased[1]. Fungal biofilters show high elimination capacities (ECs) with some hydrophobic volatile organic compounds (VOCs) due to their aerial hyphae formation, which improves the exchange area for transport, and to the surface hydrophobicity which favors the partition of these compounds[2]. García-Peña et al.[3] found an EC for toluene of around

250 g m^{-1} h^{-1} with *Paecilomyces variotii*, which is about five times higher than values usually found in bacterial biofilters. Aerial growth and surface hydrophobicity are related with the hydrophobins, small proteins (~10 kDa) with biosurfactant activity, which protect the *fungal aerial* structures both from excessive cytoplasmic water evaporation and wetting.[5] Their ability to self-assemble at hydrophilic–hydrophobic interfaces is of interest for biotechnological applications such as the immobilization of biosensors to solid surfaces and as surfactants in two-phase liquid systems.[4] Recent work by Vigueras et al.[5] showed that the fungus *Rhinocladiella similis* adapted by modifying its surface hydrophobicity and expression of hydrophobin-like molecules when grown with compounds of opposite polarities. On the other hand, Crespo et al.[6] showed that the virulence of the entomopathogenic fungus *Beauveria bassiana* increased when cultivated in media with n-hexadecane, HXD. This compound has been used as a model molecule to study aliphatic hydrocarbon biodegradation. *Paecilomyces lilacinus*, widely used against phytopathogenic nematodes, metabolizes the aromatic hydrocarbons biphenyl and dibenzofuran.[7] The objectives of this study were to investigate toluene gas phase uptake by *P. lilacinus* in biofilters and the simultaneous recovery of added value product such as biomass and hydrophobins comparing the results with production on solid or liquid culture medium with HXD.

MATERIALS AND METHODS

Toluene gas phase biofiltration

Toluene biodegradation was evaluated at 28°C and 180 rpm in 125 mL flask inoculated with mineral medium[8] containing 2 × 10^7spores of *Paecilomyces lilacinus* CBS 284.36 mL^{-1}. The flasks were sealed with Teflon valves and liquid toluene (5 μL) were added to obtain an headspace concentration around 35 g m^{-3}. Biofiltration with toluene vapors were performed at 28°C using 0.230 L cylindrical glass tubes packed with perlite. A stream of toluene vapors was mixed with moistened air and passed through the biofilter at 100 ml min^{-1} (EBRT = 1.7 min). Toluene concentration in the inlet was 5 g m^{-3} (±1 g m^{-3}). Hydrophobins were also determined in SmF and SSF microcosms with HXD (36.5 g L^{-1}). Toluene and CO_2 were monitored by gas chromatography (GC)[8]. Biomass was determined as total protein and with biomass of *P. lilacinus* grown in liquid as control. Experiments were tested by duplicate.

Extraction and purification of the hydrophobins

The hydrophobin proteins were extracted from the mycelium combining extraction with concentrated formic acid and precipitation by electrobubbling in a glass column with a Pt filament at the bottom, 300 mA were applied during 1 h and the foam was collected for freeze drying.[8] The Bradford method was used to determine total protein.

Analysis by size exclusion high performance liquid chromatography (SEC-HPLC)

SEC-HPLC was run using a gel filtration column of 5 μm Bio-Select 250 SEC (Bio-Rad). The volume injection was 20 μl. The mobile phase was NaH_2PO_4 0.05 M, Na_2HPO_4 0.05 M and NaCl 0.15 M, pH 6.8. The flow rate was 1.0 mL min^{-1} and detection was performed at 280 nm. Molecular weight standard (Bio-Rad) was used to calibrate the system.

Protein analysis by SDS-polyacrylamide gel electrophoresis (SDS-PAGE)

Protein profiles were analyzed on the basis of molecular mass by SDS-PAGE. The gels were stained with Coomassie blue R-250, using broad range standard proteins (Invitrogen).

Analysis by reversed phase high performance liquid chromatography (RP-HPLC)

Protein samples were analyzed by RP-HPLC with a 5 μm Supelcosil LC 304 column (Supelco). The volume injection was 20 μl. The mobile phase consisted of 0.1% trifluoracetic acid (TFA) and 50% (v/v) acetonitrile with a pH of 3.0. The flow rate was 0.8 mL min^{-1} and detection was performed at 280 nm.[8]

Identification of hydrophobin by peptide mass fingerprint using MALDI-TOF

Hydrophobin peptides were generated and extracted from the SDS separated proteins following in-gel trypsin digestion protocols. Peptides were analyzed on a Voyager DE-PRO MALDI-TOF as spectrometer (Applied Biosystems). The masses of monoisotopic peaks were searched against the NCBI database.[8]

Surface activity of hydrophobin on hydrophobic solid

Surface activity of hydrophobin was evaluated over a hydrophobic surface of polytetrafluoroethylene (Teflon) according to method reported by Wösten et al.[4] Surface hydrophobicity was determined with a goniometer (CAM-Micro 600, Tantec). Analyses were repeated six times in three different points of each sample.

RESULTS AND DISCUSSIONS

TOLUENE GAS PHASE BIOFILTRATION

P. lilacinus was able to grow using toluene as the sole carbon sources. An estimated Vmax of 5.15 mg toluene g biomass^{-1} h^{-1}, with a total consumption of 53% and 44.3% mineralization to CO_2 of the consumed toluene corresponding to 1.44 mg CO_2 mg toluene^{-1} were obtained by fitting the Gompertz model. Earlier reports have shown that some filamentous fungi can biodegrade toluene in gas phase.[1,3] The evolution of toluene consumption in the biofilter is shown in figure 1. The EC and CO_2 production increased from day 10 reaching values of 50 and 120 g m^{-3} h^{-1}, respectively. The maximum removal efficiency, (RE), was 53% and was reached on day 34. These results were lower than the volumetric ECs reported for *P. variotii* CBS 113409 EC 55 g m^{-3} h^{-1} (RE 99%)[9], *Exophiala oligosperma* EC 77 g m^{-3} h^{-1} (RE 99%) *E. lecanii-corni* EC 80 g m^{-3} h^{-1} (RE 95%)[10] and *P. variotii* CBS 115145 EC 200 g m^{-3} h^{-1} (RE 98%)[11]. The discrepancy between the results may be due to differences in metabolism of the microorganism involved, fungal acclimatization or biomass concentration in the biofilters. Despite the low volumetric results, the specific EC value of 13.5 mg toluene g biomass^{-1} h^{-1} obtained with *P. lilacinus* was similar to the value of 10 mg toluene g biomass^{-1} h^{-1} reported by Aizpuru et al.[11] with *P. variotii* suggesting that higher volumetric rates might be expected if more growth is obtained in the reactor.

PURIFICATION AND IDENTIFICATION OF HYDROPHOBINS

Proteins between 1.35 and 17 kDa were detected in the SEC-HPLC analysis. Figure 2 shows the analysis by SDS-PAGE and RP-HPLC of the hydrophobin

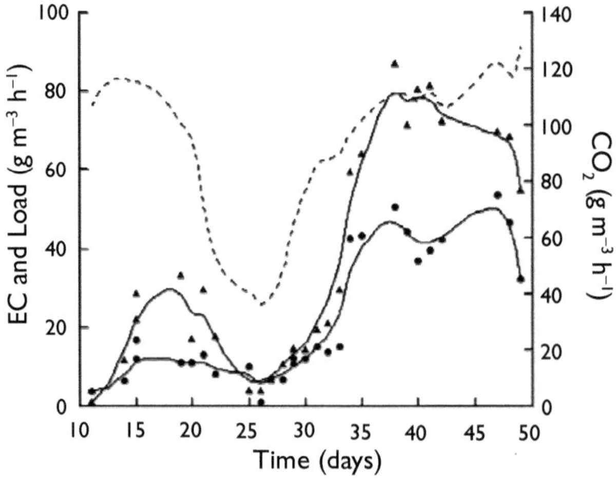

Figure 1 Time course of toluene load (dashed line), toluene EC (circles) and CO_2 production (triangles) in the biofilter inoculated with *P. lilacinus*.

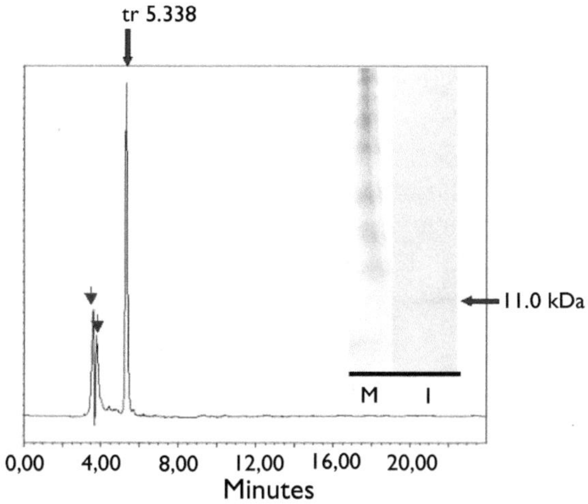

Figure 2 RP-HPLC and SDS-PAGE *Lanes* M, molecular weight standard; I, PLHYD hydrophobin purified from the mycelium of *P. lilacinus* growth in the biofilter with toluene.

extracted from the mycelium obtained in the biofilter. The molecular weight was 11 kDa which corresponds to the reported for hydrophobins, around 12 kDa[4]. The RP-HPLC shows only three peaks, the most intense (60.02% of area) was detected at 5.3 minutes which correspond to the hydrophobin. The other two peaks were not identified.

The hydrophobin final yield of 1.1 mg PLHYD g biomass^{-1} was four times lower than the 4.9 mg HFBI g biomass^{-1} reported by Askolin et al.[12] with *Trichoderma reesei* however this strain contained three copies of the *hfb1* gene. The peptide mass fingerprinting analysis of the protein of 11 kDa resulted in the identification of two peptides that presented high homology with sequences of class I hydrophobin proteins. Class I hydrophobins have a consensus patterns X_{26-85}-C-X_{5-8}-C-C-X_{17-39}-C-X_{8-23}-C-X_{5-6}-C-C-X_{6-18}-C-X_{2-13}.[4]

EXPRESSION OF HYDROPHOBIN PROTEINS IN SMF AND SSF WITH N-HEXADECANE

P lilacinus was able to grow using n-hexadecane as the sole carbon source. The *V*max of respiratory activity in SmF and SSF were 1.16 and 0.57 g CO_2 m^{-3} h^{-1} respectively. The yield in SSF, 0.8 mg PLHYD g biomass^{-1}, was lower than the obtained in the toluene biofilters, but in liquid culture medium the PLHYD was not detected (Fig. 3).

In our study the PLHYD was able to modify the hydrophobicity of Teflon lowering the contact angle of a water drop from 130.1 (±2)° to 57.0 (±5)° and tolerated hot SDS washing. These results agree with similar work reported by Wösten et al.[4]

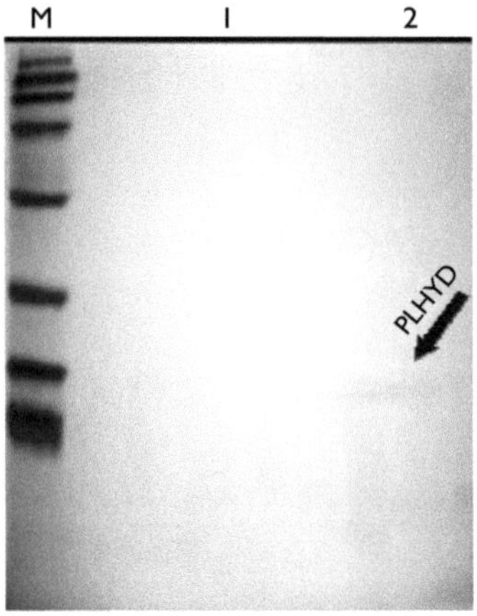

Figure 3 SDS-PAGE *Lanes* M, molecular weight standard; 1 and 2, proteins extracted from the mycelium produced in SmF and SSF with HXD respectively.

CONCLUSION

This is a first report on biodegradation of toluene by *P. lilacinus*, a nematophagous fungus, in a gas-phase biofilter and recovery of added value products such as hydrophobins proteins. Also the biosurfactant activity of hydrophobins may have an important role in the solubility and transfer of VOCs, but more work is needed in this respect.

ACKNOWLEDGEMENTS

To CONACYT, Dr. Telma Franco, the EU Alfa Polylife project and Instituto de Ciencia y Tecnología del DF, México for supporting the trip of GV.

REFERENCES

1 Kennes C and Veiga M. Fungal biocatalysts in the biofiltration of VOC-polluted air. *J Biotechnol* 113: 305–319 (2004).
2 Vergara-Fernández A, Van Haaren B and Revah S. Phase partition of gaseous hexane and surface hydrophobicity of Fusarium solani when grown in liquid and solid media with hexanol and hexane. *Biotechnol Lett* 28: 2011–2017 (2006).
3 Garcia-Peña I, Hernandez S, Favela-Torres E, Auria R and Revah S. Toluene biofiltration by the fungus Scedosporium apiospermum TB1. *Biotechnol. Bioeng* 76: 61–69 (2001).
4 Wösten H, van Wetter M, Lugones L, van der Mei H, Busscher H and Wessels J. How a fungus escapes the water to grow into the air. *Curr Biol* 9: 85–88 (1999).

5 Vigueras G, Arriaga S, Shirai K, Morales M and Revah S. Hydrophobic response of the fungus Rhinocladiella similis in the biofiltration with volatile organic compounds with different polarity. *Biotechnol Lett* 31: 1203–1209 (2009).
6 Crespo R, Júarez M, Dal Bello G, Padín S, Fernández G and Pedrini N. Increased mortality of Acanthoscelides obtectus by alkane-grown Beauveria bassiana *BioControl* 47: 685–696 (2002).
7 Gesell M, Hammer E, Mikolasch A and Schauer F. Oxidation and ring cleavage of dibenzofuran by the filamentous fungus Paecilomyces lilacinus. *Arch Microbiol* 182: 51–59 (2004).
8 Vigueras G, Shirai K, Martins D, Franco T, Fleuri L and Revah S. Toluene gas phase biofiltration by Paecilomyces lilacinus and isolation and identification of a hydrophobin protein produced thereof. *Appl Microbiol Biotechnol* 80: 147–154 (2008).
9 Estévez E, Veiga C and Kennes C. Biofiltration of waste gases with the fungi Exophiala oligosperma and Paecilomyces variotii. *Appl Microbiol Biotechnol* 67: 563–568 (2005).
10 Woertz J, Kinney K, McIntosh N and Szaniszlo P. Removal of toluene in a vapor-phase bioreactor containing the strain of the dimorphic black yeast Exophiala lecanii-corni. *Biotechnol Bioeng* 75: 550–558 (2001).
11 Aizpuru A, Dunat B, Christen P, Auria R, García-Peña I and Revah S. Fungal biofiltration of toluene on ceramic rings. *J Environ Eng* 13: 396–402 (2005).
12 Askolin S, Nakari-Setälä T and Tenkanen M. Overproduction, purification, and characterization of the Trichoderma reesei hydrophobin HFBI. *Appl Microbiol Biotechnol* 57: 124–130 (2001).

Sulfur compounds

Chapter 21

CONTROL OF SULFUR COMPOUNDS EMISSIONS

Sergio Revah
Departamento de Procesos y Tecnología, Universidad Autónoma Metropolitana-Cuajimalpa, Mexico City, Mexico

ABSTRACT

Emission of sulfur pollutants have been the main target of biological air treatment methods since its first reports in the '30s. Among these compounds hydrogen sulfide (H_2S) has been the most studied followed by carbon disulfide (CS_2) and the organic sulfides such as mercaptans, di methyl sulfide, di methyl disulfide and finally the oxidized species such as sulfur dioxide emitted from combustion processes. Research and applications have rapidly evolved in the past two decades from the traditional biofilters packed with natural supports to controlled biotrickling filters operating in a wide range of conditions adapted to the specific applications and with very short residence times.

The presentation will cover the results from different applications starting with the research and scale up of a biotrickling filter with short residence times to treat the emissions from a viscose processing plant with high H_2S and CS_2 concentrations. Industrial experience with these full scale systems (up to 760 m^3/min) suggested the necessity of reducing acid production and the strategy of maximizing elemental sulfur production was further explored.

To explore superior systems for low concentrations of sulfides, higher (around 10) pHs have been explored in the past few years. This pH improves the solubility of H_2S and some other sulfides increasing the mass transfer rates which can be the limiting step under low pollutant concentrations. This strategy requires the adaptation of a different microbial population than that which normally develops under neutral or acid conditions. Results will be presented on sulfide elimination under alkaline conditions including recent advances on the microbiology of these systems.

Chapter 22

REAL-TIME PCR AND PCR-DGGE ON THE FUNCTIONAL *soxB* GENE:

A new approach to monitor sulfo oxidizing bacteria in biotricklers treating hydrogen sulfide

Marc Jovic, Sebastien Lacroix, Camille Secher, Nathalie Brack & Anne-Sophie Lepeuple
Microbiology and Biotechnologies Program, Centre de Recherche sur l'Eau (Anjou Recherche – Veolia), Chemin de la Digue, Maisons-Laffitte Cedex, France

Isabelle Charron & Philippe Zozor
Air and Odour Program, Centre de Recherche sur l'Eau (Anjou Recherche – Veolia), Chemin de la Digue, Maisons-Laffitte Cedex, France

ABSTRACT

This study deals with the ecology of biotrickling filters dedicated to the treatment of hydrogen sulfide (H_2S).

Biotrickling filters processes are biofilm processes based on the ability of microorganisms to degrade a large variety of chemical compounds. They are widely used in environmental industries to remove H_2S. Indeed, if compared with physico-chemical purification processes, these processes have shown lower engineering and working costs.

Numerous studies based on the optimisation of physico-chemical parameters showed an efficiency enhancement of biotrickling filters. Nevertheless, only few works studied the impact of their microbial ecology. In this context, analytical tools were developed in order to quantify Sulfo-Oxidizing Bacteria (SOB) and to study their diversity in biotrickling filters.

SOB can oxidize H_2S into sulphate (SO_4^{2-}). They constitute a functional group composed of a large variety of phylum: *Chloroflexi*, *Chlorobi*, *Proteobacteria* and *Spirochaeta*. Because of their large phylogenic diversity, selective analysis of SOB using 16S rRNA approach remained impossible. It was necessary to use an alternative approach using specific functional gene to study SOB in environment.

In spite of divergences in metabolisms of SOB, the *soxB* gene, which is common to all the members of the SOB group, was identified. Indeed, it was proven that SOB always implicates a hydrolase coded by this gene to oxidize sulfur compounds. As a result, *soxB* gene was used as a molecular target to develop real time PCR (qPCR) and PCR-DGGE protocols.

Once developed, these tools were applied to the monitoring of biotrickling filters pilots treating H_2S during a six month period of time.

The biotrickling filters pilots were composed of three filtration unit filled with three different packing materials. On the one hand, the monitoring consisted in the measurement of classical physico-chemical parameters (H_2S abatement, elimination capacities, and effluent pH). On the other hand, the microbiological aspect was monitored by using molecular techniques (*soxB* qPCR and DGGE, 16r DNA qPCR and DGGE and DAPI).

The results of this monitoring, associated to the molecular tools developed, showed a variation of SOB population densities with packing materials and H_2S inlet loading. This work also demonstrated an influence of the packing material used on the SOB population diversity.

This new approach highlights the interest to study microbiological aspects of processes by using a functional gene as a molecular target. It can provide useful information and can be considered as a key tool for the improvement of H_2S treatment by biotrickling filters.

Keywords: hydrogen sulfide (H_2S), biotrickler, real-time PCR, *soxB* gene, Sulfo Oxidizing Bacteria (SOB)

Chapter 23

EXTREMELY ACIDOPHILIC SULFUR-OXIDISING BACTERIA FOR WASTEGAS TREATMENT

N.J.R. Kraakman
Bioway, The Netherlands/USA

M.J. Smeulders, M.H. Zandvoort, A. Pol, M.S.M. Jetten & H.J.M. Op den Camp
Department of Microbiology, Radboud University Nijmegen, Nijmegen, The Netherlands

INTRODUCTION

The ability of micro-organisms to be active in extreme environments offers challenges to understanding the fundamentals of microbial function and opportunities for sustainable biotechnology. One group of micro-organisms that can grow and survive under extreme acidic conditions are called acidophiles. Acidophiles have a pH optimum less than pH = 3 of which some of obligate acidophiles, unable to grow at all at neutral pH.

The search for unknown micro-organisms can also open up interesting possibilities for new biological wastegas treatment applications. In biological wastegas treatment systems treating reduced sulfur compounds like hydrogen sulfide and carbon disulfide, sulfuric acid is the final product at complete conversion. This acidic water leaves the biological wastegas treatment system when it is operated without the additional of pH neutralizing chemical such as caustic. When possible, this operation without the addition of chemicals is preferred as caustic can be relatively expensive and adds to operational concerns as it requires safety measures for proper feeding and storage.

The low pH process water can, on the other hand, cause several problems. The acidity increases the stress on many microorganisms resulting in reduced activity as well as incomplete conversion which leads to elemental sulfur accumulation. The large amounts of water required to remove the produced acid from a reactor treating highly concentrated waste gasses can also be a limiting factor for many applications. With extremely acidophilic micro-organisms, many of these problems can be solved.

In our quest to obtain new microorganisms, we enriched different bacterial strains converting reduced sulfur compounds under extreme acidic conditions. We found micro-organisms capable of degrading different reduced sulfur compounds at extremely acidic conditions ($pH < 0$).

MICROBIAL SURVIVAL MECHANISMS AT LOW pH

Acidophiles use a variety of mechanisms that enables low pH tolerance. Baker-Austin and Dopson[1] described many of these mechanisms, which mainly involve maintaining a pH gradient across the cellular membrane wall. The wide repertoire of genes for the cell wall of acidophiles is indicative of a complex structure that probably limits the influx of protons into the cell. In the highly impermeable membrane a predominance of potassium-transporting energy-generating enzymes (ATPases) is found in many acidophiles, which possibly creates the positively charged molecules inside the cell inhibiting the influx of protons. Also specific proton pumping systems to remove protons from the cell have been identified although not clearly defined yet.

Organic acids are harmful to acidophiles because they are protonated (and therefore neutrally charged) in a low pH outside the cell and can pass relatively easy into the cell (cyctoplasm) where the proton dissociates at the more pH neutral conditions lowering the pH in the cell. Many acidophiles have the capacity to degrade organic acids and it is interesting to note that the most extreme acidophiles are all heterotrophs.

OUR SEARCH FOR LIFE AT LOW pH

The aim of our project, is to find microorganisms that use carbon disulfide (CS_2) as a carbon and energy source, with a high affinity for carbon disulfide, and able to tolerate high acidity levels as extreme as 10% w/v H_2SO_4, a concentration at which it is economically viable to reuse the trickling water within certain factories emitting carbon disulfide.

In earlier work from a full-scale biotrickling reactor, treating gas streams that contain carbon disulfide and hydrogen sulfide, a sulfur-oxidizing bacterium was isolated. The bacterium, which will be described elsewhere, was found to oxidize a variety of sulfur compounds including carbon disulfide (CS_2), carbonyl sulfide (COS), hydrogen sulfide (H_2S), elemental sulfur (S^0) and thiosulfate ($S_2O_3^{2-}$). In contrast to other Thiobacilli-like organisms which can oxidize carbon disulfide, this organism was found to oxidize sulfur compounds at extreme acidic conditions with pH values below 0. Since the strain also showed degradation activity at neutral pH, the strain can be defined as an extremely acidotolerant sulfur-oxidizing bacterium. Under aerobic conditions these

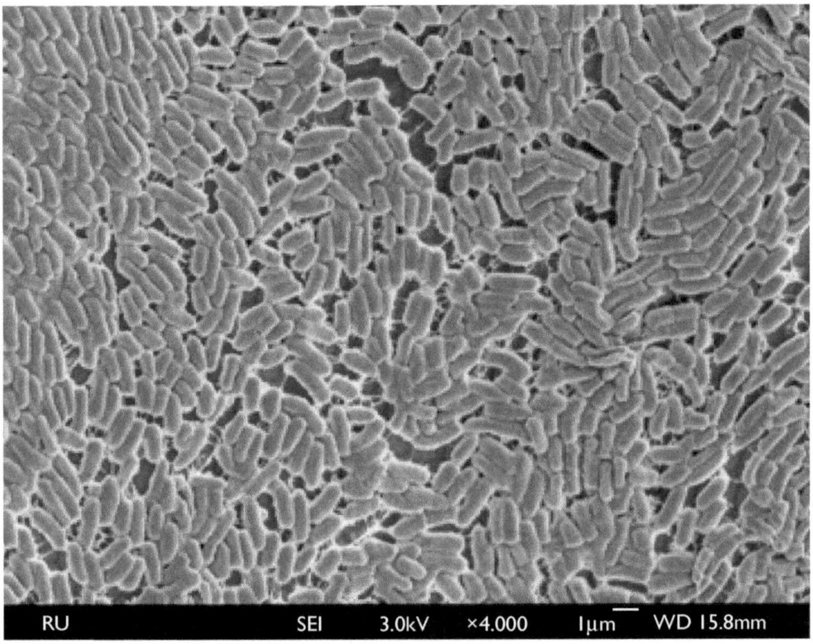

Figure 1 Extremely acidophilic bacteria.

sulfur compounds were completely oxidized to sulfate. Chemostat cultivation experiments have been performed in order to determine the kinetic parameters (molar growth yield, K_s value, maximum oxidation rate, maximum growth rate) of the isolate under various environmental conditions. Fluorescent DNA probes are being designed based on the 16S rDNA gene sequence in order to be able to monitor the presence and changes in mixed microbial communities similar to those in biotrickling filters which might contain these microbes.

In our recent quest to obtain such microorganisms that use carbon disulfide as a carbon and energy source, with a high carbon disulfide affinity, and able to tolerate high acidity levels, we enriched eleven bacterial strains from six different bioreactors treating carbon disulfide containing waste gases from five different industrial plants, and five bacterial strains from environmental samples collected at three highly acidic, sulfur rich sites (pooled hotspring samples, the Solfatara region near Naples and the Solfarata region near Rome). Although we observed substantial phenotypic differences in growth characteristics, 16S DNA analysis showed that all strains were Acidithiobacilli. Several known Acidithiobacillus strains for the ability to convert carbon disulfide were tested but found this is not a trait shared by all Acidithiobacilli. Also the tolerance for acidity levels varied largely with some strains showing very high tolerance for extreme acidic conditions.

BIOLOGICAL PROCESS CONTROL AT LOW pH

The low pH process water in a biological wastegas treatment reactor can cause several problems when treating reduced sulfur compounds: 1) a decrease in removal efficiency due to the low pH inhibiting growth and activity of the converting microorganisms, 2) increased potential of clogging of the system by elemental sulfur produced by stressed bacterial cells, 3) an increase in costs due to the large amounts of water required to remove the produced acid during treatment of highly concentrated waste gasses. Extreme acidophilic micro-organisms are able to cope much better with the stress condition of low pH and reduce or eliminate these problems.

In bench-scale reactor experiments we have seen that the biomass growth is reduced when the process conditions of the micro-organisms where more extreme acidic (data not shown here). Sulfuric acid concentration is found to be an important factor in the development of pressure drop over the reactor. Sulfuric acid concentrations of >3.5–4% (w/w) significantly limited the development of pressure drops over the reactors. At 5% sulfuric acid the pressure drop over the reactor even seems to stabilize. The expected accumulation of elemental sulfur was not observed and consequently did not cause increase in pressure drop. Actual gas retention time determination showed clearly the difference in biofilters gas distribution at different acid conditions as effect on the biomass accumulation. The reduced biomass accumulation at the more extreme acidic conditions can be explained by the reduced biomass yield when the bacteria grow under more extreme stressful conditions. An alternative explanation for the effect of sulfuric acid on pressure

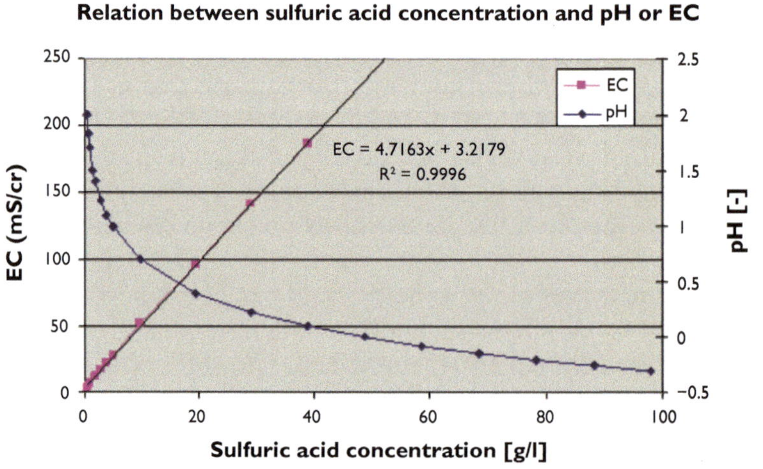

Figure 2 Relationship between acidity, pH and electrical conductivity (EC).

drop development could be the inhibition of the growth of other secondary micro-organisms that grow in the biofilm of the reduced sulfur compound degrading micro-organisms. The lower biomass growth at least provides opportunities to reduce the risk of extreme biomass growth in a bioreactor treating high loadings.

To maintain the most optimal process condition in a bioreactor, control of microbiological process parameters is essential. The pH is especially important when extreme acidic conditions are maintained using acidophilic bacteria. At extreme low acidity the pH cannot be measured accurate and the electro-conductivity (EC) can be used. The EC value increases linear with the acid concentration (see Figure 2) and is high enough to have no influence of other ions like nutrients. Using electrical conductivity as a control parameter has the advantage that frequent calibration is not required and that some EC-meters can even give a direct read-out of sulfuric acid concentration.

APPLICATIONS

Biotechnological applications using acidophiles is interesting because they can perform conversion in an extreme environment that can have important advantages.

One of the examples of using extreme acidophilic micro-organisms is the wastegas combined treatment of carbon disulfide hydrogen sulfide. Carbon disulfide emissions contribute to the greenhouse gas inventory, because carbon disulfide with its lifetime of 7–12 days is unstable and forms carbonylsulfide (COS). Carbonylsulfide may be transported to the stratosphere where its oxidation is thought to be a major source for stratospheric sulfur aerosol. This aerosol influences the earth's radiation balance and hence the climate.[2]

Traditionally, the industrial emissions of carbon disulfide have been treated by adsorption or incineration. These conventional treatment systems have relatively high operational costs and use substantial amounts of energy. They also require many safety measures for explosion-safe operation and generate waste streams (e.g., SO_2 emissions or spent activated carbon). Biological processes for waste gas treatment represent an interesting alternative and have been applied already in different industries.[3]

The sulfuric acid that is produced in a bioreactor can be neutralized by caustic in the bioreactor to form Na_2SO_4, but will require caustic which makes the process much more expensive. The produced sulfuric acid is a clear liquid, but with a concentration normally too low the applicability for re-use in a plant production process is limited. With the more extreme acidophilic micro-organisms this recycling of sulfuric acid is now more interesting and economically feasible in certain situations.

REFERENCES

1 Baker-Austin C and Dopson M. Life in acid: pH homeostatis in acidophiles. *Trends in Microbiology* **15**(4):165–171 (2007).
2 Chin M and Davis DD. Global sources and sinks of COS and CS_2 and their distributions. *Global Biogeochemical Cycles* 7:321–337 (1993).
3 Kraakman NJR and de Waal KJA. Treatment of carbon disulfide from industrial waste gas emissions: air pollutants turned into e-usable sulfuric acid. In: *Waste gas treatment for resource recovery*. Editors: Lens P.N.L., Kennes C., LeCloirec P. and Deshusses M. IWA Publishing, London (2006).

Chapter 24

BIOTRICKLING FILTERS FOR HYDROGEN SULPHIDE, METHYL MERCAPTAN, DIMETHYL SULPHIDE AND DIMETHYL DISULPHIDE REMOVAL BY *THIOBACILLUS THIOPARUS* AND *ACIDITHIOBACILLUS THIOOXIDANS*

M. Ramirez, M. Fernández, M.S. Cáceres, R.M. Pérez, J.M. Gómez & D. Cantero
Faculty of Sciences, Department of Chemical Engineering and Food Technologies, University of Cádiz, Spain

M. Fernández & R.M. Pérez
Faculty of Natural Sciences, Study Centre of Industrial Biotechnology, University of East, Santiago of Cuba, Cuba

M.S. Cáceres & G. Aroca
School of Biochemical Engineering, Pontificia Universidad Católica de Valparaíso, Valparaíso, Chile

ABSTRACT

BACKGROUND: Hydrogen sulphide (H_2S), methyl mercaptan (MM), dimethyl sulphide (DMS) and dimethyl disulphide (DMDS) are the main odorous compounds found in waste gases from wastewater treatment plants. It is known that two-stage biofiltration can be used for improving the removal efficiency of mixture compounds in gas effluent. So the performance of two biotrickling filters, the first one with *Acidithiobacillus thiooxidans* (BTA), and the second one with *Thiobacillus thioparus* (BTT) was researched.

RESULTS: For a separately reduced sulphur compounds (RSC) feeding the EC order was: DMDS>DMS>MM. The values of EC were 9.8 $g_{MM\text{-}S}$ $m^{-3}h^{-1}$ (BTT; 78% RE; EBRT 58 s), 36 $g_{DMDS\text{-}S}$ $m^{-3}h^{-1}$ (BTT; 94.4% RE, EBRT 76 s) and 57.5 $g_{H_2S\text{-}S}$ $m^{-3}h^{-1}$ (BTA; 92% RE, EBRT 59 s). For simultaneous removal in BTT, the increase of H_2S concentration from 23 to 293 ppmv (EBRT of 59 s) inhibited the RE of DMS (97–84% RE), DMDS (86–76% RE) and MM (83–67% RE). For two-stage biofiltration the RE not decresed for an increased of H_2S concentration from 75 to 432 ppmv.

CONCLUSIONS: The RSC removal can be carried out efficiently by two-stage biotrickling filters in series, the first one inoculated with *A. thiooxidans* and the second one with *T. thioparus*. The polyurethane foam as carrier for *A. thiooxidans* immobilization showed very good scale-up possibilities.

Keywords: Biotrickling filter, reduced sulphur compounds, *Thiobacillus thioparus*, *Acidithiobacillus thiooxidans*, polyurethane foam

INTRODUCTION

Several processes in sewage treatment plants generate high concentrations of malodorous components, mainly sulfur-containing compounds such as Hydrogen sulphide (H_2S) and other reduced sulphur compounds (RSC) like methyl mercaptan (MM), dimethyl sulphide (DMS), and dimethyl disulphide (DMDS),[1] because these compounds have a very low odour threshold (H_2S, 8.5–1000, MM, 0.9–8.5, DMS, 0.6–40 and DMDS, 0.1–3.6 in ppbv).[2]

The biological removal of H_2S by biofiltration have been extensively studied,[3] but not the simultaneous removal of RSC. Many microorganism have been used for the removal of simultaneous removal of RSC: *Thiobacillus*,[4–8] *Hyphomicrobium*,[9–11] *Xanthomonas*,[12] *Methylophaga sulfidovorans*,[5] and *Microbacterium sp* and *Pseudomonas*.[13,14]

The removal of RSC has been carried out by biofiltration in biofilters and biotrickling filters. Another studies show that using a two-stage biofiltration the removal efficiency of these compounds can be improved.[11,15,16] The principal advantage is that can be used different microorganism, carriers and operational conditions. The two-stage biofiltration have been researched for H_2S removal in presence of DMS,[11] MM,[16] DMS/MM,[15] DMS/DMDS,[17] but there isn't enough knowing about the simultaneous removal of MM/DMS/DMDS in presence of H_2S by two-stage biotrickling filter.

The aim of this work was the study of the simultaneous removal of a mixture of H_2S, MM, DMS and DMDS by two-stage biotrickling filters in series, the first one inoculated with *Acidithiobacillus thiooxidans* (BTA) for H_2S removal and the second one inoculated with *Thiobacillus thioparus* (BTT) for MM, DMS and DMDS removal. Other experiments were carried out to study the removal of DMDS, MM (separately) and a mixture of all RSC by BTT and H_2S removal by BTA.

EXPERIMENTAL

MICROORGANISMS AND MEDIUM PREPARATION

The original pure-culture strain of chemolithotrophic, sulfur compound oxidizer, *Thiobacillus thioparus* ATCC 23654 was obtained from the American Type Culture Collection. *T. thioparus* is a neutrophilic bacterium, able to use sulfide, sulfur, thiosulfate and tetrathionate.[18] Organic sulfur compounds like DMS, DMDS and MM also can be used as energy source by *T. thioparus*.

Acidithiobacillus thiooxidans (DSM11478) was obtained from Minas Gerais (Brazil). *A. Thiooxidans* is an acidophilic chemoautotrophic bacterium that grows in liquid medium on elemental sulfur, thiosulfate, or tetrationate.[18]

Two identical biotrickling filters was used. One biotrickling filter was inoculated with *T. thioparus* ATCC 23645 (BTT) and the another one with

A. Thiooxidans DSM11478 (BAT). The composition of basic mediums is showed in Table 1. The ATCC290:S6 medium was used in the propagation and immobilization of *T. thioparus*. The TTN+ and 0K medium was used as the recirculation medium in the biotrickling filters, TTN+ for BTT and 0K for BAT.

CHARACTERISTICS OF THE CARRIER MATERIAL

Small cubes of polyurethane foam (1 cm^3 in size) were used as the carrier. Polyurethane foam is an inert material with a very low commercial cost. The principal properties are density (20 kg m^{-3}) and porosity (96%).

A schematic of the experimental set-up of the lab scale biotrickling filters is shown in Fig. 1.

Table 1 Composition of basic mediums.

Medium	*ATCC290:S6*		*0K medium*	
Composition (g l^{-1})	KH_2PO_4	1.8	$(NH_4)_2SO_4$	3.0
	Na_2HPO_4	1.2	K_2HPO_4	0.5
	$MgSO_4 \cdot 7H_2O$	0.1	$MgSO_4 \cdot 7H_2O$	0.5
	$(NH_4)_2SO_4^a$	0.1	KCl	0.1
	$MnSO_4$	0.02	$Ca(NO_3)_2$	0.01
	$CaCl_2$	0.03		
	$FeCl_3 \cdot 6H_2O^b$	0.033		
pH		7.0		2.5
Temp. (°C)		30		30

[a]The nitrogen concentration was increased to 1 g l^{-1} for the BTT recirculation medium (referred to as TTN+).
[b]The $FeCl_3 \cdot 6H_2O$ was sterilized by filtration (0.22 mm pore size membrane).

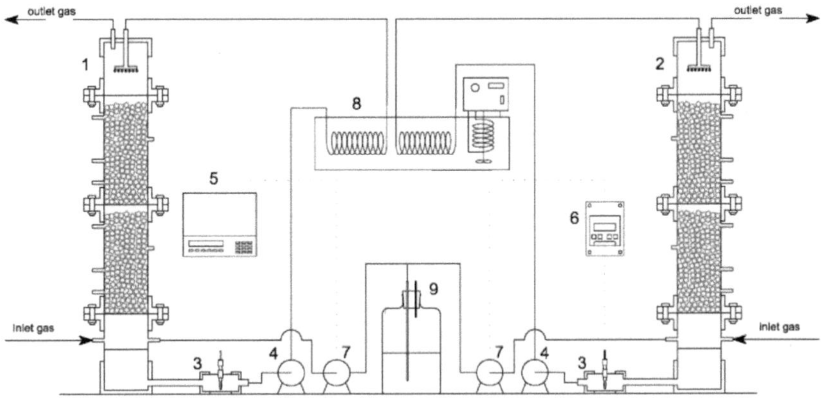

Figure 1 Schematic diagram of the laboratory-scale biotrickling filters system. 1: BTT, 2: BAT, 3: pH probe, 4: Recirculating pump, 5: pH Controller, 6: pH controller, 7: NaOH pump, 8: Termostatic bath, 9: NaOH container. Experimental set-up.

It was constituted by transparent PVC column (i.d. 0.105 m, height 1.5 m) with two stage (module height 0.25 m, total packed volume 4.33 L) joined with a screwed clamp. Five gas sampling ports were distributed along the BTF packed height at 0.12, 0.20, 0.37, 0.45 and 0.50 cm. The trickling solution was preserved at 30°C using a tubing coil submerged in a temperature controlled water bath. The pH was maintained at 7.0 in the BTT, and 2.0 in the BAT by adding sterile NaOH (2 N) using two pH controllers (Biocontroller ADI 1030, APPLIKON and PH28, CRISON) and an electrode with a sleeve diaphragm (CRISON 5221). The medium was recirculated at constant velocity of 7.85 m h^{-1} with centrifugal pump (ECOCIRC D5–8/810N).

GASEOUS GENERATING SYSTEM

A Schematic diagram of the gaseous generating system is shown in the Figure 2. Compressed air was passed through three consecutive filters, filled with silica gel, active carbon, and glass wool, respectively; after that, it was sterilized using a Millipore filter SLG05010 (0.45 µm), and humidified by passing through a column filled with distilled water. The synthetic waste gas stream was supplied to the biotrickling filters from two gas cylinders compressed H_2S (0.5vol%, balance N_2) and MM (0.1vol%, balance N_2).

The flow rates of H_2S and MM stream were regulated by mass flow rate controllers (Bronkhorst F-201C). A constant DMS and DMDS concentration stream was produced using the system reported by Smet.[19] Capillary steel tubing (O.D. 1/16", I.D.0.53 mm, AISI 316) was used to carry gaseous DMS and DMDS, by diffusion, from a steel vessel (Hoke, USA, vol. 75 mL) filled

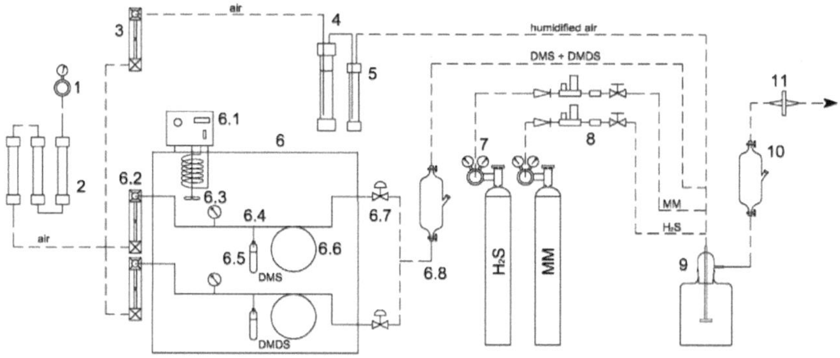

Figure 2 Schematic diagram of gaseous generating system. 1: Pressure regulator, 2: Air filters, 3: Flow meter, 4: Humidifier, 5: Water trap, 6: Gaseous DMS and DMDS generation system, 6.1: Heater pump recirculator, 6.2: Flow meters, 6.3: Pressure gauge, 6.4: T union, 6.5: Steel vessels, 6.6: Steel-tubing coils, 6.7: Needle valves and 6.8: Sampling glass bulb, 7: compressed gas cylinders of H_2S and MM, 8: Mass flow controllers, 9: Expansion tank, 10: Sampling glass bulb, 11: Membrane filter.

with the liquid; DMDS (purum, ≥98.0%, Fluka) and DMS (purum, ≥95.0%, Sigma-Aldrich). Needle valves and a 3 m long steel tubing spirals were used to apply an overpressure (~0.27 bar) on the diffusion tube to reduce variations at the final concentration of DMS and DMDS. The stream was then combined with humidified air. An expansion tank with a capacity of 2.5 L was used to homogenize the input stream. The final concentration was sampled from a glass bulb.

IMMOBILIZATION AND INITIAL CONDITIONS

The immobilization of BTT by *T. thioparus* was made in 16 Erlenmeyer flasks (vol. 1 L) containing 0.5 L of ATCC290:S6 medium, 0.1 L of inoculum and 5.4 g of foam cubes. The flasks were inoculated with a culture growing in exponential phase. The flasks were incubated in a rotary shaker (30°C, 150 rpm) monitoring the thiosulphate concentration. When thiosulphate was depleted, the mineral medium was replaced with fresh ATCC290:S6 medium. This operation was repeated thrice to obtain enough biomass attached to the polyurethane foam cubes. The BTT was packed with the colonized polyurethane foam and 1.5 L of ATCC290:S6 medium, without thiosulphate, were added as mobile phase and recirculated. After the immobilization, the study of DMS removal was carried out (89 day) by Arellano-García et al.[20]

The immobilization of BAT by *A. thiooxidans* was made with the polyurethane foam cubes of the biotrickling filter (total working volume of 1.278 L) used previously by Ramirez et al.[21] The BTA was packed with 10 g of polyurethane foam colonized and 40 g of polyurethane foam new (25 g in each stage). The biotrickling was fed with an air flow of 0.26 $m^3\ h^{-1}$ and a concentration of hydrogen sulfide between 30–50 ppmv. Fresh medium was added when the sulphate concentration reached 30 $g\ l^{-1}$ during a period of 405 days.

BIOFILTRATION EXPERIMENTS

The operational conditions are summarized in Table 2. For the evaluation of MM and DMDS removal by the biotrickling filter BTT experiments from 1 to 6 were carried out. The principal parameters studied were EBRT, inlet load and sulphate concentration. The performance of BAT was analyzed in the experiment 7 increasing the H_2S concentration. After that, mixtures of the H_2S, DMS, DMDS and MM were testing in the BTT (Exp. 8), the H_2S concentration was increased for knowing the effect in the removal efficiency of the rest of compounds. Finally, the simultaneous removal was studied by the two biotrickling filters in series, the first one BAT and the second one, the BTT (Exp. 9).

For the evaluation of MM and DMDS removal by the biotrickling filter BTT experiments from 1 to 6 were carried out. The principal parameters studied were EBRT, inlet load and sulphate concentration. The performance

Table 2 Operating conditions.

Exp.	*Compounds*	*EBRT (s)*	*Inlet load* $(gS\ m^{-3}h^{-1})$	C_{in} *(ppmv)**	$[SO_4{=}]$ $(g\ l^{-1})$	*Biotrickling filter*
1	MM	58 ± 1	1.2 to 12.1	16 to 144	<10	BTT
2	MM	31 to 208	5.1 ± 0.04	34 to 230	<10	BTT
3	MM	58 ± 1	5.0 ± 0.2	62 ± 2	3 to 23	BTT
4	DMDS	78 ± 0.6	3.0 to 37.3	25 to 312	<10	BTT
5	DMDS	78 to 15	27.7 ± 3.9	42 to 205	<10	BTT
6	DMDS	78 ± 1	24.6 ± 0.4	206 ± 3	5 to 15	BTT
7	H_2S	59 ± 1	4.2 to 102.0	54 to 1320	<30	BAT
8	H_2S		1.8 to 29.5	23 to 376		BTT
	MM	59 ± 1	2.4 ± 0.2	30 ± 2	<10	
	DMS		1.8 ± 0.3	23 ± 3		
	DMDS		7.2 ± 0.7	46 ± 5		
9	H_2S		5.9 to 33.9	75 to 432		BAT + BTT
	MM	59 ± 1	2.7 ± 0.2	34 ± 3	<10	(in series)
	DMS		2.3 ± 0.2	29 ± 2		
	DMDS		5.7 ± 0.8	36 ± 5		

*at 30°C.

of BAT was analyzed in the experiment 7 increasing the H_2S concentration. After that, mixtures of the H_2S, DMS, DMDS and MM were testing in the BTT (Exp. 8), the H_2S concentration was increased for knowing the effect in the removal efficiency of the rest of compounds. Finally, the simultaneous removal was studied by the two biotrickling filters in series, the first one BAT and the second one, the BTT (Exp. 9).

For the evaluation of MM and DMDS removal by the biotrickling filter BTT experiments from 1 to 6 were carried out. The principal parameters studied were EBRT, inlet load and sulphate concentration. The performance of BAT was analyzed in the experiment 7 increasing the H_2S concentration. After that, mixtures of the H_2S, DMS, DMDS and MM were testing in the BTT (Exp. 8), the H_2S concentration was increased for knowing the effect in the removal efficiency of the rest of compounds. Finally, the simultaneous removal was studied by the two biotrickling filters in series, the first one BAT and the second one, the BTT (Exp. 9).

Empty bed residence time, EBRT (s); removal efficiency, RE (%), load, L $(gS\ m^{-3}\ h^{-1})$ and elimination capacity, EC $(gS\ m^{-3}\ h^{-1})$ were determined using the relationships between the inlet concentration, C_{in} $(gS\ m^{-3})$, outlet concentration, C_{out} $(gS\ m^{-3})$, the gas flow rate, Q $(m^3\ h^{-1})$, and the filter bed volume, V (m^3) as follows: $EBRT = (V/Q)$; $RE = [(C_{in} - C_{out})/C_{in}] \times 100$; $L = C_{in}/V \times Q$; $EC = (C_{in} - C_{out})/V \times Q$.

ANALYTICAL TECHNIQUES

Gaseous DMS and DMDS concentrations were determined with a FID-GC (Agilent Technologies 6890), with a Porapak Q packed column (HP, 5' × 1/8", 80/100 mesh) and nitrogen (40 mL min^{-1}) as carrier gas. Temperatures of 190, 200 and 200°C were respectively used for the oven, injector and detector. The injected sample volume was 200 μL in all the cases. A specific sensor (Gasbadge Pro, ISC-Oldham) with a constant flow hand pump (ISC-Oldham) was employed to analyze the H_2S concentration in the gas phase for H_2S concentration below 500 ppmv (0–500 ppmv in 0.1 ppmv increments). For concentration bigger than 500 ppmv, the H_2S concentration was measured by iodometric tritation (iodine, $5 \cdot 10^{-3}$ N; starch indicator, 2 g l^{-1}) with a Tutweiler burette. MM gas concentration was measured by gas detector tube (Gastec 810–71, SKC). The detection ranges of the MM gas tube was 0.25–140 ppmv. Tedlar bag (10 L single PP fitting 232–08, SKC) was used for the sampling gas. Sulphate concentration was analyzed by turbidimetric method according the Standard Methods (APHA, 1998) and the thiosulphate concentration was measured by iodometric tritation acoording to Rodier.[22]

RESULTS AND DISCUSSION

EFFECT OF THE INLET LOAD OF MM AND DMDS ON THE RE

In the experiments 1 and 4 the effect of the inlet load was studied at a constant EBRT of 58 ± 1 and 78 ± 0.6 s for MM and DMDS respectively. The minimum time between each value of inlet concentration was for 6 h. Ramírez et al.[23] reported that the steady-state with *Thiobacillus thioparus* for H_2S removal was reached after 40 times the EBRT. The total time of the experiments were for 5 and 8 days for MM and DMDS respectively.

The removal of RSC has been investigated by *Thiobacillus thioparus* DW44. This reports indicated that the specific uptake rates for these compounds are H_2S > MM ≥ DMDS > DMS.[24] In this work for DMDS the critical elimination capacity was of 31.27 $g_{DMDS\text{-}S}$ $m^{-3}h^{-1}$ (RE 99.6%) and for MM the RE maximum only was of 94% for 1.14 $g_{MM\text{-}S}$ $m^{-3}h^{-1}$ (Fig. 3). Therefore, it can be concluded that the degradation rate of DMDS was better.

EFFECT OF THE EBRT ON MM AND DMS REMOVAL

In the experiments 2 and 5 the effect of EBRT was studied at two constant loads of 5.1 ± 0.04 $g_{MM\text{-}S}$ $m^{-3}h^{-1}$ and 27.7 ± 3.9 $g_{DMDS\text{-}S}$ $m^{-3}h^{-1}$. The minimum time between each value of EBRT was for 6 h. For MM removal the initial value of EBRT was 59 s (RE 95%). A decreasing in EBRT from 59 to 31 s affected the RE of MM nearly 24%. For DMDS the decay in the RE, when the EBRT was decreased from 77.5 to 27.2 $g_{DMDS\text{-}S}$ $m^{-3}h^{-1}$, was from 98 to 94%,

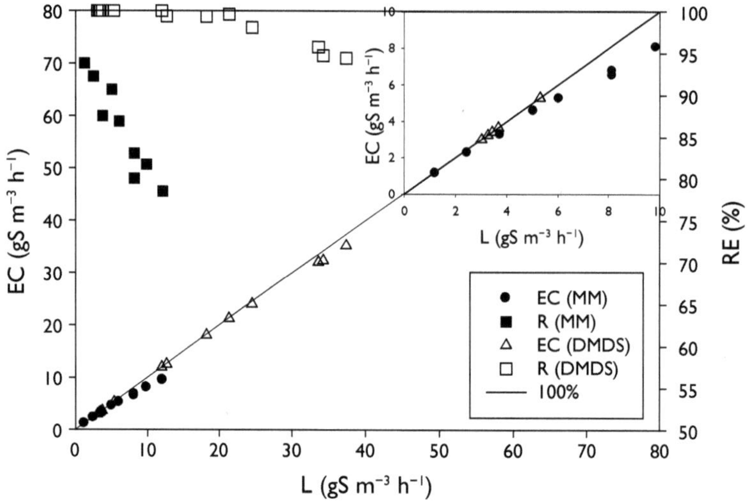

Figure 3 Elimination capacity (EC) and Removal efficiency (RE) versus Inlet load for MM and DMDS removal.

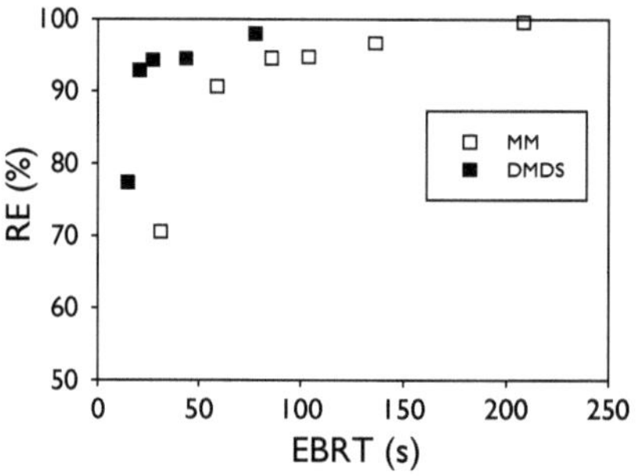

Figure 4 Removal efficiency (RE) versus EBRT for MM and DMDS removal.

even with a 5.4 times higher DMDS load (Fig. 4). For a constant EBRT over 40 s the EC were 33.56 $g_{DMDS\text{-}S}$ $m^{-3}h^{-1}$ (RE 95%), 3.56–4.58 $g_{MM\text{-}S}$ $m^{-3}h^{-1}$ (RE 71–91%) and 11.85 $g_{DMS\text{-}S}$ $m^{-3}h^{-1}$ (RE 98%) for DMDS, MM and DMS[20] respectively. Therefore, for a separately RSC feeding, the elimination capacity order was: DMDS > DMS > MM.

MM is an intermediate oxidation product in the metabolic pathway of *Thiobacillus*,[25,26] because of this, the degradation rate of MM should be better. These pathways are not consistent with the results, but the existence of another different pathways is unlikely. It is possible that another microorganism was present in the biotrickling filter. After a long operation time is possible the growth of another microorganisms in the biofilm. Sercu et al. analyzed the microbial community after 60 days of operation and the bacteria inoculated

(*Hyphomicrobium* VS) was not the dominant microorganism (biotrickling filter at pH 7).[11] The EC of each compounds depends of the community biofilm, Chung et al. reached an EC of H_2S greater than of MM working in a biofilter with *Thiobacillus sp.* HA43 and heterotrophic bacteria, they suggest the coexistence of the other H_2S-degrading microorganisms or the enhancement of the HA43 strain ability.[27]

EFFECT OF THE SULPHATE CONCENTRATION ON MM AND DMS REMOVAL

For DMDS removal, the recirculation medium was not changed during 32 days (Experiment 6). In this period the sulphate concentration increased from 5.0 to 14.8 g l^{-1}. The RE was constant (100%) up to 12 g l^{-1}, from this concentration, the RE decreased to 84%. For MM (Experiment 4) the RE was constant about 87.8 ± 2.2% in a range of sulphate concentration between 3.2–23.4 g l^{-1} (21 days of operation). As a result, it was decided to use 10 g l^{-1} as the sulfate concentration limit for the subsequent experiments.

Some authors recommend keep the sulphate concentration below 5 g l^{-1} for a mixture removal of RSC with *T. thioparus* TK-m,[4] 10 g l^{-1} or 5 g l^{-1} for DMS removal[20] or H_2S[23] respectively with *T. thioparus* ATCC 23654.

STUDY OF THE HYDROGEN SULPHIDE REMOVAL IN THE BTA

In this study the BTA was scaling-up from a previous biotrickling filter[21] inoculated with *A. thiooxidans* from a packing volume of 1.278 to 4.33 L. The scaling-up was necessary for having two identical biotrickling filers, one of them with *T. thioparus* and the another one with *A. thiooxidans*. The critical EC obtained by Ramirez et al.[21] was 58.7 $g_{H_2S\text{-}S}$ $m^{-3}h^{-1}$ (EBRT 24.4 s, RE 98%).

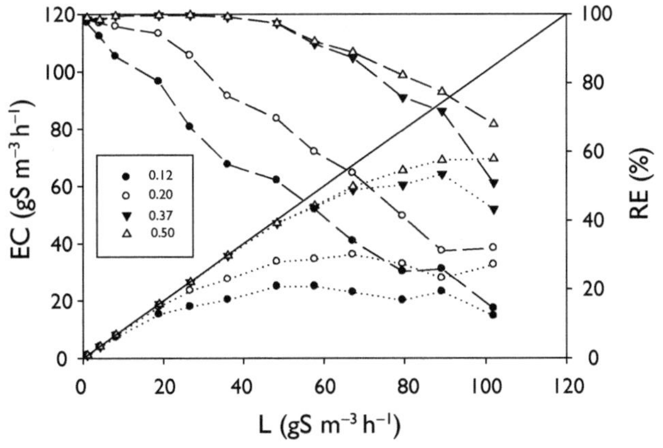

Figure 5 Elimination capacity (EC) and removal efficiency (RE) versus inlet load (L) at different bed high.

In the BAT for a load of 57.5 $g_{H_2S\text{-}S}$ $m^{-3}h^{-1}$, the removal efficiency was 92% for an EBRT of 59 s (Fig. 5, Experiment 7). The minimum time between each value of inlet concentration was 6 h for 8 days. The results, using two biotrickling filters with different size in the same conditions, were very similar. Therefore, the possibility of scaling-up of the system is easy. The maximum value of EC was 69.4 $g_{H_2S\text{-}S}$ $m^{-3}h^{-1}$ with a 68% RE.

For a similar biotrickling filter with *T. thioparus,* an EC of 14.9 $g_{H_2S\text{-}S}$ $m^{-3}h^{-1}$ (RE 99.8%) *was reached.*[23] Therefore, in this work better EC was obtained with *A. thiooxidans* . The same performance was observed using polyethylene rings as carrier by Aroca et al.[28]

STUDY OF H_2S, MM, DMS AND DMDS REMOVAL BY THE BTT

A mixture stream with the four compounds was supplied to the BTT (Experiment 8). For a constant concentration of MM, DMS and DMDS, the inlet concentration of H_2S was increased from 23 to 376 ppmv (Fig. 6). The minimum time between each value of inlet concentration was for 6 h in 6 days. When the H_2S concentration was increased, the removal efficiency was decreasing: 97 to 67% (EC 1.75–1.20 $g_{DMS\text{-}S}$ $m^{-3}h^{-1}$), 86 to 71% (EC 6.19–5.11 $g_{DMDS\text{-}S}$ $m^{-3}h^{-1}$), 83 to 33% (EC 1.96–0.79 $g_{MM\text{-}S}$ $m^{-3}h^{-1}$) for DMS, DMDS and MM respectively. So, when the H_2S was increased, the removal efficiency of RCS decreased, 30%, 15% and 50% for DMS, DMDS and MM. Therefore, the H_2S inhibited the removal of RSC.

Wani et al.[17] studied the elimination of H_2S with DMDS and DMS. They found not significant difference in the maximum removal rate of H_2S for three different biofilters. Nevertheless, the RE of DMS was <30% and the RE of DMDS was between 30–35%. RE of 100% for H_2S, MM, DMS and DMDS have been achieved in a biotrickling filter with *Pseudomonas putida* treating

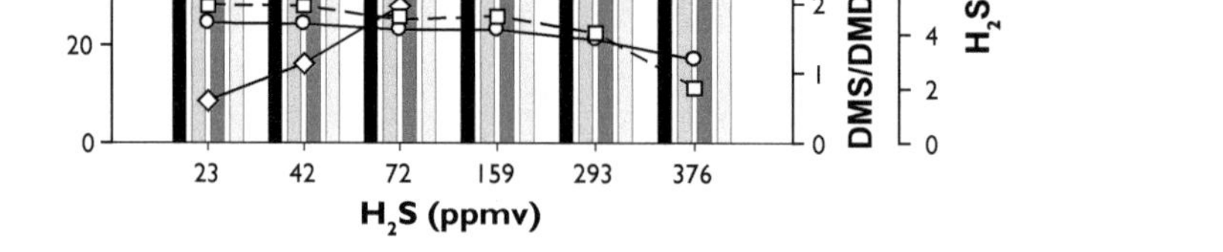

Figure 6 Elimination capacity (bars) and removal efficiency (icons) versus inlet H_2S concentration.

gas at low concentration for odour decreasing (Load H_2S 0.03–0.48 g H_2S $m^{-3}h^{-1}$).[14] However, when the H_2S concentration is increased, the RE of RSC decrease. Others authors found that in an experiment with a peat biofilter inoculated with *Hyphomicrobium sp.* I55, the removal of DMS was inhibited by the presence of H_2S and MM, but the removal of H_2S was not affected by the coexistence of DMS. Ruokojärvi et al.[29] in a biotrickling filter inoculated with activated sludge found inhibition of DMS removal when the H_2S-S concentration exceeded 200 mg m^{-3}.

STUDY OF H_2S, MM, DMS AND DMDS REMOVAL BY THE BTA AND BTT IN SERIES

The both biotrickling filters in series were fed with a four compounds stream mixture (Experiment 9). In the first biotrickling filter (BTA) the RE was approximately constant over 98.5 ± 1.4%, 84.6 ± 5.0%, 13.3 ± 5.3% and 4.2 ± 4.1% for H_2S, DMDS, DMS and MM respectively (Fig. 7a). When the RE was measured with both biotrickling filters in series, the total RE was 99.4 ± 0.3%, 92.8 ± 1.5%, 95.4 ± 1.9% and 90.7 ± 3.3% for H_2S, DMDS, DMS and MM respectively (Fig. 7b). Therefore, the BTA removed efficiently the H_2S and a high percentage of the DMDS, in this way, the global system increased the total RE. The BTT removed the almost all of the MM/DMS and the rest of DMDS. The EC in the global system were: 33.7 ± 0.1%, 5.38 ± 0.1%, 2.2 ± 0.05%, 2.4 ± 0.05% gS $m^{-3}h^{-1}$ for H_2S, DMDS, DMS and MM respectively.

Roukojärvi et al.[15] obtained an EC of 47.9, 36.6 and 2.75 gS $m^{-1}h^{-3}$ for H_2S, DMS and MM in two-stage biotrickling filters. The first biotrickling (low pH) removed approximately half of the DMS and almost all of the H_2S, and the rest of MM and DMS was oxidized in the second biotrickling filter at neutral pH.

If we compared the figures 7b and 6, it can be observed that the biotrickling filters in series are more efficient. When the H_2S was increased the RE of RSC was not decreased.

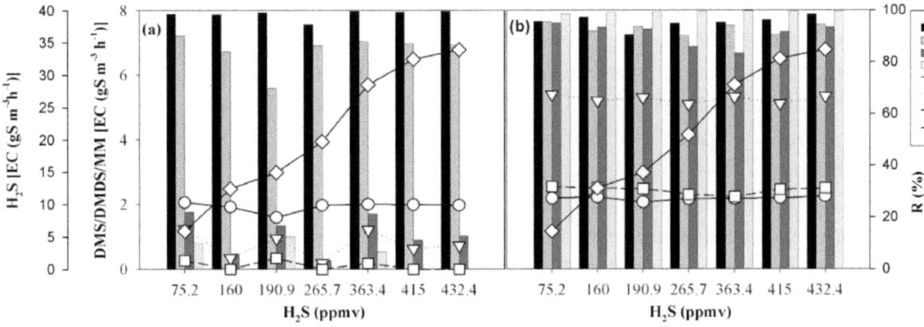

Figure 7 Elimination capacity (EC) and removal efficiency (RE) versus inlet H_2S concentration. (a) BTA; (b) BTA and BTT in series.

CONCLUSIONS

The RSC removal can be carried out efficiently by two-stage biotrickling filters in series, the first one inoculated with *A. thiooxidans* and the second one with *T. thioparus*. For inlet loads of H_2S until 33.7 $g_{H_2S\text{-}S}$ $m^{-3}h^{-1}$ the RE of the rest RSC didn't decrease in the two-stage biotrickling filter.

The polyurethane foam as carrier for *A. thiooxidans* immobilization show very good scale-up possibilities.

ACKNOWLEDGEMENTS

The authors wish to express their sincerely gratitude to the Ministerio de Ciencia y Tecnología for supporting this work through the Project CMT2006-05497 and to BIORECA network (CYTED Spain- Programa Iberoamericano de Ciencia y Tecnología para el Desarrollo) and BIOPROAM (Alfa Project UE- AML 190901/06/18414/II-O 548-FC-FA) by funding the research stays.

REFERENCES

1 O'Neill DH and Phillips VR, A review of the control of odour nuisance from livestock buildings: Part 3, properties of the odorous substances which have been identified in livestock wastes or in the air around them. *J Agr Eng Res* **53**:23 (1992).
2 Smet E, Van Langenhove H and Lens P, Treatment of waste gases contaminated with odorous sulfur compounds. *Cri Rev Env Sci Tec* **28**:89–117 (1998).
3 Mahmood Q, Zheng P, Cai J, Hayat Y, Hassan MJ, Wu DL and Hu BL, Sources of sulfide in waste streams and current biotechnologies for its removal. *J Zhejiang Univ Sci A* **8**:1126 (2007).
4 Tanji Y, Kanagawa T and Mikami E, Removal of dimethyl sulfide, methyl mercaptan, and hydrogen sulfide by immobilized *Thiobacillus thioparus* TK-m. *J Fermen Bioeng* **67**:280–285 (1989).
5 De Zwart JMM, Sluis JMR and Gijs Kuenen J, Competition for dimethyl sulfide and hydrogen sulfide by *Methylophaga sulfidovorans* and *Thiobacillus thioparus* T5 in continuous cultures. *Appl Environ Microbiol* **63**:3318–3322 (1997).
6 Cho KS, Zhang L, Hirai M and Shoda M, Removal characteristics of hydrogen sulphide and methanediol by *Thiobacillus sp.* isolated from peat in biological deodorization. *Journal of Fermentation and Bioengineering* **71**:44–49 (1991).
7 Cho KS, Hirai M and Shoda M, Enhanced removal efficiency of malodorous gases in a pilot-scale peat biofilter inoculated with *Thiobacillus thioparus* DW44. *J Fermen Bioeng* **73**:46–50 (1992).
8 Park SJ, Cho KS, Hirai M and Shoda M, Removability of malodorous gases from a night soil treatment plant by a pilot-scale peat biofilter inoculated with *Thiobacillus thioparus* DW44. *J Fermen Bioeng* **76**:55–59 (1993).
9 Cho KS, Hirai M and Shoda M, Enhanced removability of odorous sulfur-containing gases by mixed cultures of purified bacteria from peat biofilters. *J Fermen Bioeng* **73**:219 (1992).

10 Zhang L, Hirai M and Shoda M, Removal characteristics of dimethyl sulfide, methanethiol and hydrogen sulfide by *Hyphomicrobrium sp.* I55 isolated from peat biofilter. *J Fermen Bioeng* **72**:392 (1991).

11 Sercu B, Van Langenhove H, Nuñez D, Aroca G and Verstraete W, Operational and microbiological aspects of a bioaugmented two-stage biotrickling filter removing hydrogen sulfide and dimethyl sulfide. *Biotechnol Bioeng* **90**:259–269 (2005).

12 Cho KS, Hirai M and Shoda M, Enhanced removability of odorous sulfur-containing gases by mixed cultures of purified bacteria from peat biofilters. *J Fermen Bioeng* **73**:219–224 (1992).

13 Shu CH and Chen CK, Enhanced removal of dimethyl sulfide from a synthetic waste gas stream using a bioreactor inoculated with *Microbacterium sp.* NTUT26 and *Pseudomonas putida. J Ind Microbiol Biot* **36**:95 (2009).

14 Ho KL, Chung YC, Lin YH and Tseng CP, Microbial populations analysis and field application of biofilter for the removal of volatile-sulfur compounds from swine wastewater treatment system. *J Hazard Mater* **152**:580 (2008).

15 Ruokojärvi A, Ruuskanen J, Martikainen PJ and Olkkonen M, Oxidation of gas mixtures containing dimethyl sulfide, hydrogen sulfide, and methanethiol using a two-stage biotrickling filter. *J Air Waste Manage Assoc* **51**:11–16 (2001).

16 Pinjing H, Liming S, Zhiwen Y and Guojian L, Removal of hydrogen sulfide and methyl mercaptan by a packed tower with immobilized micro-organism beads. *Water Sci Technol* **44**:327–333 (2001).

17 Wani AH, Lau AK and Branion RMR, Biofiltration control of pulping odors – hydrogen sulfide: performance, macrokinetics and coexistence effects of organo-sulfur species. *J Chem Technol Biot* **74**:9–16 (1999).

18 Brenner DJ, Krieg NR and Staley JT, *Bergey's Manual® of Systematic Bacteriology.* Springer US (2006).

19 Smet E, *Biofiltration of volatile organic sulfur compounds.* Appendix A, Ph.D. diss., Ghent University (1996).

20 Arellano-García L, Revah S, Ramirez M, Gomez JM and Cantero D, Dimethyl sulfide degradation using immobilized *Thiobacillus thioparus* in a biotrickling filter. *Environ Technol* (2009). IN PRESS

21 Ramirez M, Gomez JM, Cantero D, Paca J, Halecky M, Kozliak E.I and Sobotka M, Hydrogen Sulphide removal from air by *Acidithiobacillus thiooxidans* in a Trickle bed reactor. *Folia Microbiol* 54 (2009). IN PRESS

22 Rodier J, *Análisis de las aguas.* Ediciones Omega, S.A. Barcelona (1998).

23 Ramirez M, Gomez JM, Aroca G and Cantero D, Removal of Hydrogen sulfide by immobilized *Thiobacillus thioparus* in a biotrickling filter packed with polyurethane foam. *Bioresource Technology*:doi:10.1016/j.biortech.2009.1005.1022 (2009).

24 Cho KS, Hirai M and Shoda M, Degradation characteristics of hydrogen sulfide, methanethiol, dimethyl sulfide and dimethyl disulfide by *Thiobacillus thioparus* DW44 isolated from peat biofilter. *J Fermen Bioeng* **71**:384–389 (1991).

25 Kelly DP and Smith NA, Organic sulfur compounds in the environment. *Adv Microb Ecol* **11**:345 (1990).

26 Visscher PT and Taylor BF, A new mechanism for the aerobic catabolism of dimethyl sulfide. *Appl Environ Microb* **59**:3784–3789 (1993).

27 Cho KS, Zhang L, Hirai M and Shoda M, Removal characteristics of hydrogen sulphide and methanediol by *Thiobacillus sp.* isolated from peat in biological deodorization. *J Fermen Bioeng* **71**:44 (1991).
28 Aroca G, Urrutia H, Núñez D, Oyarzún P, Arancibia A and Guerrero K, Comparison on the removal of hydrogen sulfide in biotrickling filters inoculated with *Thiobacillus thioparus* and *Acidithiobacillus thiooxidans*. *Electron J Biotech* **10**:514 (2007).
29 Ruokojärvi A, Aatamila M, Hartikainen T, Olkkonen M, Salmi J, Ruuskanen J and Martikainen PJ, Removal of dimethyl sulphide from off-gas mixtures containing hydrogen sulphide and methanethiol by a biotrickling filter. *Environ Technol* **21**:1173 (2000).

Chapter 25

THE DEVELOPMENT OF BIOLOGICAL DESULFURIZATION FOR POLLUTED AIR AND GAS STREAMS

Robert Lems & Erwin H.M. Dirkse
DMT, Joure, The Netherlands

ABSTRACT

Hydrogen Sulfide (H_2S) has been a problematic gas compound for centuries. In this paper the development of H_2S treatment technologies from past to future is demonstrated with field data from industrial and pilot plants. DMT has designed and constructed biological desulfurization systems for treatment of both air and biogas. The development starts with bio-filters followed by lava rock- and bio-trickling filters (BTF). The newest bio-reactors are now highly sophisticated biological scrubbers. The internal processes are pH, temperature, nutrient and oxygen controlled. During an extensive research period over the last 15 years, the elimination capacity (EC) has been multiplied 10 to 50 times from minimum 10 to maximum 100–500 gH_2S/m^3 * hr with removal efficiencies over 99%. Optimum process conditions and maximum removal efficiencies have been determined by variation of all relevant process conditions. Even higher ECs can be expected in the near future due to the development of new highly structured packing materials.

The newest generation of biological H_2S removal systems have overcome the disadvantage of the first bio-filters while maintaining the advantage of a biological system. This makes the bio-reactor today the best technology to treat H_2S polluted gas/air streams even compared to a chemical system.

Keywords: Bio-filters, Biogas, Biological desulfurization, Bio-trickling filters, DMT BioSulfurex®, Elimination Capacity, H_2S, H_2S oxidation, Hydrogen Sulfide, Lava rock filters

BACKGROUND

Hydrogen Sulfide (H_2S) has been a problematic gas compound for centuries. The smell of rotten eggs is not only a nuisance for people but can even be toxic at low concentrations. In biogas the acid gas can cause corrosion of piping and engines preventing gas utilization without pre-treatment. In the past the main treatment for H_2S were chemical plants based on water scrubbing with caustic (NaOH) or dry processes which binds H_2S to iron or activated carbon. Both systems consume high amounts of chemicals and produce chemical waste.

The first steps towards biological treatment were already made in 1923 but it took until the 60s before the bio-filtration was a more or less accepted system to treat mal-odours of air containing H_2S, mercaptanes and other pollutants in air.[1] From the 1960s till present the biological treatment of odours and H_2S has been greatly improve to increase elimination capacity (EC), decrease footprint and increase operational stability.[2] In principle the bio-reactor is developing into a chemical scrubber without the disadvantages like chemical consumption and waste.

This is demonstrated in the next chapters by research and field data from different DMT plants like bio-filters, lava rock filters, bio-trickling filters, Odourex® and BioSulfurex®. Also the first results of the next generation, the BioSulfurex® HSC, are presented as a biological scrubber which even surpasses the characteristics of a chemical scrubber.

PRINCIPLE DESIGN BIO-REACTOR FOR AIR/BIOGAS DESULFURIZATION

DMT has developed several processes for the biological removal of H_2S and other components like mercaptanes and VOC's from air and biogas.[3] The design principle is the same for each plant[2] (Fig. 1). Water is supplied to a system containing more or less specific micro-organisms. The systems contain a packing material which acts as carrier to which the micro-organisms can

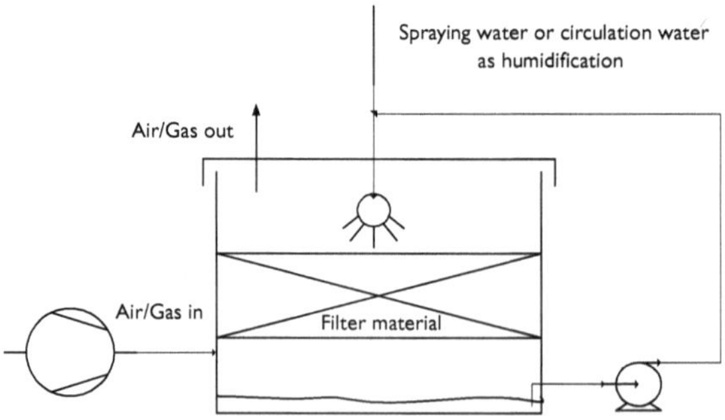

Figure 1 Simplified pfd of a biological air/gas treatment system.

attach. The air/gas stream flows through the packing material while the bacteria degrade the pollutants. Nutrients are available in the carrier material or are supplied separated and dosed into the water circulation stream. The main H_2S degrading bacteria is *Acidithiobacillus thiooxidans*. The *Acidithiobacillus thiooxidans* converts H_2S into sulfuric acid or sulfur in the presence of oxygen (equation 1 and 2). If there are oxygen limitations the reaction will stop at equation 1 and elemental sulfur will be formed.[4]

$$H_2S + ½O_2 \rightarrow S + 2H_2O \tag{1}$$

$$S + 1½O_2 + H_2O \rightarrow H_2SO_4 \tag{2}$$

As like all biological processes the activity depends on environmental parameters like temperature, pH, substrate concentration, nutrient availability, oxygen concentration etc. A lot of field and laboratory research has given indications for these relations.[5,6] E.g. the optimal pH is about 2–3.5 while the bacteria are still very much active at a pH of <0.5[7] Further more the optimal temperature is around 25–35°C with minimum activity below 15°C.[8] Oxygen concentrations down to <0.5% in the air/gas stream still show high elimination capacities.[7]

Besides the bacteriological activity also the density of the population influences the maximum possible EC and removal efficiency. The population density is influenced by the specific surface of the filter media at field conditions. The development of biological desulfurization is a continuous research of increasing the activity by optimizing process parameters and at the same time increasing the population density without clogging the bio-reactor. Also the reactor design is developing from high footprint and low height to small footprint with higher towers.

BIO-FILTERS

The first bio-filters were large beds containing soil and/or compost and could only remove small concentrations of a large variety of (organic) compounds like H_2S, mercaptanes, ammonia and VOC (Fig. 2a). Different filter materials have been developed to improve these bio-filters. Pure material or mixtures of compost, manure, bark, wood and coconut fibres have been used to increase the capacity of the bio-filters and decrease the footprint and pressure loss. At the same time it was tried to get a better control over the environmental parameters (See effect of no control, Fig. 2b) of the filters. Bio-filters were roofed and equipped with a pre-air humidifier. Process parameters like humidity, temperature, pressure loss etc. are measured with weight- humidity-, temperature- and pressure transmitters which control a spraying system and indicate maintenance intervals for the filter bed (Fig. 2c). More or less inert materials like coconut fibres were introduced to extent the life time of the filters.

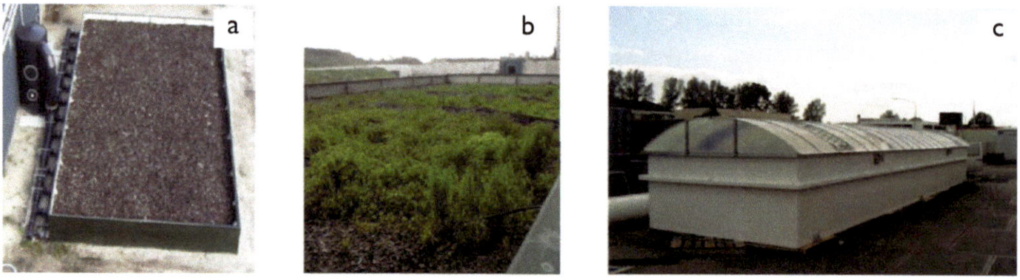

Figure 2 a) Old fashioned bio-filter, b) "Vegetable" garden on the old fashioned filter, c) Modern, parameter controlled bio-filter.

Today there are containerized systems that can be stacked to minimize footprint, but still the filters are usually only equipped to treat "small" waste streams with a high variety of low concentrations of pollutants (e.g., H_2S 10–50 ppm). The filters remain large (EC max 10 $gH_2S/m^3 * hr$) and have a large pressure drop (10 mbar/m). The filter material acidifies and is degraded by the bacteria so that is has to be replaced every 3–5 years.[9,10]

BIO-TRICKLING FILTERS

A second step in the development of biological treatment of H_2S rich air streams was the introduction of lava rock bio-trickling filters (BTF) (Fig. 3a). The organic filter material of the bio-filters was replaced by inorganic lava rock. The high specific surface of the lava rock increased the population density of the (H_2S degrading) bacteria reducing the total required filter volume. At the same time higher filters were possible due to the decrease in pressure drop reducing the footprint significant. Using bio-trickling filters the process conditions for the bacteria can be better monitored and controlled due to the smaller installations and by using a water circulation system (Fig. 3b). This resulted in higher removal efficiencies and EC's up to 20–60 $gH_2S/m^3 * h$.[9,10]

Also the first multi stage biological treatment systems were introduced, which consist of two lava rock series if necessary followed by a conventional bio-filter. In this way different environmental systems (e.g., pH) can be created to degrade different pollutants by specialized bacteria (Fig. 3c). However the lava rock is still difficult to handle, has a large weight and is susceptible to clogging especially at high H_2S concentrations and can therefore only be used for air with maximum H_2S concentrations of 100–500 ppm.

ODOUREX®

The next generation bio-(trickling) filters is making use of random structured plastic filter materials like Pall Rings and/or foam. This resulted in higher

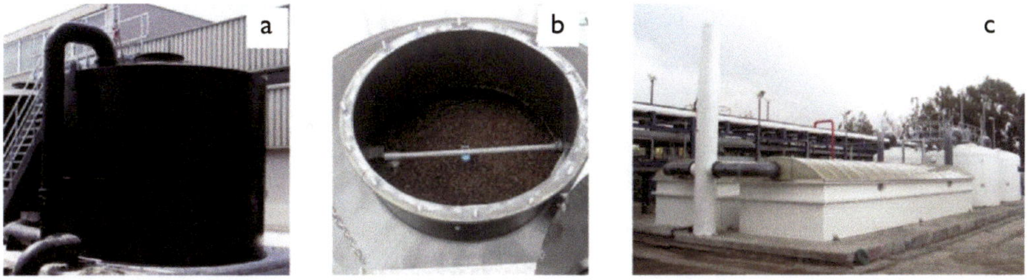

Figure 3 a) Single stage lava filter, b) Spraying system on the lava rocks, c) Multi stage biological system.

Figure 4 a) Pilot for comparison of trickling filter materials, b) Top, DMT-HS media, *Middle* 1″ Pall Ring, *Bottom* lava rock, c) DMT multi-stage Odourex® system.

possible EC's (foam) less clogging (Pall Ring) and a better process control (pH, T, pressure etc) due to higher circulation capacities. Although foam material tends to clog at high concentrations, the higher EC makes it possible to treat most polluted air streams on a small foot print.[11,12] Although with Pall Rings the increase in EC is less, they can be used as filter material to also treat H_2S from biogas, due to less chance of clogging. DMT has executed a pilot plant to compare the different packing materials (Fig. 5c). This installation treated a polluted air stream (H_2S:150–300 ppm) from the WWTP of the paper factories around Eerbeek (NL). Lava rock was tested and compared with Pall Rings and a newly developed highly structured packing material: DMT-HS packing (See figure 4; Left and middle).

The results of the pilot plant demonstrated that the Pall Rings had about a two times higher EC (10–20 $gH_2S/m^3 * hr$ for lava compared to 40–50 $gH_2S/m^3 * hr$ for Pall Rings at pilot field conditions) than lava rock but at the same time a lower removal efficiency (typical about 10 ppm H_2S out compared to <2 ppm for lava rock). However the DMT-HS had elimination capacities up to 100 $gH_2S/m^3 * hr$ with still removal efficiencies up to 90–98% which means less than 2 ppm out (See figure 5; Left). Also mercaptanes, alcohols and aldhydes were removed to a significant extent (50–90%). Both Pall Ring and

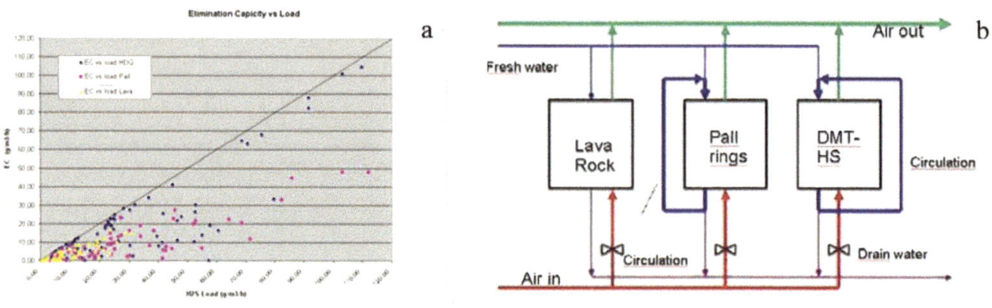

Figure 5 a) H_2S loading and EC for pilot tests of lava rock, Pall Ring and DMT-HS media, b) PFD of the pilot.

DMT-HS had a minimum pressure loss (1–5 mbar) compared to lava rock and were "self-cleaning" after artificial clogging. DMT developed the Odourex® (Fig. 4c) system with the results of this pilot.[8,13] This is a multi stage biological air treatment system which consists out of only one column and is able to treat H_2S, mercaptanes, NH_4 and VOC on a minimal footprint with minimal operational interventions.

BIOSULFUREX®

H_2S is not only a problem for air streams but it is also present in gases like biogas or flare gas. Before the gas can be used in a gas engine the H_2S has to be removed.[14] This is usually done by a chemical system due to the ultra high concentrations (up to 5% Vol.) and the lack of oxygen. But with the upcoming of Pall Rings as filter material for bio-reactors it is also possible to treat H_2S polluted gas stream with a bio-reactor. A small air stream is introduced into the system for the essential oxygen (see figure 6; Left and middle).

The first bio-reactors for biogas were built with an EC of 25–40 $gH_2S/m^3 * hr$. H_2S removal efficiencies of 99% are observed (figure 6; Right). Due to the small pressure loss the bio-reactors can be built on a relative small footprint. The control of the supplied oxygen makes it possible to predetermine the ratio between conversion of H_2S to sulfuric acid and elemental sulfur.

DMT has increased the bacteriological activity by optimization of the process parameters and a strict control over these parameters. In this way elimination capacities of 80 $gH_2S/m^3 * hr$ have been reached without limitations on further improvements (See figure 7; Left).[7,15] The newest reactors are designed at a pH of 0.5–1.5 at a constant temperature of 30°C and a strict oxygen control of 0.5–1% at the outgoing biogas. The formed elemental sulfur is about 30%. Nutrient dosing is calculated and monitored to reach a maximum EC without inducing excessive growth which would clog the bio-reactor.

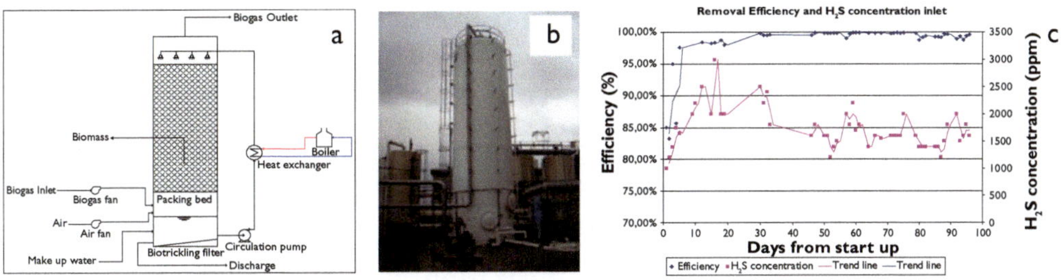

Figure 6 a) PFD BioSulfurex®, b) Photo BioSulfurex®, c) H_2S concentration in and out from start up BioSulfurex® full-scale.

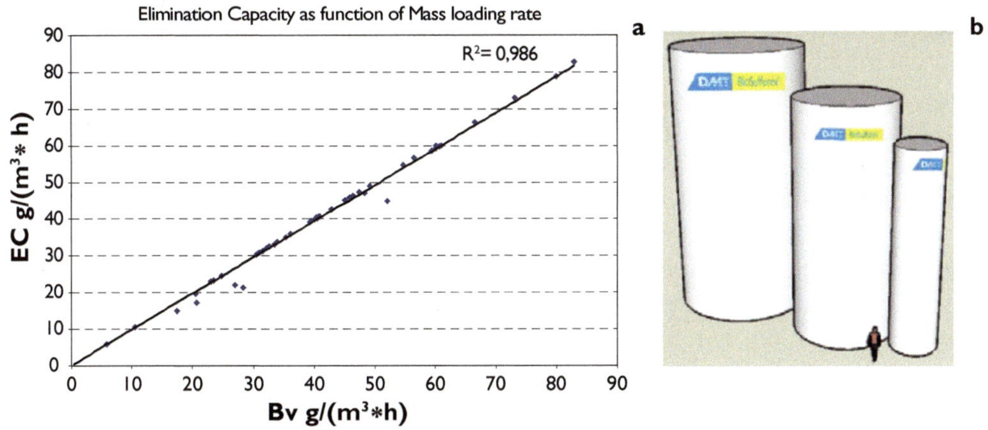

Figure 7 a) Graph of H_2S load and EC for the BioSulfurex® in Israel, b) Schematic to scale drawing of the development of the BioSulfurex® with variable EC's from 40–80–500 g $H_2S/m^3 * h$.

In both Israel and Greece full scale plants for biological desulfurization have been extensively monitored.[7,15] The reactors are about 4 meter in diameter and 14 meter in height and have sample points over the length of the bio-reactor. They are designed to treat up to 2000 m^3/hr of biogas with 4000ppm of H_2S. The results of the start up of the bio-reactor in Israel are displayed in Fig. 6c. The plants has been running at temperatures of 28–42°C. After 1 month almost 100% removal efficiency has been reached. The plant has been running at different H_2S loadings (variations in flow and H_2S concentration) resulting in the EC/Bv plot of Fig. 7a. The straight line indicates that the maximum EC (elimination capacity) will be minimal 80 $g/m^3 * hr$ but probably can be much higher. Both plants have been running on a pH range of 0.5–3 which had a minimum effect on the efficiency. The oxygen concentration varied from 0.5–2.5% resulting in a slight decrease in water demand at lower oxygen concentrations. This is due to extra formation of elemental sulfur and less sulfuric acid which increased from 30% up to almost 40%. Both plants have shown to be very robust in operation even at fluctuations in H_2S loading and gas stops of several hours (due to engine failure). Even after a cleaning stop for 24 hrs the removal efficiency was back to 99% within 1 day.[15]

BIOSULFUREX® HSC

The current biological biogas desulfurization resembles a chemical system already quite closely. To overcome the last obstacle for biological treatment versus chemical treatment the EC has to be further improved and the clogging problems have to be further minimized/eliminated. Paques B.V.[16] has overcome these problems by combining chemical scrubbing with biological conversion of H_2S to elemental sulfur in two steps. This results in a combination of a chemical scrubber and a bio-reactor. The chemical scrubber uses caustic (NaOH) which is recovered in the biological stage at which elemental sulfur is formed at oxygen limiting conditions. The Paques system is less sensitive to clogging but still needs a minimum amount of chemicals and has relative a higher footprint due to the combination of two systems.

The future processes which DMT is already developing are using the highly structured packing material (e.g., DMT-HS packing) which will decrease the bio-reactor size to a footprint similar to a chemical scrubber (Fig. 7b). Results from the "air" pilot in Eerbeek suggest that ECs of 500–800 $gH_2S/m^3 * hr$ are possible (6–8× improvement for DMT-HS compared to Pall Rings). Also research from Marc Fortuny et al. demonstrate that EC's of minimal 250 $gH_2S/m^3 * hr$ are possible at conditions which are still limited in carbon source, oxygen, pH and temperature. Marc Fortuny et al.[17] also demonstrated that at oxygen limitations about 94–95% of the H_2S is converted to elemental sulfur instead of sulfuric acid.

The BioSulfurex® HSC has several advantages due to the use of the DMT-HS packing material. The elimination capacity can be greatly increased and at the same time the gas load increased due to the low pressure drop. This results in a very small bio-reactor with a minimum foot print (See figure 7: right, right bio-reactor) which is almost comparable to a chemical system. At the same time the system is self-cleaning and there is no effect of clogging due to sludge and elemental sulfur. This makes it possible to add a minimal amount of oxygen and converting all H_2S to elemental sulfur. The sulfur can be collected and re-used in various processes (e.g., fertilizer). At the same time minimal to no water consumption is required due to the operation at low pH, condensation from the biogas and formation of elemental sulfur instead of sulfuric acid. Compared to the Paques system no chemicals are needed and only small waste stream of acid water is produced which can be directly discharged in a WWTP. The smaller design also reduces power consumption and the gas quality is guaranteed due to minimal air injection.

Table 1 The development of biological H_2S treatment techniques compared to a chemical scrubber.

Technique	ΔP	A	EC	EBCT	Weight	Robust	Rem. Eff	Oper. costs	Chemicals/ waste	Treatment of biogas
Chemical	+	++	++	++	++	++	+	+/-	-	++
Bio-filter	-	-	-	-	-	-	+/-	+	+	-
Lava rock	+/-	+/-	+/-	+/-	-	+/-	+/-	+	++	-
Pall Rings	+	+	+-	+/-	++	+/-	+/-	++	++	+
DMT-HS	+	++	+	+	++	+	+	++	++	++

CONCLUSIONS

The biological treatment of H_2S has greatly improved from the first bio-filters with a maximum EC of 10 $gH_2S/m^3 * hr$ with a large footprint and minimal process control to highly sophisticated bio-trickling filters with EC's up to 80–100 $gH_2S/m^3 * hr$. The current bio-reactors have a significant reduced footprint, optimal control over the process parameters and can also be used for ultra high H_2S concentrations as found in biogas. The next generation of bio-reactors will overcome the last disadvantages of a biological scrubber by further improvement of the EC to 500 $gH_2S/m^3 * hr$ or more and eliminate the chance of clogging (Table 1) resulting in small, slim bio-reactors with automatic control, no chemical demand and minimal to no operational interventions.

REFERENCES

1 http://inventors.about.com/od/bstartinventions/a/biofiltration.htm
2 S. Joseph et al. Biofiltration for air pollution control, 5–15, 2008.
3 http://www.dmt-et.nl
4 F.P. van der Zee et al. Sulfide removal by moderate oxygenation of anaerobic sludge environments, *Bioresource Technology*, 98: 518–524 (2007).
5 M. Silver and O. Dinardo. Factors affecting oxidation of thiosalts by thiobacilli, *Applied and environmental microbiology*, Vol. 41, No. 6: 1301–1309 (1981).
6 Germán Aroca et al. Comparison on the removal of hydrogen sulfide in biotrickling filters inoculated with thiobacillus thioparus and acidithiobacillus thiooxidans, *Electronic journal of biotechnology*, Vol. 10, No. 4 (2007).
7 E. Bruné. *Biological biogas desulfurization at Emek Hefer*, internal graduation report, 2007.
8 J. Edens. Evaluation report biological odour removal techniques/research odour removal at test installation Eerbeek, internal evaluation report 2009.
9 Evaluation of Biological air treatment systems at WWTPs (Dutch), 2003.
10 Various internal inspection reports.
11 G.R. Moosavi et al. Biotechnology advances in treatment of air streams containing H_2S, *Journal of Biological Sciences*, 5 (2): 170–175 (2005).

12 D. Gabriel and M.A. Deshusses. Retrofitting existing chemical scrubbers to biotrickling filters for H_2S emission control, *Proceedings of the national academy of sciences*, Vol. 100, No. 11, 6380–6312 (2003).
13 Dirkse, E.H.M. *Odour control on Waste Water Treatment Plants and Pumping Stations using DMT Biotrickling Filtration,* DMT environmental Technology. 12th European Biosolids Conference, Manchester (2007).
14 Dirkse, E.H.M. *Biogas Desulphurisation using the DMT multiple stage Sulfurex® process,* 11th European Biosolids Conference, Manchester (2006).
15 P. van der Linden. *Optimization of the BioSulfurex®,* internal graduation report, (2009).
16 http://www.paques.nl
17 M. Fortuny et al. Biological sweetening of energy gases mimics in biotrickling filters, *Chemosphere*, 71, 10–17 (2008).

Chapter 26

OUTLOOK FOR WASTE GAS TREATMENT – SCIENTIFIC PERSPECTIVE

Yves Andrès
GEPEA, UMR CNRS 6144, Ecole des Mines de Nantes, Nantes Cedex, France

ABSTRACT

Air pollution is one of the major problems to be solved over this century. So in recent years, a strong technological base has been developed for air quality control. The driving force has been the public awareness of the need for human health and environmental protection. Consequently, environmental legislation and regulations have been initiated by governments to control air quality. Removing air pollutants from industrial gaseous emissions is achieved by different physical and chemical ways like the transfer and/or concentration of compounds in a liquid phase (condensation, absorption) or onto a solid (adsorption) or by the destruction of molecules by thermal and catalytic incineration. However, more recently new treatment alternatives have been developed, involving the biological degradation of pollutants present in the gaseous phase. Thus, these microbial systems can remove a large number of molecules, especially volatile organic compounds (VOC) and odorous molecules. Such systems present several advantages, among which is their high efficiency and low cost. Moreover, the treatment of polluted air in bioreactors allows the complete degradation or transformation of the contaminants. As a result of microbial metabolism, the pollutants are degraded into carbon dioxide and water or converted into cell biomass.

Biological units commonly used for air management are based on three different technologies: biofilter, trickling filter and bioscrubber. Recently, two other types of bioreactors have been developed: membrane bioreactor and two-phase partitioning bioreactor. These processes can be distinguished either by the mobility of the micro-organisms and the liquid phase (as in bioscrubber) or only the liquid phase (as in trickling filter). In biofilters and trickling filters, the micro-organisms are fixed on a support or packing material. Waste gas treatment performances of such processes are very interesting when operating parameters, such as waste gas loads for biofiltration and washing solution loads in the case of bioscrubbing and trickling filtering, are well controlled.

The objective of this presentation is to focus on the scientific new development of reactor design, enhancement of pollutant transfer, nutrient contribution and the packing material effects, microbial activities measurement and population characterization studies.

Poster presentations

Chapter 27

DEVELOPMENT OF AN ORIGINAL APPROACH TO EVALUATE EFFECTS OF SURFACTANTS, BIOMASS AND POLLUTANTS ON THE SCALING-UP OF A TWO-PHASE PARTITIONING BIOREACTOR

Jean-Marc Aldric, Sébastien Gillet, Frank Delvigne & Philippe Thonart
Centre Wallon de Biologie Industrielle, Unité de Bio-industries,
Faculté des Sciences Agronomiques de Gembloux, Gembloux, Belgium

BACKGROUND

Two-phase partitioning bioreactors (TPPBs) are considered as a new technology for xenobiotic degradation in gaseous effluents. However, there is still a need for more knowledge on how to design and scale up TPPBs. The partitioning of the two phases remains a misunderstood way of research. In particular, the impact of pollutant (IPB), biomass and surfactant extract needs to be better evaluated.

RESULTS

An adaptated scale-down apparatus has been developed in order to quantify the speed of phase partitioning (SPP) into a plug flow section. Firstly, it was shown that IPB doesn't destabilize more significantly the stability of the system. Secondly 0.5 $g \cdot L^{-1}$ and 0.05 $g \cdot L^{-1}$ of respectively biomass and surfactant extract, separately or in mixture, were sufficient to ensure a good stability of the two-phase system. Finally, a 100 m^3 limit of scaling-up was suggested on the basis of the circulation time comparison.

CONCLUSIONS

The scaling up of an aqueous/silicone-oil two-phase partitioning bioreactor was found to be definitely conceivable when the presence of biotic compounds were considered. However, further considerations are needed to verify our assumptions, in particular by taking into account the velocity field pattern in full-scale bioreactors and reproduce it in lab-scale apparatus.

Chapter 28

BIO ALCOHOL PRODUCTION FROM AGRICULTURAL RESIDUES

Iran Alemzadeh & Manouchehr Vossoughi
Department of Chemical and Petroleum Engineering, Sharif University of Technology, Tehran, Iran

ABSTRACT

The bioconversion of abundant and renewable cellulosic biomass into ethanol as an alternative to petroleum is gaining importance due to the realization of diminishing natural oil and gas resources. Agricultural and foresty plant residues are an abundant and renewable source of sugar substrates that could be fermented to ethanol. A thermochemical treatment of biomass in which both cellulose and hemicellulose are hydrolyzed to soluble sugar is necessary before yeast fermentation. After thermochemical treatment, cellulase enzymes must be introduced in the system to hydrolyze any remaining cellulose. The simultaneous saccharification and fermentation (SSF), is a method which converts agricultural residue to ethanol. In this investigation, pretreatment of rice husk was effected under, acid, alkaline and mixed treatment, which alkaline pretreatment showed higher biomass degradation. Production of enzyme complex (cellulase) was studied by fungal strain cultivation on pretreatment biomass,using Trichoderma reesei on alkaline pretreated rice husk. The biomass was then subjected to an SSF process, using the cellulase enzymes complex extracted from fungal culture and Saccharomyces cerevisiae yeast cells in batch scale fermentation sysrem. Ethanol production was evaluated by changing various parameters such as: substrate and enzyme concentration, corn steep liquor addition and inoculum percent under SSF conditions.

Keywords: Bio alcohol, SSF, Cellulose and hemicellulose, Corn steep liquor

INTRODUCTION

Fuel ethanol production from lignocellulosic materials such as forages and agricultural residues offer many potential economic and environmental benefits.[1–7] Several methods are developed to make the production of ethanol based on lignocellulosic biomass technologically feasible but the challenge of developing and commercializing a cost effective process remains a major problem in utilizing lignocellulosic materials as fermentation substrate due to their resistance to hydrolysis. Various lignocellulosic material such as woody, low grade wheat flour and herbaceous material have been assayed.[8,9] A wide variety of pretreatment have been applied to improve the surface area and porosity of such materials as well as their accessibility for the enzymes.[2] No single method has been found as a widespread commercial application. The ammonia fiber explosion (AFEX) process reported as a pretreatment method for improving the reactivity of lignocellulosic materials.[3,4] Hydrolysis step using HCl at different concentrations and temperature in the production of alcohol from wheat straw was studied.[5] Steaming method described by many investigators for upgrading the nutritive value of forage and also fiber material would be liable to biochemical or microbial processes.[7] Among the systems used for ethanol production from cellulose, SSF (Simultaneous Saccharification and Fermentation) method has attracted many investigators' attentions due to its high yield in producing ethanol and lactic acid by eliminating end product inhibition and its special characteristic that does not need separate reactors in Saccharification and fermentation.[7,10] Besides fed_batch cultivation of *Mucor* for ethanol production using lignocellulosic hydrolysate was also studied.[11]

In this investigation, ethanol production by pretreated cellulosic material (rice husk) was done in the presence of Saccharomyces cerevisiae by SSF method.

MATERIALS AND METHODS

MICROORGANISMS

Microorganisms: Trichoderma reesee, T. Reesee, PTCC 5142 which was stocked on PDA slants and Saccharomyces cerevisiae, S. cervisiae PTCC, was cultivated on YMPG (yeast extract, malt extract, pepton and glucose) agar, and incubated at 30°C.

SUBSTRATE AND PRETREATMENT

Dry rice husk was selected as substrate in this investigation. Different treatments; alkaline and alkaline, acid (mixed) treatment were effected. For alkaline treatment, 30 grams of milled substrate (0.2 mm mesh) were soaked in 360 ml 1% NaOH and were autoclaved for 1 hour at 15 psi and 120°. Washing the treated substrate 3 times with distilled water so that pH reaches neutral. After the substrate is dried for 10 hr at 50°C. For mixed treatment, beside alkaline treatment, acid treatment with H_2SO_4 1% and autoclaving for 20 min at 15 psi and 120°C. Washing the substrate 3 times with distilled water and drying under the same conditions for alkaline treatment.

YEAST PRE-CULTURE

Pre-culture contains 250–500 ml medium with the composition g/l: glucose, 10; peptone, 5; yeast extract 3; malt extract 3, and distilled water up to 1 lit. Sterilization was done at 120°C. Inoculate were effected by transferring organisms maintaind on YMPG slant agar into the pre-culture. Growth was carried out at 30°C on 200 rpm shaker for 24 hr.

ENZYME

Crude cellulase from T. reesee PTCC 5142 were utilized. Enzyme was produced according to the method described by many authors.[12,13] Filter paper activity (FPA) or fpu, filter paper unit was determined according to the method described by Mandels et al. 1976.[14] One unit of enzyme activity is the amount of enzyme that liberates 1 μ mole of reducing sugar per minute under standard assay.

For preparation of crude cellulase complex, culture medium were prepared with the following composition (g/l): Pre-treated substrate, 50; $CaCl_2$, $6H_2O$, 0.0037; $(NH_4)_2SO_4$, 11.7; corn steep liquor (CSL) 1.5; $Tween_{80}$, 0.2 and distilled water up to 1 lit. pH was adjusted to 4.8. After sterilization, 5% inoculums from spore suspension of T. reesee on PDA slant in 5 ml physiological serum was transferred and flasks were placed in a model rotary shaker at 200 rpm and 28°C, shaking. Fermentation was conducted in duplicates for enzyme production. Samples from the fermentation batches were centrifuged with a rotary model centrifuge at 2000 g for 10 min at 4°C. Crude enzyme produced was stored under refrigeration temperature.

SIMULTANEOUS SACCHARIFICATION AND FERMENTATION (SSF) PROCESS

SSF reaction mixture consisted of alkaline pre-treated rice husk as substrate, 10% yeast inoculums, YPM medium and crude cellulase. Fermentation medium, YMP medium contained (g%): Yeast extract, 0.3; malt extract, 0.3; peptone, 0.5; pre-treated substrate, 4; and distilled water up to 100 ml. After sterilization, cellulase 15 fpu/g substrate and 10% yeast pre-culture inoculums were added the SSF mixture, its pH then was adjusted to 5 and mixtures were incubated on a rotary shaker at 150 rpm and 37°C. Aliquots (2 ml) were withdrawn aseptically and assayed for ethanol content.

Ethanol determination was estimated by GC method using a flame ionization detector and stainless column (2 m length, 3 mm i.d.) packed with propak Q (50–80 mesh) the column oven was operated isothermally at 150°C and the detector and injector were kept at 170°C. Nitrogen was used as the carrier gas at 30 ml/min flow rate and the combustion gas was a mixture of hydrogen and air. All experiments were conducted in duplicates and average values reported.

Reducing sugar produced during enzymatic hydrolysis was determined by DNS method.[15]

RESULTS AND DISCUSSION

EFFECT OF PRETREATMENT ON FPA (FILTER PAPER ACTIVITY)

Different rice husk pretreatments were studied to determine the most suitable and possible method. Alkaline, alkaline and acid (mixed) treatment and the results were compared with untreated substrate. At different fermentation time, in the growth medium of Trichoderma reesee, samples were collected and reducing sugar was measured and FPA was determined. Table 1 depicts effect of pre-treatment on FPA at different fermentation time.

The results presented in Table 1 showed that alkaline pre-treatment conducted higher activity FPA 4.3, after 10 days, compared to mixed treatment 3.16 after one week and untreated 1.17, after 5 days. Because of the aforementioned reason alkaline pre-treatment was chosen for SSF process.

Table 1 Effect of different rice husk pretreatments on activity (U/ml) at various fermentation times by Trichoderma reesee.

Time (days)	*Mixed treatment*	*Alkaline treatment*	*Untreated*
1	0.11	0.54	0.32
2	0.80	1.22	0.54
3	1.86	1.62	0.64
4	2.70	1.86	0.86
5	3.10	2.83	1.17
6	3.10	2.83	0.76
7	3.12	3.17	0.73
8	3.11	3.70	0.71
9	3.10	3.90	0.70
10	3.10	4.30	0.69
11	3.09	4.10	0.67

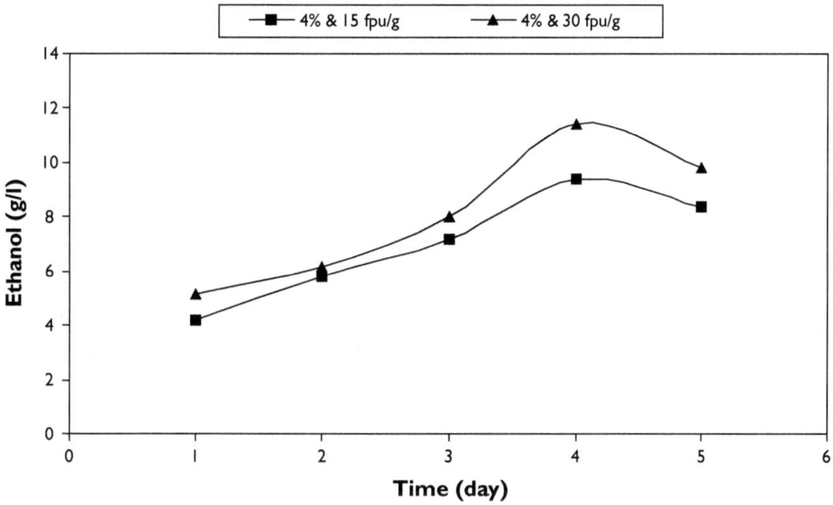

Figure 1 Effect of enzyme (crude cellulases) 15 and 30 fpu/g concentration in the medium containing 4% substrate on ethanol production (g/l).

EFFECT OF ENZYME CONTENT

To identify a suitable enzyme concentration in ethanol production, in ymp medium, two different enzyme concentrations 15 and 30 fpu/g with 4% pre-treated substrate were used. With increasing fpu cellulase in SSF the amount of liberated reducing sugar increases which results in higher conversion of substrate to ethanol (Fig. 1).

EFFECT OF SUBSTRATE CONCENTRATION

Pretreated rice husk by alkaline method was utilized as substrate in 4 and 6% (w/w) and 15 fpu/g cellulase were added in SSF fermentation medium. Ethanol production was measured with time (Fig. 2). The results presented in Fig. 2 shows that maximum reaction time was observed after 4 days in the presence of 6% substrate.

EFFECT OF CSL (CORN STEEP LIQUOR)

CSL could be added to the fermentation medium as an enriched source of nitrogen, organic carbon and other ingredients. 2% CSL was added to YMP medium with 4% pre-treated substrate and 15 fpu cellulase, and ethanol production was measured with time. The result is shown in Fig. 3. Maximum ethanol production was observed after 4 days in the absence of CSL which could be due to its inhibitory effect on ethanol production.

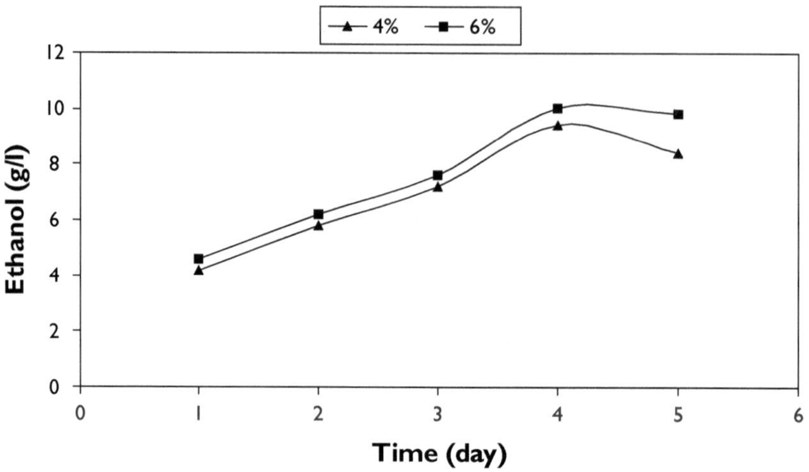

Figure 2 Effect of % pre-treated substrate on the ethanol production.

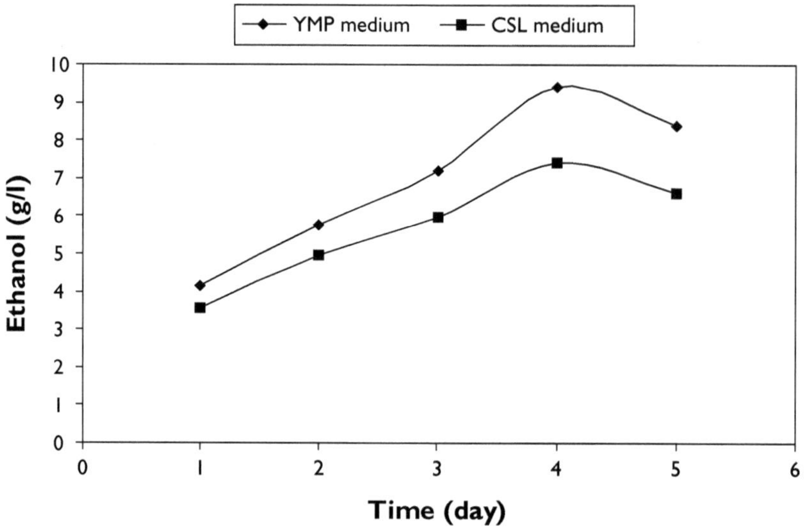

Figure 3 Effect of 2% CSL addition to YMP medium in SSF on ethanol production.

INOCULATION EFFECT

To determine effect of % of yeast inoculation on ethanol production efficiency, two inoculation percents 10 and 20% of yeast pre culture were added to SSF culture with 4% pre-treated substrate and ethanol production were measured with time (Fig. 4). The result showed that increasing inoculation percent for 10 and 20 had not significant influence on ethanol production.

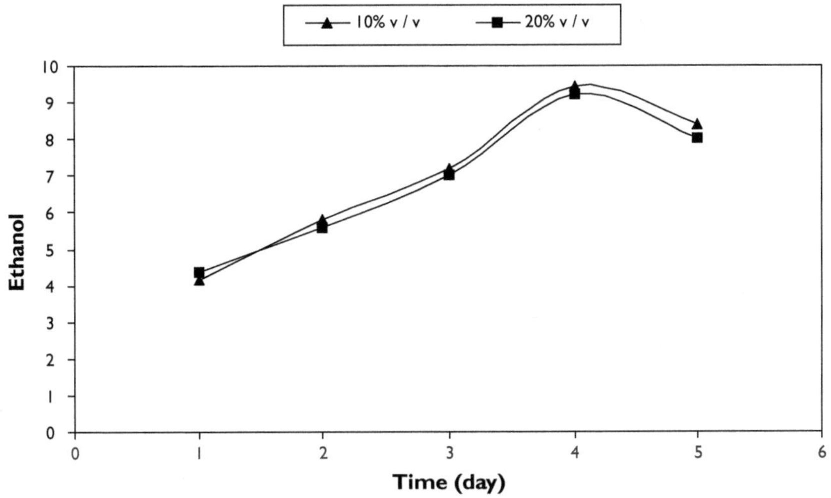

Figure 4 Effect of % inoculation on ethanol production with time.

CONCLUSIONS

Ethanol production in an SSF process with alkaline method pretreated substrate, rice husk, using crude cellulase from T. reesee, and Saccharomyces yeast cells at 37°C, pH 5 and 150 rpm were effected. The parameters determined were: Enzyme concentration, 30 fpu; substrate concentration, 6%; inoculation percent, 10; and csl showed inhibitory effect on ethanol production. Here the maximum ethanol reached was 12 g/lit. Compared to the other investigator worked on different species the results are reliable, however here crude enzyme, which its lower price can be counted as a conspicuous benefit, is used for hydrolyzing.

REFERENCES

1 Gerhartz W. Enzyme in industry; Germany: VCH, pp. 77–84 (1990).
2 Rani KS, Swamy MW and Seenayya G. Production of ethanol from various pure and natural cellulosic biomass by Clostridium thermocellum. *Process Biochem* **26**: 435–440 (1998).
3 Dale B and Moreira MJ. A Freeze exoplosion technique for increasing cellulose hydrolysis. *Biotechnol Bioeng Symp* **12**: 31–43 (1982).
4 Belkacemi K, Turcootte G, Hhalleux DD and Savoie P. Ethanol production from AFEX- treated forages and agricultural residues. Vol. 70–72 *Applied biochemistry and biotechnology* **70–72**: 441–462 (1998).
5 Jimenez L and Ferrer JL. Saccharification of wgeat straw for the production of alcohols. *Process Biochem* **26**: 153–156 (1999).
6 Fujita Y, Ito J, Ueda M, Fukuda H and Kondo A. Synergistic saccarification and direct fermentation to ethanol, of amorphous cellulose by use of an engineered

yeast strain Codisplaying three type of cellulytic enzyme. *Applied and Environmental Microbiology* **70**: 1207–1212 (2004).
7 Krishna SH, Parasanthi K, Ghowdary GV and Ayyanna C. Simultaneous saccharification and fermentation of pretreated sugar cane leaves to ethanol. *Process Biochem* **33**: 825–830 (1998).
8 Ballesteros M, Oliva JM, Negro MJ, Manzanares P and Ballesteros I. Ethanol from lignocelluloses materials by a simultaneous saccharification and fermentation process (SFS) with Kluyvermyces marxianus. *J Process biochem* **39**: 1843–1848 (2004).
9 Neves MAD, Kimura T, Shimizu N and Shiiba K. Production of Alcohol by Simultaneous Saccharification and fermentation of Low-grade Wheat Flour, Braz. *Arch Biol Technol* **49**: 481–490 (2006).
10 Patel MA, Ou MS, Ingram LO and Shanmugam KT. Simultaneous Saccharification and Co_Fermentation of Crystalline cellulose and sugar cane bagasse Hemicellulose Hydrolysate to lactate by a Thermotolerant Acidophilic Bacillus sp. *Biotechnology prog* **21**: 1453–1460 (2005).
11 Karimi K, Brandberg T, Edebo L and Taherzadeh MJ. Fed- Batch cultivation of Mucor indicus in dilute-acid lignocellulosic hydrolyzate for ethanol production. *Biotechnology Lett* **27**: 1395–1400 (2005).
12 Helle SS, Duff SJB and Cooper DG. Effect of surfactants on cellulose hydrolysis. *Biotechnol Bioeng* **42**: 611–617 (1993).
13 Ooshima H, Sakata M and Hareno Y. Enhancement of enzymatic hydrolysis of cellulose by surfactant. *Biotechnol Bioeng* **28**: 1727–1734 (1986).
14 Mandels M, Andreotti RE and Roche C. Measurement of saccharifying cellulose. *Biotechnol Bioeng Symp* **6**: 21–23 (1976).
15 Miller GL. Use of dinitro salicylic acid reagent for determination of reducing sugar. *Anal Chem* **31**: 426–428 (1959).

Chapter 29

GHG REDUCTION POTENTIAL IN THE SEAFOOD INDUSTRY:

A case study of seafood processing industry in Thailand

Awais A. Baloch & Ajit P. Annachhatre
Environmental Engineering and Management, Asian Institute of Technology, Klong Luang, Pathumthani, Thailand

INTRODUCTION

Seafood processing industries, in general, consume large quantities of water during processing activities and generate the wastewater with high organic and nutrient loads which ultimately becomes a source of methane (CH_4) and nitrous oxide (N_2O) emissions. Intergovernmental Panel on Climate Change (IPCC) guidelines for national GHG inventories[1] have illustrated the methods for calculation of CH_4 and N_2O emissions from the wastewater based on degradable organic matter and nitrogen components respectively. One of the seafood processing industries located in Samut Sakhon province, Bangkok was selected as a case study with average production capacity of 35–38 tons per day. The main objective of this study was to recommend a practical method for reducing GHG emissions by collecting methane from anaerobic process. Furthermore, methane utilization scenarios such as flaring and steam generation are also suggested. Reduction in GHG emissions for each scenario is also estimated.

METHODOLOGY

BASELINE STUDY

Description of existing wastewater treatment process

The quantity of wastewater from the industry varies between 1200–1400 m^3 per day, and treated in a wastewater treatment plant (WWTP) comprising of rotating drum screens, equalization tank, upflow anaerobic units, followed by activated sludge process. The final effluent comes to the polishing ponds before discharge to the nearby Tha Chin Klong river. The upflow anaerobic filter treats wastewater with COD 3000–5000 mg/l[2] without capturing any biogas from it.

Project study

It is suggested that the biogas escaping from upflow anaerobic filter be collected and utilized in the plant. Hence two scenarios (Figure 1) are suggested for biogas utilization as project activity; such as, scenario 1- only flaring and scenario 2- using biogas with partial saving of diesel oil which will reduce equivalent emissions from replacement of fossil fuel (i.e., diesel oil).

The proposed method for capturing biogas from anaerobic filter consists of flexible high density polyethylene (HDPE) cover with gas collection assembly and all other necessary equipments.

Estimation of GHG emissions

Average values of some important parameters, which were used for the estimation of GHG emissions, are given in Table 1. All calculations are based on the values of those parameters.

Total Baseline emissions are the sum of all emissions arising from anaerobic filter for wastewater treatment, sludge treatment, and fossil fuel burning as fuel for heating and electricity replacement purposes. These emissions are calculated as mentioned in methodologies AMS.III.H combination of

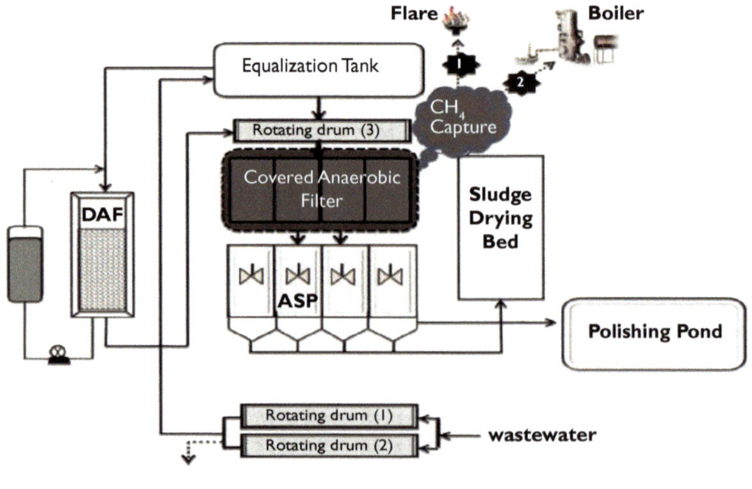

Figure 1 Scenarios for biogas utilization.

Table 1 Average values of the parameters used.

Parameters	*Description*	*Average values*
Working days per year	Total operating days of the industry in a year	300 days/year
Flow (Q_{ww})	Flow of wastewater	1300 m^3/day = 390,000 m^3/year
$COD_{removed}$	Chemical oxygen demand removed by anaerobic filter	2,400 mg/L = 0.0024 t/m^3
COD_{out}	Chemical oxygen demand out from anaerobic filter	450 mg/L = 0.00045 t/m^3
$TKN_{removed}$	Total *Kjeldahl* Nitrogen removed by Activated sludge process	310 mg/L
$B_{0,ww}$	Methane producing capacity	0.25 kg CH_4/kg COD
$MCF_{ww,treatment,i}$	Methane correction factor	0.8 (in case of this study)
GWP	Methane global warming potential [tCO_2/tCH_4]	21 (for CH_4) 298 (for N_2O)
Fuel used	Total fuel used in a year	504 t/year
Energy value of fuel	Energy value of diesel oil	0.043 TJ/ton
$N_{effluent}$	Nitrogen in the effluent of N_2O from the wastewater treatment	120.9 tN/year
$EF_{effluent}$	Emission factor for nitrous oxide emissions	0.005 tN_2O-N/tN
CEF_{fuel}	Carbon emission factor of the fuel	74.1 tCO_2/TJ
$MCF_{ww,final,i}$	Methane correction factor based on type of treatment	0.1 (anaerobic treatment)
CFE_{ww}	Capture and flare efficiency of methane recovery in the WWTP	0.9 (anaerobic treatment)
$[CH_4]_{y,ww,treated}$	Dissolved methane in the treated wastewater	0.0001 tCH_4/m^3 (anaerobic treatment)

version 6 and 9, and AMS.I.D version 13 respectively for small scale CDM project activities.[3]

Project activity emissions: The total amounts of GHG emissions due to the project activities are estimated by referring to IPCC, 2006 National GHG Inventories.[1]

Reduction in GHG emissions: Based on the estimation of emissions from two scenarios, the amount of possible reductions was estimated for each scenario.

Table 2 Comparison of two scenarios for biogas utilization.

Scenarios	*Baseline emissions (tCO_2/yr)*	*Project emissions (tCO_2/yr)*	*Emission reductions (tCO_2/yr)*	*Savings*
1. Flaring	5,819.20	2,909.26	2,909.94	–
2. Replacement of diesel oil	5,819.20	2,211.085	3,608.115	About 43.5% of diesel oil consumption can be reduced

RESULTS AND DISCUSSIONS

The results of emissions reductions for each scenario are presented in Table 2.

The emission reductions are estimated as 2,909.64 tCO_2/year and 3,608.12 tCO_2/year for scenario 1 and 2 respectively. Scenario 2 proves to be economically viable than scenario 1.

Food industry wastewaters are often treated in anaerobic ponds[4] which are incapable of trapping methane emissions. On the other hand, high-rate anaerobic processes[5] are capable of entrapping methane emissions from organic wastes. N_2O emissions from nitrogenous wastewaters from food industries[6] can further be trapped by employing novel biological nitrogen removal processes[7] such as partial nitrification[8] followed by anammox process.[2,9] Thus, high rate anaerobic process followed by partial nitrification and anammox process are recommended for effective GHG mitigation from sea food industry wastewater.[10]

CONCLUSION

Current wastewater treatment practices in seafood industry release significant amount of GHG emissions into the environment due to untapped methane emissions from anaerobic processes. This study conducted on a seafood processing unit from Thailand brings out that methane from anaerobic treatment of wastewater can be effectively collected and either flared or utilized in plant for steam generation, thus replacing diesel oil. Replacing diesel oil with the collected biogas was found to be more effective scenario as compared to flaring of biogas. In this case about 3,608 tCO_2 equivalent emissions per year can be reduced, and captured biogas can replace 43.5% of diesel oil used for the boiler.

REFERENCES

1 IPCC (2006), *IPCC Guidelines for National Greenhouse Gas Inventories*, Prepared by the National Greenhouse Gas Inventories Programme, Eggleston HS, Buendia L, Miwa K, Ngara T and Tanabe K (eds). IGES, Japan.

2 Lamsam A (2008), The anammox process for nitrogen removal from seafood processing industrial wastewater. (Doctoral dissertation No. EV-08-8, *Asian Institute of Technology*, 2008). Bangkok: Asian Institute of Technology.

3 UNFCCC (2005), Retrieved on 3 April, 2009 from websites: http://unfccc.int/kyoto_protocol/items/2830.php and http://cdm.unfccc.int/index.html

4 Rajbhandari BK and Annachhatre AP (2004), Anaerobic ponds treatment of starch wastewater: Case study in Thailand, *Bioresource Technology*, 95(4), 135–143.

5 Amornkaew A and Annachhatre AP (2001), Upflow Anaerobic Sludge Blanket Treatment of Starch wastewater containing Cyanide, *Water Environment Research* 73(5), 622–632.

6 Park, JW (2005), Surimi and Surimi Seafood, Second Edition, CRC Press. 4–5: 286–289.

7 Khin T and Annachhatre AP (2004), Novel microbial nitrogen removal processes, *Biotechnology Advances*, 22, 519–532.

8 Sinha B and Annachhatre AP (2007), Partial nitrification – operational parameters and microorganisms involved. *Reviews in Environmental Science and Biotechnology*, 6(4), 285–313.

9 van Dongern U, Jetten MSM and van Loosdrecht MCM (2001), Combined Sharon/Anammox process: a sustainable method for N-removal from sludge water. Delft, Netherlands: IWA Publishing.

10 Lamsam A, Laohaprapanon S and Annachhatre AP (2008), Combined activated sludge with partial nitrification (AS/PN) and anammox processes for treatment of seafood processing wastewater. 43(10), 1198–1208.

Chapter 30

SEWAGE BIOGAS TREATMENT:
Relevant options based on the biogas end-use

L. Bouchy, N. de Arespacochaga & JL. Cortina
CETaqua, Paseo de los Tilos, Barcelona, Spain

E. Larrotcha
Agbar, Paseo de los Tilos, Barcelona, Spain

M. Sanchez
Emuasa, Plaza Circular, Murcia, Spain

C. Peregrina
Suez Environnement, Le Pecq, France

ABSTRACT

Within the framework of sustainable development, today burning biogas via a flare cannot be considered a long-term solution. Moreover, the ever-increasing energy costs constitute another important driver for wastewater treatment activities. Furthermore, although cogeneration has been widely developed, other emerging more promising biogas end-use routes must now also be considered. Within this context, the goal of this study is to assess opportunities for different biogas end-uses, in particular: fuel cells, conventional cogeneration and car motors. The focus of this poster is the biogas treatment, and the objective is to establish a "mapping" of adequate technologies available and their relevance depending on the biogas end-use, based on technical, economic and environmental criteria. This study was carried out by literature research, and carrying out audits of industrial or pilot plants.

Keywords: Biogas treatment, biogas use, technical assessment, economic assessment, environmental impact

BACKGROUND

Within the framework of sustainable development, today burning biogas via a flare cannot be considered a long-term solution.[1,2] Moreover, the ever-increasing energy costs constitute another important driver for wastewater treatment activities. Within this context, the goal of this study is to assess opportunities for different biogas end-uses. Furthermore, although cogeneration has been widely developed and used for many years, other emerging more promising biogas end-use routes must now also be considered: for instance fuel cells are rapidly developing and promising both in terms of technical efficiency (overall efficiency over 75%, greater energy densities) and in terms of environmental impact (lower emissions, lower noise impact).[3,4] Another promising route is biogas utilisation as a fuel for vehicles, which requires thorough biogas conditioning as CO_2 must be removed. This product is called biomethane and its use in engines allows for lower global and local emissions in comparison to traditional vehicle fuels. The main drawbacks of these emerging technologies to this day is the price (prices of fuel cells can reach 10.000 €/kW), and higher biogas quality requirements. These statements also imply that in order to assess these biogas end-use options, it is necessary to take into account both the biogas treatment and its end-use, and also to consider technical, economic and environmental issues.

The focus of this study is set on the application of these technologies to sewage biogas, and in particular the most adapted biogas treatment technology which is the topic of this poster. The objective is to establish a "mapping" of adequate technologies available and their relevance depending on the biogas end-use.

METHOD

This study was focused mainly on H_2S and siloxanes (main contaminants of energy conversion equipment),[1] as well as CO_2. Dust, NH_3 and water removal technologies are widely commercialised and controlled and will not be described here. Table 1 shows the main technologies considered:

This poster is focused on the biogas treatment (however in order to assess thoroughly each option, it will obviously be necessary to consider the whole energy production line, including the energy conversion system). In order to compare different options, several criteria were considered.

- Technical: removal rate (%), capacity (g/kg of media or chemical), limits (inlet and outlet), pre-conditioning requirements
- Economic: capital and operating costs (including consumables, electricity, man power)

Table 1 Main treatment technologies considered.

H_2S	*Siloxanes*	*CO_2*
Iron sponge	Activated carbon/graphite	Chemical scrubbing
Activated carbon	Biological treatments	Adsorption
Biological treatments	Freezing	Freezing
Chemical scrubbing		Membranes

- Environmental: impact on climate change due to the use of consumables and energy.

In order to be able to take into account both commercialised and developing technologies, and to both pinpoint technical specific technical issues and highlight operating limitations, this study was carried out by 3 means: literature research, contacts with treatment technologies providers and carrying out audits for data collection from industrial or pilot plants.

Figure 1 shows a pilot plant on a wastewater treatment plant (Murcia, Spain) where biogas is used as car fuel, and is to be used shortly as fuel for a fuel cell.

RESULTS

Due to development of biogas use over the past few years, much development has taken place, not only in terms of energy conversion technologies, but also in terms of biogas treatment. Evolutions have allowed to adapt existing technologies to biogas: for example scientific works, currently being applied on the field, have allowed to determine adequate conditions to avoid elemental sulphur accumulation in biotrickling filters and therefore ease their operation.[5] Improvements have also been made to increase the efficiency of the technologies or adapt them to treat emerging compounds (e.g., siloxanes)[6] to

Figure 1 Biogas treatment plant of Murcia operated by Emuasa (Spain).

comply with the stringent inlet requirements of emerging energy conversion technologies. Moreover, many mature technologies which have shown important drawbacks have developed "advanced" systems to counteract these drawbacks. For example, scrubbers are acknowledged to show high chemical and energy costs: the advanced systems put emphasis on chemical regeneration (e.g., the Thiopaq system).

A first pre-selection of adapted technologies has been established for each of the target compounds, based on the 3 performance criteria mentioned above. Moreover, for each technology, the application field may differ. For example, concerning H_2S removal, several technologies are considered most relevant: iron sponge, biotrickling filters and advanced scrubbers (with regeneration). However, activated carbon is a necessary technology to be used for final polishing to achieve stringent inlet requirements such as for fuel cells. These results and references will be further described and discussed in the full poster.

REFERENCES

1 Abatzoglu N and Boivin S. A review of biogas purification processes. *Biofuels, bioproducts and biorefining*. 3: 42–71(2009).
2 Truong LV and Abatzoglou N. A H_2S reactive adsorption process for the purification of biogas prior to its use as a bioenergy vector. *Journal of Biomass and Bioenergy* 29: 142–151(2005).
3 Pipatmanomai S, Kaewluan S and Vitidsant T. Economic assessment of biogas-to-electricity generation system with H_2S removal by activated carbon in small pig farm. *Applied Energy, article in press* (2008).
4 Trogisch S, Hoffmann J and Daza L. Operation of molten carbonate fuel cells with different biogas sources: A challenging approach for field trials. *Journal of Power Sources* 145: 632–638(2005).
5 Fortuny M, Baeza JA, Gamisans X, Casas C, Lafuente J, Deshusses MA and Gabriel D. Biological sweetening of energy gases mimics in biothricking filters. *Chemosphere* 71: 10–17(2008).
6 Popat SC and Deshusses MA. Biological removal of siloxanes from landfill and digester gases: opportunities and challenges. *Environmental Science and Technology* (2008).

Chapter 31

SENSITIVITY ANALYSIS OF THE MODEL THAT DESCRIBES THE BIOFILTRATION OF A MIXTURE OF H_2S AND DMS

G. Aroca, J. Silva, M. Morales, M.S. Cáceres & R. San Martín
School of Biochemical Engineering, Pontificia Universidad Católica de Valparaíso, Valparaíso, Chile

ABSTRACT

Sensitivity analysis was performed for the parameters involved in the modelling of a biotrickling filter for biooxidation of a mixture of DMS and H_2S. The Thiele module values are extremely low and the system is sensitive to its value, showing that the system is controlled by mass transfer. An important effect was observed for the value of the inhibition constant used in the modelling. Almost no variation was predicted in the removal capacity of the biofilter for different residence times when using low Peclet numbers.

INTRODUCTION

Volatile reduced sulphur compounds (VRSC) generate a high environmental impact in the vicinity of their sources. Several technologies have been proposed for their elimination using biological systems, being more difficult to eliminate all the compounds when they are in a mixture. To solve this problem two-stage systems have been proposed.[1,2] In the first reactor H_2S is biooxidized and in the second the rest of the VRSC mixture, avoiding the inhibition effects of H_2S over the biooxidation of the other compounds. In the systems described the complete oxidation of H_2S must be performed in the first reactor, if some H_2S pass though out the first reactor it have a negative effect on the bio-oxidation of the other VRSC present in the mixture in the second bioreactor. This situation was modeled and validated[3] using experimental results. The biotrickling filter was modeled as a fixed bed of packing material which supports the growth of micro-organisms (*T. Thioparus*) as a biofilm. The model describes the mass transfer of the compounds from gas to liquid, and the biooxidation of the compounds by the biofilm attached to the support. In the system studied H_2S is preferentially biooxidized over DMS,[4,5,6] this effect was observed experimentally. The model assumes the kinetic of a competitive inhibition for the biooxidation of DMS in the presence of H_2S.

The aim of this work is to present a sensitivity analisys of the parameters involved in the modelling of a biotrickling filter for bioxidation of a mixture of DMS and H_2S. The parameter studied are: the Peclet number, the Thiele number, the inhibition constant.

MATHEMATICAL MODEL

The model developed describes the mass transfer gas to liquid and the biodegradation of the compounds in the mixture by a biofilm attached to the support (*T. Thioparus*). The model also takes into account the inhibition of H_2S on the DMS bioxidation using a Monod-type kinetics.

EQUATIONS

General Mass balance for compound i in the gaseous phase:

$$\frac{1}{Pe}\cdot\frac{\partial^2 C_{gi}}{\partial\zeta^2}-\frac{\partial C_{gi}}{\partial\zeta}+v_b\cdot Ti\cdot\left.\frac{\partial C_{bi}}{\partial\psi}\right|_{\psi=0}=0 \tag{1}$$

where C_g is the dimensionless concentration in the gas phase, Pe is the Péclet number, ζ is the dimensionless axial co-ordinate along the bed height, v_b is the specific volume, Ti is the residence to diffusion time ratio, C_b is the dimensionless concentration in the biofilm and ψ is the dimensionless spatial co-ordinate in the biofilm.

General mass balance for the i compound in the liquid phase dimensionless

$$\frac{\partial^2 C_{bi}}{\partial\psi^2}-Th_i^2\cdot\kappa_i=0 \tag{2}$$

where Th is the Thiele modulus and κ is the dimensionless specific degradation rate. Kinetic equation for the dimensionless rate of H_2S bioxidation:

$$\kappa_{H_2S}=\frac{1}{Y_{x/H_2S}}\cdot\frac{C_{bH_2S}}{\sigma_i\cdot\left(1+\frac{C_{bH_2S}}{\iota_i}\right)+C_{bH_2S}} \tag{3}$$

where Y is the yield coefficient of biomass (g biomass g^{-1} substrate), σ is the dimensionless Monod constant and ι is the dimensionless inhibition constant. Kinetic equation for the dimensionless rate of DMS bioxidation:

$$\kappa_{DMS} = \frac{1}{Y_{x/DMS}} \cdot \frac{C_{bDMS}}{\sigma \cdot \left(1 + \frac{C_{bH_2S}}{\iota_i}\right) + C_{bDMS}} \tag{4}$$

The boundary conditions are:

$$\begin{aligned}
&\zeta = 0 \quad 0 \le \psi \le 1 \quad -\frac{\partial C_g}{\partial \zeta} + \upsilon_b \cdot Ti \cdot \left.\frac{\partial C_b}{\partial \psi}\right|_{\psi=O} = 0 \\
&\zeta = 1 \quad 0 \le \psi \le 1 \quad -\frac{\partial C_g}{\partial \zeta} + \upsilon_b \cdot Ti \cdot \left.\frac{\partial C_b}{\partial \psi}\right|_{\psi=O} = 0 \\
&\psi = 0 \quad 1 \le \zeta \le 0 \quad C_b = \frac{C_g}{m} \\
&\psi = 1 \quad 1 \le \zeta \le 0\!: \quad \frac{\partial C_{bi}}{\partial \psi} = 0
\end{aligned} \tag{5}$$

where m is the partition coefficient.

RESULTS

Figure 1 shows the effect of the Péclet number. In the range studied the *Pe* has not influence on the removal capacity due to the high dispersion that have the

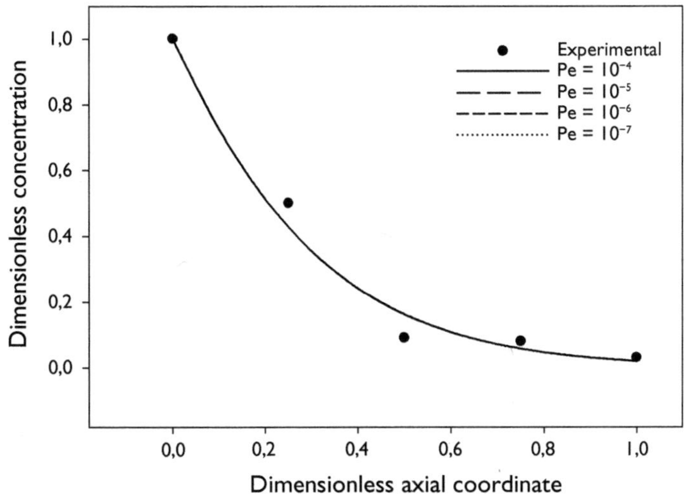

Figure 1 Effect of Peclet number in the removal of H_2S in the biofilter.

contaminant in the system beside the turbulent flow regime. The results of the simulation for different *Pe* values and the experimental validation shows that the system behaves as a plug flow reactor. Figure 2 shows that the model has a high sensitivity to the Thiele modulus which indicates a high influence of the diffusion in the liquid phase on the kinetics of bioxidation.

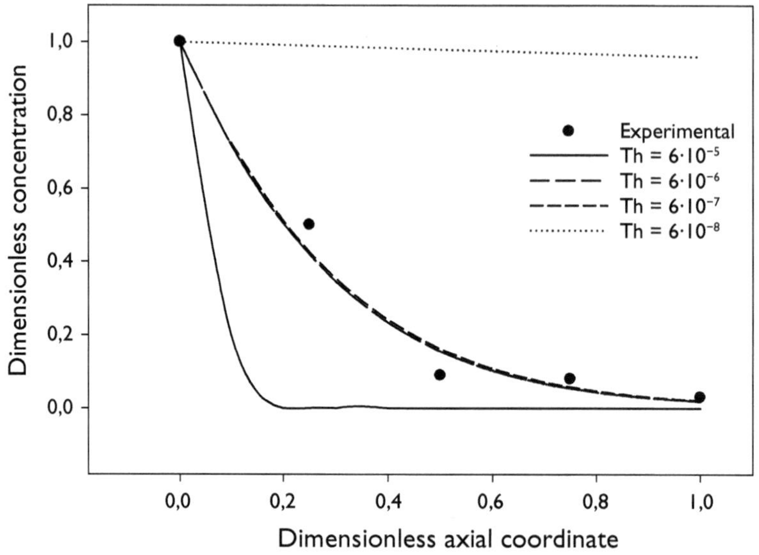

Figure 2 Effect of Thiele module in the removal of H_2S in the biofilter.

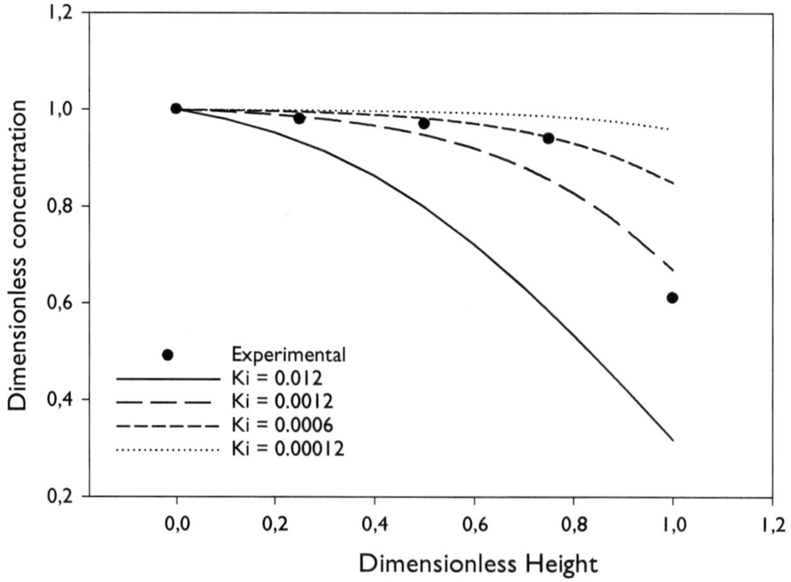

Figure 3 Effect of inhibition constant in the removal of DMS in the biofilter.

Some experiments[2] shown that H_2S is preferentially bioxidazed than DMS, suppressing the biooxidation of the second when there is presence of the first. This fact can be modeled as a competitive inhibition. Figure 3 shows the important effect of the value of the inhibition constant in the simulation of the performance of the biofilter.

CONCLUSIONS

Due to the high dispersion of H_2S in the gas phase, the Peclet number has no effect on the biooxidation. The Thiele module values are extremely low and the system is sensitive to its value, showing that the system is controled by mass transfer. The inhibition constant is a parameter that can be difficult to determine experimentally and its influence is very important in the simulation.

REFERENCES

1 Ruokojärvi, A., J. Ruuskanen, P.J. Martikainen and M. Olkkonen. Oxidation of gas mixtures containing dimethyl sulfide, hydrogen sulfide, and methanethiol using a two-stage biotrickling filter. *J. AWMA*, 51: 11–16 (2001).

2 Sercu, B., D. Nuñez, H. van Langenhove, G. Aroca and W. Verstraete. Operacional and Microbiological Aspects of a Bioaugmented Two-stage Biotrickling Filter Removing Hydrogen Sulphide and Dimethylsulfide. *Biotechnology and Bioengineering* 90: 259–269 (2005).

3 Aroca, G., J. Silva, M. Morales, M. Cáceres and R. San Martín. Biofiltration of volatile reduced sulfur compounds in biotrickling filters: Prediction through mathematical modeling and simulation. *Conference on Biofiltration for Air Pollution Control.* 22–24 October, Long Beach, Florida (2008).

4 Zhang, L., M. Hirai and M. Shoda. Removal characteristics of dimethyl sulfide, methanethiol and hydrogen sulfide by *Hyphomicrobium* sp. I55 isolated from peat biofilter *J. Ferment. Bioeng.* 72(5): 392–396 (1991).

5 Cho, K.S., Hirai M and Shoda M. (1991). Degradation characteristics of hydrogen sulfide, methanethiol, dimethyl sulfide and dimethyl disulfide by Thiobacillus thioparus DW44 isolated from peat biofilter. *J Ferment. Bioeng* 71(6): 384–389.

6 Wani, A.H., A.K. Lau and R.M.R. Branion, (1999). Biofiltration control of pulping odors–hydrogen sulfide: performance, macrokinetics and coexistence effects of organo-sulfur species. *J. Chem. Technol. Biotechnol.* 74: 9–16.

7 Devinny, J.S. and J. Ramesh. A phenomenological review of biofilter models (2005) *Chemical Engineering Journal*, 113(2–3), pp. 187–196.

Chapter 32

EXPERIMENTAL DETERMINATION OF KINETIC PARAMETERS IN BIOFILTERS TREATING METHANOL

A. Avalos Ramirez, J.P. Jones & M. Heitz
Faculty of Engineering, Department of Chemical Engineering, Université de Sherbrooke, Sherbrooke (Québec), Canada

ABSTRACT

The present study was designed to determine experimentally in a biofilter and in a biotrickling filter a) the biodegradation rate of methanol at different operating conditions, and b) the specific growth rate at those operating conditions. The pollutant biodegradation rate was determined in a laboratory scale biofilter and a laboratory scale biotrickling filter. Models were fitted to the experimental data of pollutant biodegradation occurring in both types of biofilters. The models tested were based on the Michaelis-Menten and Haldane models.

The specific growth rate in the biofilter was calculated using the exponential growth model, expressed in terms of the biomass produced. In the case of the biotrickling filter, the specific microbial growth rate was calculated using an expression which links the specific biodegradation rate of the pollutant with the biomass yield coefficient.

INTRODUCTION

Treatment of polluted air using biological processes, such as biofiltration with and without percolation, is an environmental friendly option to control the emissions of volatile organic compounds (VOCs) to the atmosphere. The performance of these bioprocesses is evaluated in terms of the VOCs' biodegradation and their conversion to biomass and carbon dioxide.

Methanol is widely used in industry where it is often eventually released to atmosphere. In Canada, the total emissions of methanol were 13.5 ktons year^{-1} in 2008.[1] Methanol is toxic to humans and therefore its emissions must be well controlled in order to avoid health problems.[2] Methanol is biodegradable and this property leads to the use of biofiltration for the control of its release.

The estimation of methanol biodegradation and the kinetic parameters of the bioreaction led to the correct estimation and application of the biofiltration for control emissions of this pollutant. The specific growth rate (μ) has often been predicted by using the Monod model employed for population growth studies.[3] The μ is a biological parameter which is affected by the operating conditions, such as substrate concentration and residence time. The aim of this study is to analyze the performance and the microbial growth of a biofilter and a biotrickling filter treating methanol when both biorectors are operated under similar operating conditions.

MATERIALS AND METHODS

The present research is divided into two phases, consisting of a study on the effect of the nitrogen inlet load, type of packing material and methanol concentration on: 1) the elimination capacity of a biofilter (BF) and a biotrickling filter (BTF) treating methanol; and, 2) the specific growth rate for both types of bioreactors.

The BF and BTF were constructed of Plexiglas cylinders with an internal diameter of 0.15 m and a total bed height of 1 m divided into three sections of 0.33 m. The experimental setup (Figure 1) consisted of a bubbler

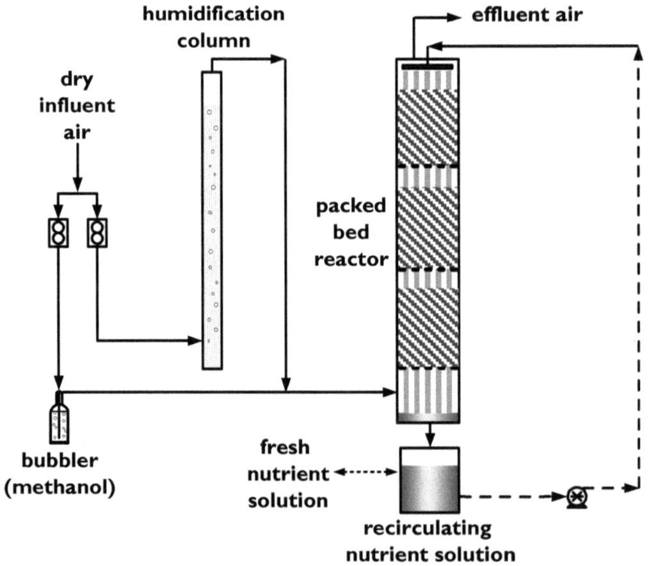

Figure 1 Schematic representation of biofilter and biotrickling filter. For a biofilter, the holding tank and the recirculation pipeline of nutrient solution (– – –) disappeared.

containing methanol, a humidification column and a packed bed with an empty bed volume of 0.018 m^3. Two types of packing materials were tested: 1) clay spheres (Hydroton®, Ökotau GmbH, Germany) of 0.012 m diameter, 310 m^2 m^{-3} specific surface and 40% void space, and 2) non-porous polypropylene (PP) spheres (Jaeger Tri-Packs®, Fabco Plastics, Toronto, Canada) of 0.025 m diameter, 280 m^2 m^{-3} specific surface and 90% void space. The nutrient liquid solution was fed into the BF and the BTF countercurrent to the air flow. The BTF included also a holding tank and a recycling pump.

The concentrations of methanol were measured continuously with a total hydrocarbon analyzer (Horiba FIA-510, Horiba Instruments Inc., Irvine, CA, USA). The following equations provide a good fit for μ for the BF and the BTF, repectively:[4,5]

$$\text{In}(B) = \text{In}(B_0) + \mu\, t \tag{1}$$

$$Y_B = \frac{\mu(B\ \ \text{EBRT})}{(C_{\text{in}} - C_{\text{out}})} \tag{2}$$

where, μ is the specific growth rate (h^{-1}), B_0 is the initial biomass (g), B the final biomass (g), t the elapsed time (h), Y_B the biomass yield coefficient (dimensionless), EBRT the empty bed residence time (h), C_{in} and C_{out} the inlet and outlet methanol concentrations (g m^{-3}), respectively.

RESULTS AND DISCUSSION

EFFECT OF NITROGEN INLET LOAD AND TYPE OF PACKING MATERIAL ON THE ELIMINATION CAPACITY

Figure 2 shows the elimination capacity (EC) of two biofilters (BF) and a biotrickling filter (BTF) as a function of the nitrogen inlet load (N-IL). One BF was packed with clay spheres (clay-BF), and the other BF and the BTF with PP spheres (PP-BF and PP-BTF, respectively). The three biofilters were operated at different nitrogen concentrations in nutrient solution, for a nitrogen inlet load in the range from 0.2 to 3.7 gN m^{-3} h^{-1}. Figure 2 shows that the clay-BF presented a higher performance than the PP-BF; for example, at a N-IL of 3.7 gN m^{-3} h^{-1}, the EC of the clay-BF had increased an up to 65 g m^{-3} h^{-1}, whereas the PP-BF stayed at around 27.5 g m^{-3} h^{-1}. For this reason, the clay-BF was selected for the study of the effect of methanol on EC. At the highest N-IL tested, the clay-BF had the same performance than the PP-BTF, whose EC oscillated between 60 and 70 g m^{-3} h^{-1}, in the whole range of N-IL tested. As observed by Avalos Ramirez *et al.*[6] for a high water soluble pollutant like methanol, the continuous trickling in a BTF contributes with the removal efficiency. This explains why the PP-BTF presented higher EC than the clay-BF at N-IL smaller than 2 gN m^{-3} h^{-1}.

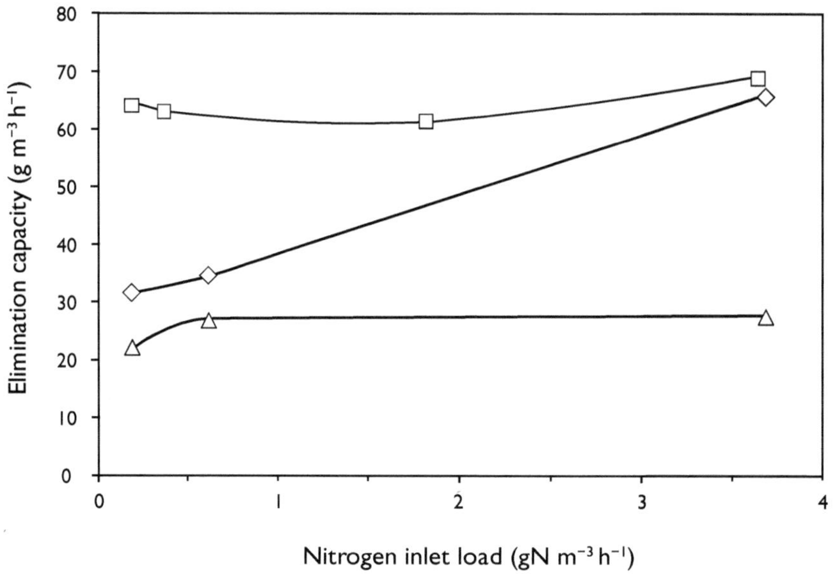

Figure 2 Effect of the nitrogen inlet load on the elimination capacity of the biotrickling filter packed with polypropylene spheres (–□–), the biofilter packed with clay spheres (–◇–), and the biofilter packed with polypropylene spheres (–Δ–).

EFFECT OF METHANOL CONCENTRATION ON THE ELIMINATION CAPACITY

Figure 3 shows the effect of the methanol concentration (logarithmic mean of inlet and outlet concentrations, C_{ln}-MetOH) on EC for the clay-BF and PP-BTF. Figure 3 shows that the C_{ln}-MetOH varied from 0.1 to 2.2 g m^{-3} for the PP-BTF and from 0.3 to 3.1 g m^{-3} for the clay-BF. For both biofilters, EC increased with C_{ln}-MetOH. In the whole ranges of C_{ln}-MetOH, EC increased from 28 to 72 g m^{-3} h^{-1} for the clay-BF and from 12 to 170 g m^{-3} h^{-1} for the PP-BTF. Mixed order models (Michaelis-Menten and Haldane) were fitted to the experimental data of EC. In the case of the PP-BTF, the Michaelis-Menten model fitted the EC data well. The kinetic parameters were obtained by non-linear regression and the fitted equation is shown in Figure 3 by a solid line, the Michaelis-Menten equation representing the EC for the PP-BTF is as follows:

$$EC = \frac{2237\ C_{In}^{gas}}{24.05 + C_{In}^{gas}} \tag{3}$$

In the case of the clay-BF, the Haldane model fitted to the EC. The kinetic parameters were obtained by non-linear regression and the fitted equation is

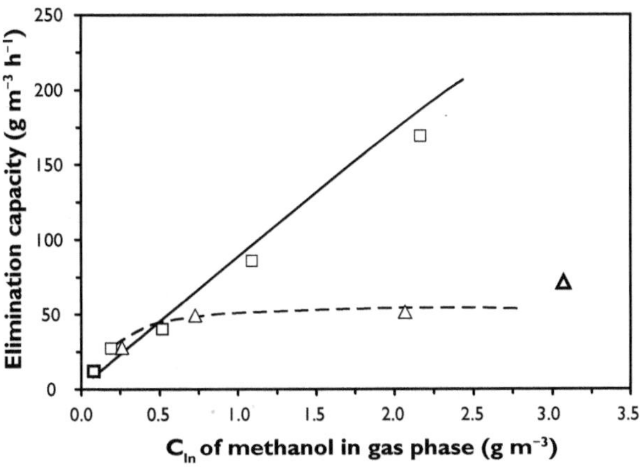

Figure 3 Effect of the methanol concentration (logarithmic mean) on elimination capacity of the biotrickling filter packed with polypropylene spheres (–□–) and the biofilter packed with polypropylene spheres (–Δ–).

shown in Figure 3 by a dashed line, the Haldane equation representing the EC for the clay-BTF is as follows:

$$EC = \frac{67.8\ C_{In}^{gas}}{0.263 + C_{In}^{gas} + \frac{(C_{In}^{gas})^2}{16.86}} \qquad (4)$$

EFFECT OF OPERATING CONDITIONS ON THE SPECIFIC GROWTH RATE

The specific growth rate (μ) was calculated for all experiments performed with the clay-BF, PP-BF and PP-BTF. Table 1 shows μ for three biofilters at the different operating conditions tested. For both BFs, μ oscillated in a range of values from 0.0008 h^{-1} (the smallest level of μ for the PP-BF) to 0.0095 h^{-1} (the highest level of μ for the clay-BF). In contrast to both BFs, μ of PP-BTF was constant for each study. For the study about the effect of N-IL, μ was 0.05 h^{-1}, and for the effect of C_{In}-MetOH, μ was 0.04 h^{-1}. These results confirm the EC of PP-BTF observed in Figures 2 and 3, which was always higher than the EC of clay-BF and PP-BF.

CONCLUSIONS

The treatment of methanol by biofiltration with and without percolation was studied. The kinetic parameters of methanol biodegradation and growth rate

Table 1 Effect of operating conditions on the specific growth rate for the biofilters tested in this study.

a) Effect of nitrogen inlet load

Type of bioreactor	*Nitrogen inlet load ($gN\ m^{-3}\ h^{-1}$)*	*Packing material*	*Specific growth rate μ (h^{-1})*
Biofilter	0.2	Polypropylene	0.0008
Biofilter	0.6	Polypropylene	0.0015
Biofilter	3.7	Polypropylene	0.0006
Biofilter	0.2	Clay	0.0005
Biofilter	0.6	Clay	0.0015
Biofilter	3.7	Clay	0.0016
Biotrickling filter	0.2 to 3.6	Polypropylene	0.05

b) Effect of methanol concentration

Type of bioreactor	*Methanol concentration, C_{In} ($g\ m^{-3}$)*	*Packing material*	*Specific growth rate μ (h^{-1})*
Biofilter	0.8	Clay	na
Biofilter	1.3	Clay	0.0013
Biofilter	2.7	Clay	0.0095
Biofilter	4.3	Clay	0.0053
Biotrickling filter	0.8 to 5.3	Polypropylene	0.04

na = non available.

in both types of biofilters were determined at different nitrogen inlet loads, packing materials and methanol concentrations. The kinetic parameters were a function of operating conditions. For biofilters packed with different packing material, the clay spheres presented better performance than polypropylene spheres. However, the biotrickling filter packed with polypropylene spheres performed better and was more stable than the biofilters, with an average EC 65 $g\ m^{-3}\ h^{-1}$ when nitrogen inlet load was varied.

A Michalis-Menten type model provided a very good fit to the elimination capacity of the biotrickling filter, whereas a Haldane type model provided a better fit to the elimination capacity of the biofilter. In the biofilter, the specific growth rate was function of operating conditions varying from 0.0008 to 0.0095 h^{-1}, whereas the specific growth rate of the biotrickling filter was higher than its level observed in the biofilters and nearly constant for each study: 0.05 h^{-1} for the study of the nitrogen effect and 0.04 h^{-1} for the study of methanol concentration effect.

ACKNOWLEDGEMENTS

The authors express their sincere acknowledgements to Natural Sciences and Engineering Research Council of Canada for the financial support.

REFERENCES

1 NPRI – National Pollutant Release Inventory (2009) NPRI, Methanol 67-56-1, Releases 2008. Environnement Canada, QC, Canada.
2 Monod, A., Doussin, J.F., Chebbi, A. and Carlier, P. (1998). Transformations chimiques des COV dans la troposphère. Impact sur la qualité de l'air. In: Le Cloirec, P. (Ed.). Les composés organiques volatils (COV) dans l'environnement. Technique & Documentation, Paris, France, pp. 119–162.
3 Liu. Y. Overview of some theoretical approaches for derivation of the Monod equation. *Applied Microbiology and Biotechnology* **73**: 1241–1250 (2007).
4 Jorio, H. and Heitz, M. Traitement de l'air par biofiltration. *Canadian Journal of Civil Engineering* **26**: 402–424 (1999).
5 Avalos Ramirez, A., Bénard, S., Giroir-Fendler, A., Jones, J.P. and Heitz, M. Kinetics of microbial growth and biodegradation of methanol and toluene in biofilters and an analysis of the energetic indicators. *Journal of Biotechnology* **138**: 88–95 (2008).
6 Avalos Ramirez, A., Bénard, S., Giroir-Fendler, A., Jones, J.P. and Heitz, M. (accepted for publication on September 2009). Treatment of air polluted with methanol vapours in biofilters with and without percolation. *J. Environ. Eng. Sci. – TBA Special Issue*. Manuscript number CJCE-26-02-09-3.

Chapter 33

COMPARATIVE STUDY OF PACKING MATERIALS FOR H_2S REMOVAL BY BIOFILTRATION

I. Charron, P. Zozor & C. Renner
Air & Odors Program – Centre de Recherche sur l'Eau, VEOLIA Environnement, Maisons Laffitte Cedex, France

M. Jovic
Microbiologie & Biotechnologies Program – Centre de Recherche sur l'Eau, VEOLIA Environnement, Maisons Laffitte Cedex, France

ABSTRACT

Biofiltration is an air treatment process widely used to remove odorous pollutants from air. Based on the ability of micro-organisms to degrade pollutants, biofilters are particularly suitable for the treatment of hydrogen sulphide (H_2S) emitted in wastewater treatment plant atmosphere. Well designed and operated, they permit to obtain very good H_2S removals with low working costs. However, some problems exist and some improvements are still necessary: for example, biofilters are often badly watered on sites or packing materials become compact with time resulting in higher pressure drop, preferential air paths.

In this context, this study focuses on the optimization of H_2S treatment by biofiltration. The main objectives are to select new biofiltration packing materials, to determine the best operating conditions and to monitor the biomass colonization in order to better understand the biological mechanisms involved.

To achieve these targets, comparison of several packing materials has first been carried out. Experiments were performed on a laboratory testing device composed of three biofiltration units. Each unit was made of a 0.2 m diameter column and a 20 L water tank. The packing height was 1 meter. The water was re-circulated over the biofilter in order to spray it in continuous at 35 L/h. Four air velocities (200 m/h, 500 m/h, 700 m/h and 1000 m/h) were successively tested, corresponding to empty bed residence time (EBRT) from 18 seconds to 3.6 seconds. The inlet H_2S concentration was kept constant at 15 mg/m^3. Pressure drops, H_2S removal and the water chemical composition (pH, sulfates, sulfites and sulfurs) were followed during all the tests.

H_2S removal rates were also measured at several packing heights (12 cm, 37 cm, 62 cm and 87 cm) in order to study the elimination profile along the column. Last, some microbiologic samples were taken at each air velocity when removal was stable in the purpose to analyze the microbial community (total biomass count by DAPI and sulfato-oxidizing bacteria identification and quantification by qPCR).

Results obtained with the three first packing materials studied (polyurethane foam, plastic support and waste treatment residue) are quite interesting. Indeed, various H_2S removals have been obtained according to the packing material used and the air velocity applied:

- For an air velocity of 200 m/h (EBRT = 18 s), removal superior to 99% has been measured with all the materials after four weeks of working; however, a difference of behavior has been noted with the biofilter filled with the waste treatment residue, which allowed to reach 99% H_2S removal as early as the first day due to its adsorption abilities.
- When the air velocity was raised from 200 m/h to 500 m/h (EBRT = 7 s), biofilters with plastic and waste treatment residue have maintained a 99% removal after only few days of adaptation while removal on polyurethane foam has stayed at 95%.
- At 700 m/h (EBRT = 5 s), only the biofilter filled with the waste treatment residue has maintained a high removal (>95%), others materials peaking at 90%.

Other differences in material behavior have also been shown, especially in terms of pressure drop and responses to incidents like spraying stops and water acidification. For example, pressure drops measured on waste treatment residue were ten times higher than those measured on polyurethane foam and plastic but stayed acceptable (20 mm $H_2O \cdot m^{-1}$ at 500 m/h and 80 mm $H_2O \cdot m^{-1}$ at 500 m/h). Biofilter filled with polyurethane foam also showed a bad resistance to watering stops in comparison to others. Generally speaking, importance of a good spraying and an adequate pH (superior to 6) were demonstrated. Variations in the sulfato-oxidizing bacteria density and diversity were also observed with the packing material.

To pursue this study, three other packing materials are currently in tests (two ceramics and a volcanic stone). The best materials amongst the six will then be chosen in order to study accurately the influence of working parameters (i.e., inlet H_2S concentration, water flow rate) and to determine the best operating conditions.

Chapter 34

BIODEGRADATION OF BTEX IN A TWO PHASE PARTITIONING SYSTEM – USE OF A BIODEGRADABLE SOLVENT

Rebiha Chikh & Annabelle Couvert
Ecole Nationale Supérieure de Chimie de Rennes, CNRS, UMR 6226, Rennes Cedex, France
Université européenne de Bretagne

Hamid Aït Amar
Laboratoire de Génie des Procédés et Environnement, Alger, Algeria

Abdeltif Amrane
Ecole Nationale Supérieure de Chimie de Rennes, CNRS, UMR 6226, Rennes Cedex, France
Université européenne de Bretagne

ABSTRACT

A biodegradable solvent was used in a two-phase aqueous-organic system for hydrophobic VOC (volatile organic compounds) elimination by activated sludge. Experiments, conducted with hexadecane as solvent and toluene as VOC, showed that toluene was removed simultaneously with part of the organic solvent elimination; cells were readily able to completely consume the toluene available at low and even at high concentrations within 48 h, even if they were not previously pre-adapted to these substrates.

Keywords: Absorption, Hydrophobic VOC, Hexadecane, Biodegradation

INTRODUCTION

Bioremediation appears to be an economical, energy-efficient and environmentally sound approach for the removal of VOC,[1] it is also the cleanest one in the case of benzene, toluene, ethyl benzene and xylene (BTEX) because they can be completely mineralized and do not generate secondary effluent problems.[2] In spite of this, the use of biological elimination methods for VOC is limited by the difficulty of achieving substrate absorption in water at the appropriate concentration; since most of the VOC are hydrophobic, absorption leads to too low substrate concentrations and cells may be starved.[3]

To overcome this problem, two-phase partitioning bioreactors can be used, it consists of the addition of water immiscible, biocompatible and non-biodegradable organic solvent to improve hydrophobic substrates mass transfer and to reduce their toxicity and inhibitory effect on the microorganisms present in the aqueous phase. Such a system is self regulating in the sense that the metabolic activity of the cells determines the rate of transfer of the xenobiotics. However, its application is limited to individual or very limited mixtures of VOC and pure strains of micro-organisms to avoid assimilation of the organic solvent.[4]

In this work, another alternative is tested; it consists of the use of a biodegradable solvent as organic phase. By this way, two objectives may be achieved: i) the biodegradation of VOC using an organic solvent at no cost since it is an industrial waste and ii) the elimination of the organic solvent harmful for the environment. As a biodegradable organic solvent for toluene elimination, the model solvent was hexadecane, which represents a considerable fraction in worn oil lubricate car engine and may be easily assimilate by a large number of microorganisms.[5]

MATERIALS AND METHODS

EXPERIMENTATION SETUP

Experiments were conducted in 250 mL Erlenmeyer flasks closed with cellulosic stoppers for oxygen supply. Due to the biodegradability and the negligible solubility of hexadecane in the aqueous phase, it was not possible to obtain a uniform sample from the two-phase system in view of analysis and hence, a sacrificial Erlenmeyer technique was used in all experiments. For this, a given number of 250 mL Erlenmeyer flasks were set up identically; each Erlenmeyer contained exactly the same amount of hexadecane, nutrient media, activated sludge and toluene, which were added at the beginning of the experiment. In each experiment, a blank test was also carried out to determine the fraction of VOC lost by possible leaks or absorption on the cellulosic stoppers. The pH of the media was maintained constant throughout experiments at approximately 7 by adding a 10 N NaOH solution. Cultures were stirred at 300 rpm and maintained at a temperature of 25°C. Samples were taken from the three phases (gas, organic and aqueous phases) and were analyzed to quantify the remaining toluene.

ANALYTICAL METHODS

Toluene concentrations in the gaseous phase were measured by gas chromatography by a Thermo Focus gas chromatograph with flame ionizer detector. In the organic phase, toluene was quantified using a Perkin-Elmer gas chromatograph also equipped with a flame ionizer detector. In the aqueous phase, it was analyzed by gas chromatography technique coupled with mass spectrometry with head space; for this purpose, a Perkin-Elmer Clarus 500 GS/MS was used.

The remaining hexadecane was evaluated with extraction by hexane (25% v/v). The extract was analyzed on a Perkin-Elmer gas chromatograph using a flame ionization detector.

The biomass growth was quantified by means of turbidimetric measurements at 600 nm (thermospectronic Helios λ Spectrophotometer, Bioblock, Illkirch, France), which was related to dry matter (DM) through a calibration curve.

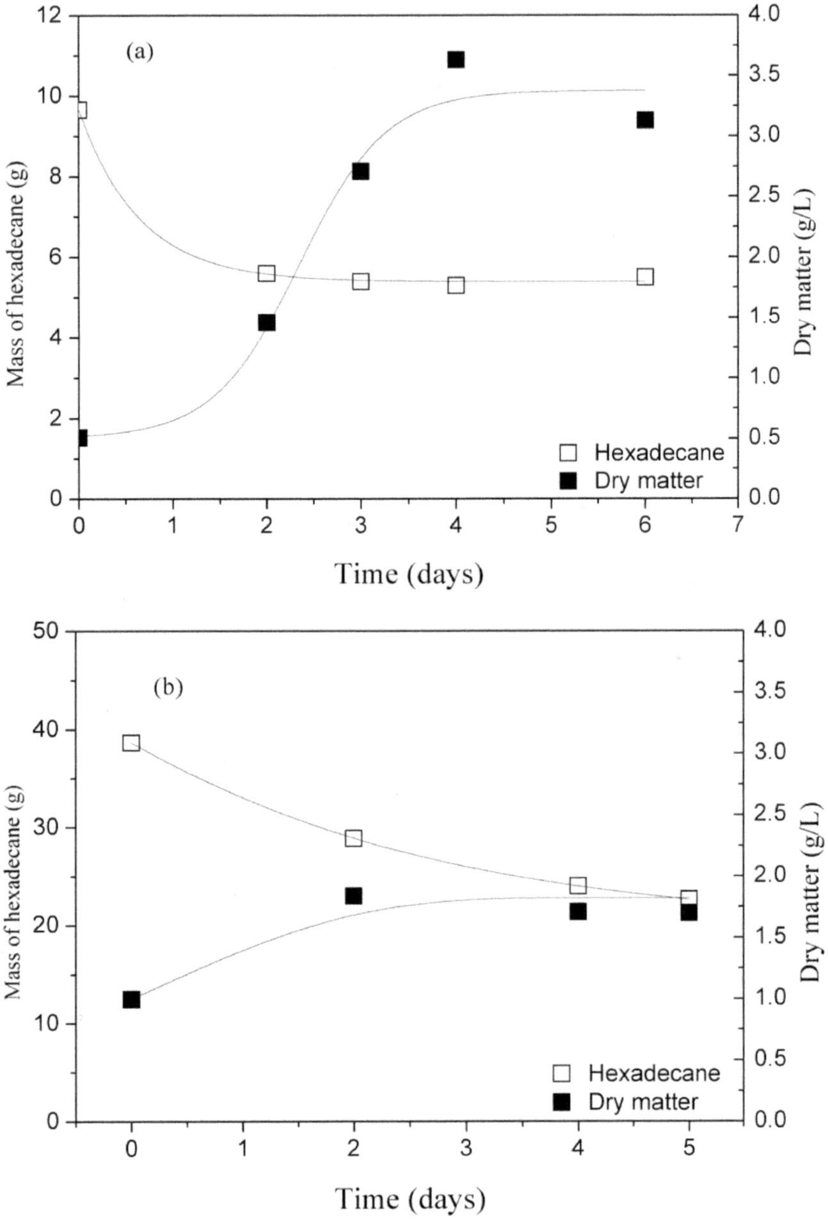

Figure 1 Biodegradation of hexadecane in water: (a): at 5% v/v ratio; (b): at 10% v/v ratio.

RESULTS AND DISCUSSION

The results obtained showed that activated sludge was able to degrade hexadecane even if there was not any pre-adaptation phase to this substrate (Fig.1). Indeed, a lot of microorganisms may metabolize paraffinic hydrocarbon like

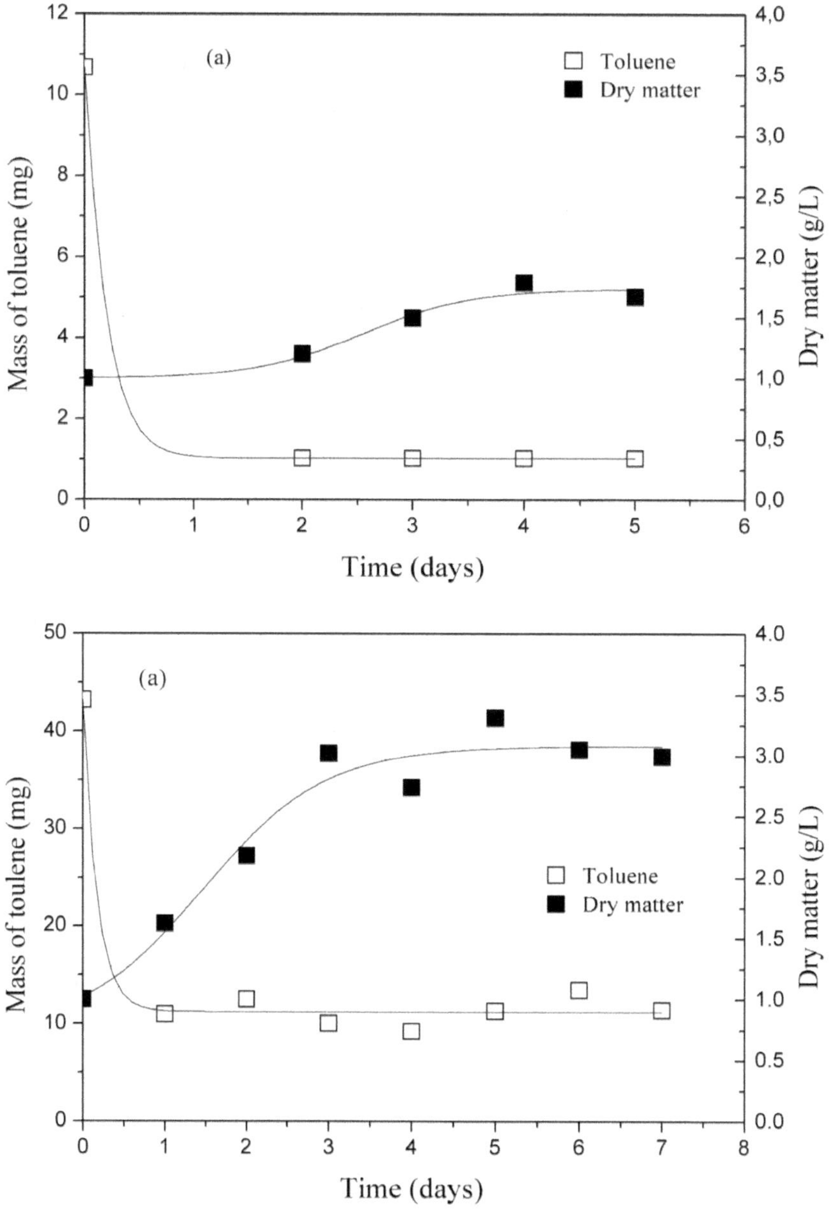

Figure 2 Biodegradation of toluene: (a): in a 10% v/v hexadecane/water. Mixture with 0.2 g/L of toluene in hexadecane; (b): in 5% v/v; hexadecane/water mixture with 4.3 g/L of toluene in hexadecane.

n-hexadecane, such as *Pseudomonas aeruginosa* which are the most commonly isolated in petroleum soils and groundwater.[6] Figure 1b shows that even at 10% hexadecane a significant growth was observed. However, an inhibitory effect can be noted at the comparison of figures 1 a and b, since the biomass concentration recorded for 5% hexadecane was significantly higher than that recorded for 10% solvent, approximately 3.5 and 1.5 g/L respectively. It is in agreement with the observations made by other authors[5] in their study of biodegradation of hexadecane by *Aspergillus niger*; they found that biomass production was proportional to the initial hexadecane concentration until 60 g/L (8% v/v approximately); while beyond this value the biomass production decreases.

The results recorded for toluene consumption are displayed in Figures 2. As observed and confirming the above results, higher final biomass concentrations were recorded for the lower solvent ratio and the difference was significant from the first culture sampling (Figs. 2 a and b). This result was not surprising since 10% hexadecane corresponded to a water activity of approximately 0.95, as shown for some hydrocarbons,[7–8] which can be considered to be inhibitory, at least partially, for microorganism growth, since most of bacteria, even gram-positive ones, are severely inhibited for water activity values below 0.95.[9]

In both cases, namely for low (0.2 g/L) and high (4.3 g/L) concentrations of toluene in the organic phase, constant residual toluene concentrations were achieved within 48 h, 90 and 74% of the pollutant was assimilated respectively (Figs. 2 and 3a). Darracq et al. also found a constant level of the cumulated toluene consumption after 2–3 days using DEHA as an organic phase.[10]

CONCLUSIONS

The feasibility of the considered process was shown, namely the simultaneous biodegradation of a hydrophobic VOC and the absorbent (organic) phase. Consequently, after appropriate experimental validation, this process could found practical applications and reduce exploitation costs of two-phase partitioning bioreactors. However, more investigations must be conducted to determine the kinetics of elimination of each substrate in order to optimize the operating parameters, so that organic solvent and toluene should be consumed with the same efficiency.

REFERENCES

1 Lapertot M, Seignez C, Ebrahimi S, Delorme S and Peringer P. Mass production of bacterial communities adapted to the degradation of volatile organic compounds (TEX). *Biodegradation* **18**: 343–350 (2007).

2 Yeom SH and Daugulis AJ. Benzene biodegradation in a two-phase partitioning bioreactor by *Alcaligenes xylosoxidans Y234*. *Process Biochem* **36**: 765–772 (2001b).

3 Collins LD and Daugulis AJ. Benzene/toluene/*p*-xylene degradation. Part I. Solvent selection and toluene degradation in a two-phase partitioning bioreactor. *Appl Microbiol Biotechnol* **52**: 354–359 (1999).
4 Littlejohns J and Daugulis A. Response of a solid-liquid two phase partitioning bioreactor to transient BTEX loadings. *Chemosphere* **73**: 1453–1460(2008).
5 Volke-Sepulveda TL, Guitierrez-Rojas M and Favela-Torres E, Biodegradation of hexadecane in liquid and solid-state fermentation by *Aspergillus niger*. *Bioresour Technol* **87**: 81–86 (2003).
6 Chayabutra C and Ju LK. Degradation of n-hexadecane and its metabolites by *Pseudomonas aeruginosa* under microaerobic an anaerobic denitrifying conditions. *App Environ Microbiol* **66**: 413–498 (2000).
7 Bell G, Janssen AEM and Halling PJ. Water activity fails to predict critical hydration level for enzyme activity in polar organic solvents: Interconversion of water concentrations and activities. *Enzyme Microb Technol* **20**: 471–477 (1997).
8 Hallsworth JE. Ethanol-induced water stress in yeast. *J Ferment Bioeng* **85**: 125–137 (1998).
9 Prescott L, Harley J and Klein D. *Microbiologie*. 2nd French ed. De Boeck Université (2003).
10 Darracq G, Couvert A, Couriol C, Amrane A and Le Cloirec P. Absorption and biodegradation of hydrophobic volatile organic compounds: determination of Henry's constants and biodegradation levels. *Water Sci Technol* **59**: 1315–1322 (2009).

Chapter 35

BIOFILTER PILOT PLANT TO REDUCE VOCs POLLUTION OF INDUSTRIAL EMISSIONS

Marcello Civilini
Food Science Department, University of Udine, Udine, Italy

Giovanni Cortella
Department of Energy Technologies, University of Udine, Udine, Italy

INTRODUCTION

Since the entry into force of new European legislation on emissions, many companies producing furniture don't satisfy the European Union rules on VOCs. Even if economy collapse reduced the industrial productivity and consequently their environmental impact, some companies from north east Italy joined a project aimed to apply the biofiltration and to integrate processes improvement, with the purpose of the emission reduction and achievement of target threshold limits.

In fact, in these two years, various solutions have been applied to different industrial characteristics. Essentially, alternative varnishing systems, such as water-based varnish for VOCs removal were adopted. Currently, the reliability of such products is not satisfactory, and they are also uneconomical in the chair sector. This is not due to their cost but to the higher costs of the work cycle combined with results that are poorer than those obtained using the basic products. Among the most efficient systems for VOCs' removal, activated carbon systems and post-combustion technologies can guarantee a satisfactory level of removal. Often, their applications are limited because of high start-up and managements costs. The biological removal of VOCs contained in gas emissions from chair varnishing is a viable alternative.

WOOD INDUSTRIAL DISTRICT

Six industries joined a project aimed to apply biofiltration and to integrate and improve their processes with the purpose of emission reduction and achievement of target threshold limits. Several steps were programmed to scale up our previous lab experiments (Civilini, 2006).

The solvent management plan of three out of six companies was followed with the purpose of gaining knowledge of the mass balance and similarity of each company in order to generalize the results obtained from the first prototype of the biofilter to the other three.

The painting cycles employed in the monitored companies consist essentially in spraying techniques involving an anthropomorphic robotic sprayer and surface painting both of which are connected to an area made for withering of the painted piece.

Table 1 Solvent composition (%) in varnish (Average calculated from safety data sheets).

		1330-20-7	108-88-3	123-86-4	110-19-0	100-41-4	108-65-6	108-10-1	%	78-93-3	108-94-1	%	141-78-6	%
CODE	KG	XIL	TOL	AnB	AiB	EtBz	A1m2m	MIBK		MEK	CEone		AE	
TZ58	**12**	37,5		1,5		6,25	6,25	6,25	**57,75**					
TZ47	**28**	37,5			6,25	6,25				17,5	1,5	**69**		
time (min)	**82**													
Varnish 1	Kg/min	*0,549*	*0,000*	*0,007*	*0,064*	*0,091*	*0,027*	*0,027*		*0,179*	*0,015*			
CODE	KG	XIL	TOL	AnB	AiB	EtBz	A1m2m	MIBK		MEK	CEone		AE	
TZ58	**4,5**	37,5		1,5		6,25	6,25	6,25	**57,75**					
TZ47	**10,5**	37,5			6,25	6,25				17,5	1,5	**69**		
Time (min)	**37**													
Varnish 2	Kg/min	*0,456*	*0,000*	*0,005*	*0,053*	*0,076*	*0,023*	*0,023*		*0,149*	*0,013*			
CODE	KG	XIL	TOL	AnB	AiB	EtBz	A1m2m	MIBK		MEK	CEone		AE	
TU98	**20**	17,5	6,25	6,25	6,25	6,25	6,25	1,25	**50**					
TU00	**80**	17,5		6,25	6,25	6,25				17,5	6,25	**60**		
Time (min)	**194**													
Varnish 3	Kg/min	*0,271*	*0,019*	*0,097*	*0,097*	*0,097*	*0,019*	*0,004*		*0,216*	*0,077*		*0,000*	
CODE	KG	XIL	TOL	AnB	AiB	EtBz	A1m2m	MIBK		MEK	CEone		AE	
TZ58	**20**	37,5		1,5		6,25	6,25	6,25	**57,75**					
DIL 89	**6**			100								**100**		
THO7	**10**		6,25	17,5						37,5			6,25	**67,5**
time (min)	**45**													
Varnish 4	Kg/min	*0,500*	*0,042*	*0,537*		*0,083*	*0,083*	*0,083*		*0,250*			*0,042*	
CODE	KG	XIL	TOL	AnB	AiB	EtBz	A1m2m	MIBK		MEK	CEone		AE	
TU98	**10**	17,5	6,25	6,25	6,25	6,25	6,25	1,25	**50**					
DIL 89	**2,5**			100								**100**		
THO7	**5**		6,25	17,5						37,5			6,25	**67,5**
Time (min)	**16**													
Varnish 5	Kg/min	*0,328*	*0,059*	*0,750*		*0,117*	*0,117*	*0,023*		*0,352*			*0,059*	
CODE	KG	XIL	TOL	AnB	AiB	EtBz	A1m2m	MIBK		MEK	CEone		AE	
TZ58	**10**	37,5		6,25		6,25	6,25	6,25	**62,5**					
DIL 89	**2,5**			100								**100**		
THO7	**5**		6,25	17,5						37,5			6,25	**67,5**
time (min)	**10**													
Varnish 6	Kg/min	*1,125*	*0,094*	*1,200*		*0,188*	*0,188*	*0,188*		*0,563*			*0,094*	

Table 1 shows the first step of the approach used to determine the VOCs mass balance of a chair painting cycle by an anthropomorphic robot. The average VOCs values shown in the table were used in order to correlate the calculated consumed quantitative with the data obtained instrumentally. Then, the VOCs values (kg min^{-1}) of each varnish mixture, were shared in the air volumes of the interested exhaust fans for its interval time of use.

A series of diagrams were obtained in order to characterize the behaviours of total VOCs and each main ones found both in Safety Data Sheets and gas-chromatographic analysis.

Following a series of analytic procedures and converting the PID traces to real instantaneous values (Figure 1), it has been possible to carry out realistic weekly simulations of the monitored processes. At different interval times, conversion factors were calculated which made possible to quantify the VOCs polluting emissions of each productive cycle.

As an example, in Table 2 VOCs final values and a comparison between estimated and measured consumption of robot painting cycle are reported. The sum between "measured consumption and emissions" of the robot cabin aspiration and withering zone would have to be as near as possible to the values of the calculated consumption [average of the VOCs intervals declared on the safety sheets compared with the operating air flow rate and time].

For every company, correlations were performed among real data obtained from the analytical (instrumental) and volumetric (air) monitoring of the main emission points of the productive chain with the theoretical removal capacity of a 50 m^3 septum. Weekly simulations were made with instantaneous VOC

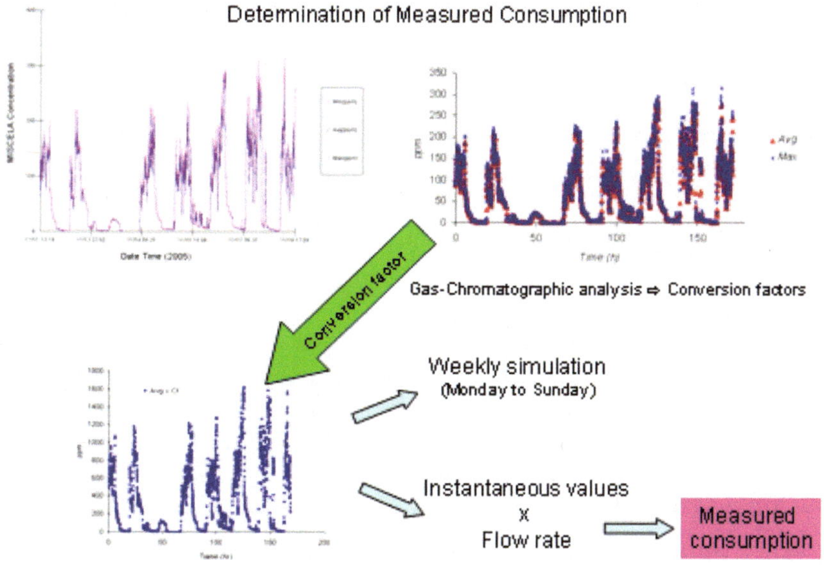

Figure 1 Procedures to evaluate mass balance.

Table 2 Robot Painting Cycle mass balance.

	XIL	*TOL*	*AnB*	*AiB*	*EtBz*	*A1m2m*	*MIBK*	*MEK*	*CEone*	*AE*	*DA*	*Total*
ROBOT PAINTING CYCLE												
Varnish 1	15,00		0,18	1,75	2,50	0,75	0,75	4,90	0,42			
Varnish 2	5,63		0,07	0,66	0,94	0,28	0,28	1,84	0,16			
Varnish 3	17,50	1,25	6,25	6,25	6,25	1,25	0,25	14,00	5,00			
Varnish 4	7,50	0,63	8,05		1,25	1,25	1,25	3,75		0,63		
Varnish 5	1,75	0,94	4,00	0,63	0,63	0,63	0,13	1,88		0,31		
Varnish 6	3,75	0,31	4,00		0,63	0,63	0,63	1,88		0,31		
Estimated consumption kg	*51,13*	*3,13*	*22,55*	*9,28*	*12,19*	*4,78*	*3,28*	*28,24*	*5,58*	*1,25*		*141,39*
Measured consumption kg	45,45	1,93	15,56	8,58	11,15	3,76	2,24	25,04	5,42	0,73		*119,85*
WITHERING ZONE												
Measured emissions kg	5,07	1,60	2,26	1,70			1,17	2,14			1,18	*13,93*

XYL – xylene isomers (o – ortho, m – meta, p – para), EtBZ – ethyl benzene, TOL – toluene, BA – butyl acetate (n – normal, i – iso), MPA methoxy propyl acetate, iPO – isopropanol, EA – ethyl acetate, DA – diaceton alcohol, MEK – methyl ethyl ketone, MIBK – methyl isobutyl ketone, A1m2m – Acetate 1 methyl 2 methoxy ethylene, CEone – Cyclo exanone.

estimation every three minutes and were elaborated for each emission point (Civilini, 2008).

To control the monitored emissions by means of biofiltration plants and following the biofiltration Best Available Techniques (BAT) (DGR Lombardy, 1.8.03, n. 7/13943, Table 1), an indicative empty bed contact time of 30 seconds should be respected. The resulting biofilter size should be:

Company 1 Vm = 786 m^3
Company 2 Vm = 390 m^3
Company 3 Vm = 422 m^3

Two important considerations must be made in order to judge such results: the first is that the residence time is quite high, thus needing a huge filter volume that possibly is not available for all; the second consideration is that the system does not take into account various VOCs loads.

Essentially for these reasons, and in order to verify the possibility to apply VOCs reduction by biofiltration, a prediction model was developed based, among other things, on the VOCs microbial removal ability (Civilini, 2008). Based on such data, it results that a VOC load of about 500 kg/week may be controlled by 50 m^3 of septum.

BIOFILTER

All six manufacturers agreed to build a prototype, designed to verify the possibility of application of microbial removal ability. The initial investment, like the starter and management costs, was subdivided equally.

The prototype of biofilter was built to control a fraction of the emissions from Company 1. It has a working volume of 25 m^3 and a max capacity to treat 16000 mc/hr of exhausted air coming from the working on and painting of the wooden components. The withering zone emissions were chosen to feed the biofilter with the polluted air. The uniformly and relative high concentration of these VOCs emissions may help our studies in this start up period. In Table 3 and Figure 2 are represented the mass VOCs consumption and relative emission concentrations during a regular productive week, respectively. A system of pipes and valves on vent holes was designed to allow for a possible reduction of the VOCs' concentration at the inlet of the biofilter through dilution with clean air. Air flow rate was regulated controlling the velocity of the air extractor by means of an electric inverter.

To control fluid dynamics behaviour and on line concentrations, a series of instrumentations were set up for inlet and outlet lines.

In particular, pressure values (and consequently the pressure drop) were monitored with pressure transducers (Kimo Instruments, CP101-AO, scale –500/+1000 Pa and CP102-AO, scale –500/+1000 mm H_2O for inlet and outlet points respectively) (Figure 3); air velocity at the outlet point was measured

Table 3 Weekly VOCs speciation of withering zone emissions.

							XYL		
	EA	*MEK*	*iPO*	*iBA*	*TOL*	*nBA*	*isom*	*MPA*	*DA*
Solvent composition (%)	1,11	0,20	0,48	8,44	10,47	9,87	46,89	21,24	1.29
Total (kg/week)	1,25	0,23	0,54	9,54	11,84	11,16	53,01	24,01	1,46

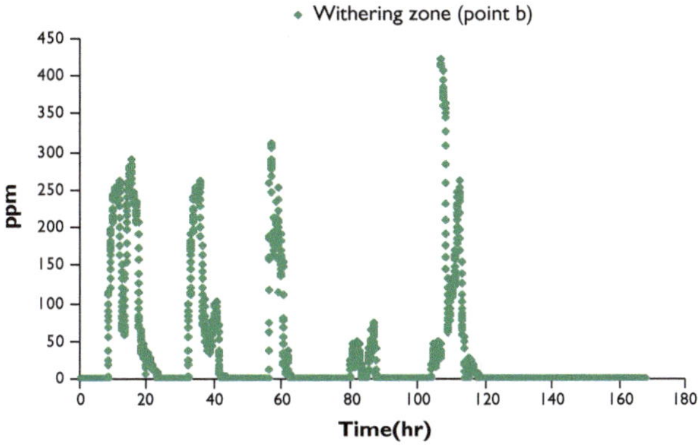

Figure 2 Weakly VOCs concentration of withering zone emissions.

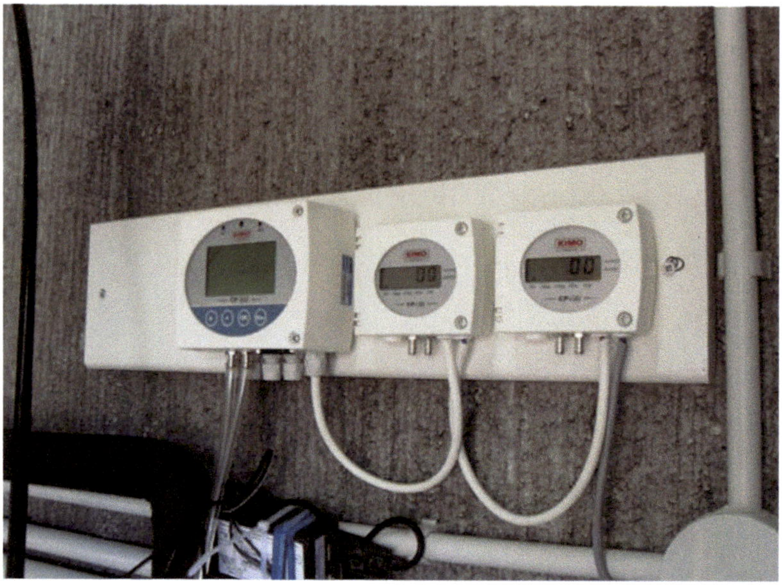

Figure 3 Instrumentation for pressure and air velocity values determination.

with a differential pressure transducer CP304 scale –10000/+10000 Pa connected with two DEBIMO pitot tubes sensors.

CEE CEN264 n.326 and UNI EN 13526 standards to determine TOC, VOC and CH_4 in emissions were applied to monitor continuously the biofilter performance. By means of alternate (3 min^{-1}) online sampling between inlet and outlet air flows, air samples were continuously withdrawn (0.6 L min^{-1}) by a F.I.D. analyzer (SIEMENS mod. FIDAMAT 6). In discontinuous mode, coconut charcoal tubes (Anasorb CSC, SKC Inc.) were positioned on the outlet pump flow for 20 min to confirm VOC concentrations by gas-chromatographic (GC) analysis. The GC instrument was an Autosystem XL Perkin- Elmer gas chromatograph fitted with a 30-m Supelcowax 10 column (0.53 mm, 1 µm film) (Supelco) equipped with a flame-ionization detector (FID). Compounds were identified and separated using the internal standard method under the following conditions: nitrogen as carrier gas, split mode injection (250°C), column temperature 60°C, held isothermally for 10 min, and programmed to rise to 210°C at 4°C min^{-1}.

RESULTS

The first action was to start up the biofilter without any matrices. This phase has been necessary to optimize the equilibrium flow conditions between the biofilter and its extractor fan, the blower and the line (diameter 0,63 m) connections to the withering zone. As a starting parameter, a flow rate of 5700 m_3/h was chosen for the further tests. After having overcome some technical problems, a correct pressure equilibrium was reached, that guaranteed the absence of dilution due to unforeseen clean air infiltration. The most important parameter indicating a right behaviour of the system was the overlapping measurement of inlet and outlet VOC concentrations (Figure 4).

Then, a partial quantity of prepared substrate was filled up in the biofilter. All conditions were optimized and the fan switched on. The results of these first months were very positive. Observing the average of 8 hours of working periods and the respectively successive 8 hours of pause, a series of data were obtained (Table 4).

The on-line monitoring, whose samples are alternated between inlet and outlet concentrations, confirms these good results in all observed period. During the high VOCs load, the septum is sufficiently able to contain the outlet concentration. When the inlet VOCs concentration was between 600 and 700 ppm, the removal efficiency was about 65–70%, whereas at inlet VOCs concentration of 700–800 ppm was about 60–65%.

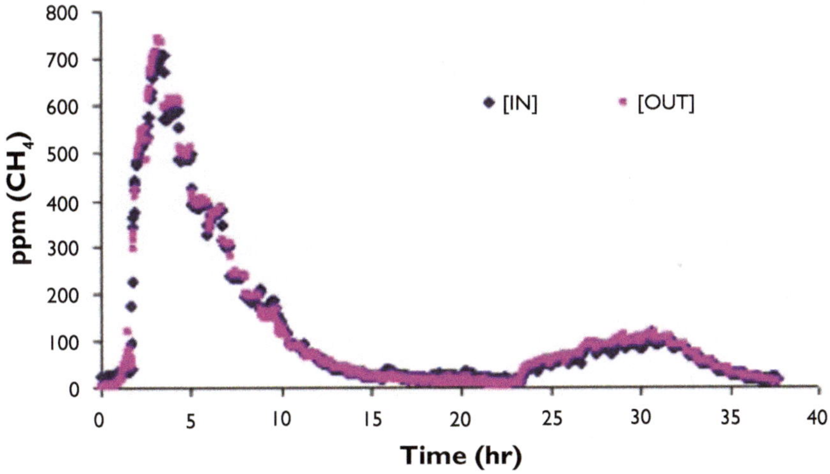

Figure 4 Two days of equilibrium of between inlet and outlet VOC concentrations.

Table 4 Average values of VOCs Removal Efficiency (%) at different Inlet concentrations (ppmCH_4).

≤20 ppm		*≥20 ppm, <50 ppm*		*≥50 ppm, <100 ppm*		*≥100 ppm, <300 ppm*		*≥300 ppm, <500 ppm*	
Min	Max	Min	Max	Min	Max	Min	Max	Min	Max
93,51	98,22	47,54	99,27	71,27	86,53	52,24	81,31	78,30	80,02

The speciation of compounds may explain the behaviour obtained during these months.

REFERENCES

1 Civilini M., Lolini S., Giacomello S. (2008). Predictive Applications of Biofiltration on VOCs.
2 Industrial Emissions. USC-UAM 2008 Conference on Biofiltration for Air Pollution Control.
3 Long Beach, CA, USA. October 22–24, 2008 Proceedings: p. 119–128.
4 Civilini M. (2006). Multiple microbial activities for Volatile Organic Compounds reduction by biofiltration. *J. Air & Waste Manage. Assoc.* 56: 922–930.

Chapter 36

SILICONE OIL:
An effective absorbent for hydrophobic Volatile Organic Compounds (VOC) removal

G. Darracq, A. Couvert, C. Couriol, A. Amrane & P. Le Cloirec
Ecole Nationale Supérieure de Chimie de Rennes, CNRS, Rennes Cedex, France and Université européenne de Bretagne

D. Thomas
Faculté Polytechnique de Mons, Rue de l'Epargne, Mons, Belgique

E. Dumont & Y. Andrès
Ecole des Mines de Nantes, La Chantrerie, Nantes Cedex, France

ABSTRACT

Hydrophobic VOC (like toluene, dimethylsulphide (DMS) and dimethyldisulphide (DMDS) are poorly soluble in water and classical air treatment processes are not efficient. An alternative technique involving an absorption step in an organic solvent followed by a biodegradation phase was proposed. The solvent must fulfil several characteristics, and a previous study allowed to select polydimethylsiloxane (or PDMS, *i.e.*, silicon oil). The aim of this paper was to determine some of its characteristics like absorption capacity and velocity performances (Henry's constant, diffusivity and mass transfer coefficient), and to verify its non-biodegradability.

For the three VOC, Henry's constants in silicone oil were very low. Diffusivities values were found in the range 10^{-10} – 10^{-11} $m^2 \cdot s^{-1}$ and mass transfer coefficients did not show significant differences between the values in pure water and pure silicone oil, in the range 1.0×10^{-3} to 4.0×10^{-3} s^{-1} for all the considered VOC. These absorption performances warranted its use during the first step of the process. Silicone oil was also found weakly biodegradable, owing to the COD on BOD_5 ratio (2.30), and no toxic for microorganisms confirming its relevance as an absorption phase for the considered integrated process.

Keywords: Absorption, Silicone oil (PDMS), Hydrophobic VOC, Mass transfer coefficient, Diffusivities, Biodegradation

INTRODUCTION

Hydrophilic compounds can usually be removed by chemical scrubbing, biological treatment (biofilter), or thermal oxidation. But these processes are not really efficient for hydrophobic compounds like toluene, dimethylsulphide (DMS) and dimethyldisulphide (DMDS) that have low olfactory thresholds. An alternative process, which consisted in the coupling of an absorption step involving an organic solvent[1] and a biodegradation step, has been developed. The aim was to achieve the consumption of the VOC and the regeneration of the absorbent phase. In 1991, Bruce and Daugulis set up a method to choose the best organic phase for implementation in a multiphase bioreactor.[2] The solvent must fulfil several characteristics: high absorption capacity and velocity performances, a low emulsion-forming tendency, a low solubility in water, good chemical and thermal stabilities, non-biodegradability, biocompatibility and of course, cheapness. Silicone oil (polydimethylsiloxane or PDMS) was used by several authors[3,4] as a pollutant reservoir and seemed to be suitable. Hayachi *et al.*[5] and Matsumoto *et al.*[6] showed that the partitioning coefficient octanol/water (log Kow) allowed to check the toxicity of a solvent towards microorganisms. Since log Kow of PDMS is 4.25, this solvent could be considered as biocompatible. The aim of this work was to measure several parameters such as dimensionless Henry's constants (H), diffusivities (D_L) and global mass transfer coefficients (K_La) for each VOC in order to characterise the absorption performances of silicone oil.

MATERIALS AND METHODS

HENRY'S CONSTANT AND GLOBAL MASS TRANSFER COEFFICIENT

In 2003, Roustan described the physical absorption between a gaseous compound and a liquid phase[7] (Eq. 1);

$$dN = K_L a dV (C_L^E - C_L) \tag{1}$$

where dN (mol · s^{-1}): transferred amount, dV (m^3): volume element, K_L (m · s^{-1}): global liquid mass transfer coefficient, a (m^2 · m^{-3}): volumetric interfacial area, C_L: pollutant liquid concentration, C_L^E (mol · m^{-3}): pollutant liquid concentration at the equilibrium with the gas phase given by Eq. 2;

$$C_L^E = p/H'' = \frac{C_G RT}{H''} \tag{2}$$

where p (atm): VOC partial pressure, H'' (atm · m^3 · mol^{-1}): Henry's constant. Dimensionless Henry's constants H were obtained using a static method during three days at 25°C. The VOC concentration in the gas phase (C_G) was determined by gas chromatography. For the $K_L a$ measurements, the absorption of toluene, DMS and DMDS in water and in silicone oil was independently studied at a constant gas flow rate (1 $m^3 \cdot h^{-1}$). A dynamic method[8] was used to determine the volumetric mass transfer coefficients of pollutants during absorption at 25°C.

DIFFUSIVITY

A specific mathematical model was developed to simulate the VOC absorption into viscous solvents in a vertical wetted-wall column with a co-current gas-liquid downflow. Considering a plane flow, the liquid stream was considered as laminar and exempt of ripples for Reynolds numbers lower than 4, requiring a low liquid flow rate if the viscosity was not so high.[9]

BIODEGRADABILITY

Biodegradability was deduced from BOD_5 measurements, carried out in Oxitop IS6 (WTW, Alès, France). Activated sludge from a wastewater treatment plant was used to inoculate samples and the initial bacteria concentration was 0.5 $g \cdot L^{-1}$.

RESULTS AND DISCUSSION

HENRY'S CONSTANT

The three selected VOC were preferentially absorbed in silicone oil. The experimental value was then compared to the theoretical value calculated by means of Eq. 3.

$$\frac{1}{H''_{mixture}} = \frac{1}{H''_{so}} x + \frac{1}{H''_{w}} (1 - x) \tag{3}$$

where H''_{SO}: Henry's constant of COV in pure PDMS, H''_w: Henry's constant of COV in pure water and x: volumetric fraction of PDMS in emulsion. Results are reported in Tab. 1 for toluene. Since the relative error was inferior to 15%, Henry's constant of toluene` in an oil/water emulsion could be estimated by Eq. 3.

In the case of toluene, H''_{so} value appeared very low if compared to H''_w value; hence the term $1/H''_w$ can be neglected towards $1/H''_{so}$ and Eq. 3 led to:

$$H''_{mixture} \cong \frac{H''_{so}}{x} \tag{4}$$

Table 1 Henry's constants of toluene in silicone oil/water emulsion. Comparison between experimental and calculated values.

% PDMS	H" exp ($Pa \cdot m^3 \cdot mol^{-1}$)	R^2	H" calculated ($Pa \cdot m^3 \cdot mol^{-1}$)	Relative error (%)
0%	–	–	609[10]	–
9%	18.7	0.994	19.5	4
24%	6.8	0.999	7.5	10
49%	3.3	0.998	3.7	11
85%	1.8	0.998	2.1	14
100%	1.6	0.999	1.8[11]	14

Table 2 Henry's constants of DMDS in silicone oil/water emulsion.

% PDMS	H" exp ($Pa \cdot m^3 \cdot mol^{-1}$)	R^2	H" calculated ($Pa \cdot m^3 \cdot mol^{-1}$)	Relative error (%)
0%	119	0.993	111[12]	7
8%	27.1	0.997	20.8	30
24%	8.0	0.997	8.2	3
48%	4.4	0.999	4.3	2
84%	2.6	0.996	2.5	6
100%	2.3	0.995	2.1[13]	8

According to Eq. 4, $H"_{mixture}$ increases when the volumetric fraction of PDMS in emulsion decreases. For DMDS, results reported in Tab. 2 lead to the same conclusions.

GLOBAL MASS TRANSFER COEFFICIENT

The results in Tab. 4 show that there was no significant difference between the values of $K_L \cdot a$ in pure water and pure silicone oil. However and owing to the higher affinity of pollutants for silicone oil than for water, higher $K_L \cdot a$ values were expected for silicone oil than for water.

For DMS, experimental results (Tab. 3) differed from those calculated.

The dynamic viscosity of the silicone oil used, which was 5 times higher than that of water, could account for this unexpected result. Nevertheless, the use of silicone oil enabling to absorb large amounts of pollutant should be beneficial to increase the mass transfer rate.

MODELLING OF VOC ABSORPTION INTO SILICON OIL FOR THE ESTIMATION OF THE LIQUID DIFFUSIVITIES

The liquid diffusivities resulting from these absorption tests achieved in the wetted wall column using the experimental procedure described above were in the range of 10^{-11}–10^{-10} $m^2 \cdot s^{-1}$. Absorption tests results and diffusivities values are given in Tab. 5.

BIODEGRADABILITY

The results given in Tab. 6 are the average of two experiments with two control flasks (glutamic acid and glucose) or two experiments with four samples (silicone oil in water). However, the increase of the BOD_5 in the flasks containing silicone oil proved an assimilation partially of this compound by bacteria. Silicone oil was therefore (partially) biodegradable and non-toxic

Table 3 Henry's constants of DMS in silicone oil/water emulsion.

% PDMS	*H"exp* *($Pa \cdot m^3 \cdot mol^{-1}$)*	*R^2*	*H"calculated* *($Pa \cdot m^3 \cdot mol^{-1}$)*	*Relative error (%)*
0%	124	0.998	143[14]	13
8%	184	0.986	65	203
24%	123	0.999	29	331
48%	70	0.997	16	342
84%	46	0.995	10	379
100%	42	0.982	8	413

Table 4 Global mass transfer coefficients of VOC in PDMS.

Pollutant	*$K_L \cdot a$ in pure water (s^{-1})*	*$K_L \cdot a$ in pure silicone oil (s^{-1})*
Toluene	1.4×10^{-3}	1.6×10^{-3}
DMS	4.0×10^{-3}	4.0×10^{-3}
DMDS	1.8×10^{-3}	1.6×10^{-3}

Table 5 Liquid diffusivities of VOC in PDMS.

	C_{Gin} ($g \cdot Nm^{-3}$)	*C_{Gout} ($g \cdot Nm^{-3}$)*	*D_L (25°C) ($m^2 \cdot s^{-1}$)*
Toluene	1.630	1.286	5.65×10^{-10}
DMS	2.615	2.486	Not given because of very high *H"*
DMDS	3.277	3.026	8.72×10^{-11}

Table 6 COD and BOD_5 for PDMS in water (≅ 80 mg · L^{-1}).

Samples	*COD* ($mgO_2 \cdot L^{-1}$)	BOD_5 ($mgO_2 \cdot L^{-1}$)	$\frac{COD}{BOD_5}$
Control flask (Gutamic Acid and Glucose)	–	205	–
PDMS 1	116	51	2.3
PDMS 2	117	54	2.3

towards micro-organisms. The ratio between COD and BOD_5 was close to 2.3, confirming that PDMS was weakly biodegradable.

CONCLUSIONS

As a conclusion, silicone oil seems fulfilling most of the requested characteristics and will be subsequently tested for hydrophobic VOC removal in an alternative process.

For the three VOC, the Henry's constant in pure water were significantly higher than in pure PDMS, leading to ratio of 400, 53 and 3 for toluene, DMDS and DMS, respectively. PDMS appeared to be a good absorbent since it solubilises efficiently DMDS and toluene. with the results obtained for DMS were a little bit less interesting, most likely due to its high volatility and then, to losses during experiments. This could explain the non significant results recorded for various emulsion ratio, while the Henry's constant in pure PDMS was high. The results obtained for diffusivities in PDMS showed that D_L values can be determined for two VOC and were in the range of 5.65×10^{-10} and 8.72×10^{-11} $m^2 \cdot s^{-1}$ for toluene and DMDS respectively. For DMS, the liquid diffusivity in PDMS could not be determined, owing to the high Henry's constant value. Results for global mass transfer coefficient did not show a significant gap between the absorption in water and in PDMS. The rough estimate for the three selected VOC was between 1.6×10^{-3} and 4.0×10^{-3} s^{-1}. Finally, the low biodegradability of silicone oil and its absence of toxicity justify its use as an absorbent phase for the considered integrated process.

REFERENCES

1 Daugulis AJ and Boudreau NJ. Removal and destruction of high concentrations of gaseous toluene in a two-phase partitioning bioreactor by *Alcaligenes xylosoxidans*. *Biotechnol Lett* **25**: 1421–1424 (2003).
2 Bruce LJ and Daugulis AJ. Solvent selection strategies for extractive Biocatalysis. *Biotechnol Prog* **7**: 116–124 (1991).

3 Ascon-Cabrera MA and Lebeault JM. Cell hydrophobicity influencing the Activity/Stability of Xenobiotic-degradind microorganisms in a continuous biphasic aqueuous-organic system. *J Ferment Bioeng* **80(3)**: 270–275 (1995).
4 Bouchez M, Blanchet D and Vandecasteele JP. Substrate availabiltiy in phenanthrene biodegradation: transfer mechanism and influence on metabolism. *App Microb Biotechnol* **43**: 952–960 (1995).
5 Hayachi S, Kobayashi T and Honda H. Simple and Rapid Cell Growth Assay Using Tetrazolium Violet Coloring Method for Screening of Organic Solvent Tolerant Bacteria. *J Biosc Bioeng* **96(4)**: 360–363 (2003).
6 Matsumoto M, Mochiduki K and Kondo K. Toxicity of Ionic Liquids and Organic Solvents to Lactic Acid-Producing Bacteria. *J Biosc Bioeng* **98(5)**: 344–347 (2004).
7 Roustan M. Lavoisier TEC&DOC. Paris (2003).
8 Dumont E, Andrès Y and Le Cloirec P. Mass transfer coefficients of styrene and oxygen into silicone oil emulsions in a bubble reactor. *Chem Eng Sci* **61(17)**: 5612–5619 (2006).
9 Bird RB, Stewart WE and Lightfoot EN. Transport phenomena, John Wiley & sons, New York, 7th edn (1966).
10 Staudinger J and *Roberts* PV. A critical review of Henry's law constants for environmental applications. *Crit Rev Environ Sci Technol* **26**: 205–297 (1996).
11 Heymes F, Manno-Demoustier P, Charbit F, Fanlo JL and Moulin P. A new efficient absorption liquid to treat exhaust air loaded with toluene. *Chem Eng J* **115**: 225–231 (2006).
12 Hine J and Weimar Jr RD. Carbon basicity. *J Am Chem Soc* **87(15)**: 3387–3396 (1965).
13 De Guardia A, Thèse Université de Rennes 1 (1994).
14 Vuong MD, Couvert A, Couriol C, Amrane A, Le Cloirec P and Renner C. Determination of the Henry's constant and the mass transfer rate of VOCs in solvents. *Chem Eng J* **150**: 426–430 (2009).

Chapter 37

STUDY OF THE BIOMASS ACCUMULATION FROM INOCULATION TO CLOGGING IN A BIOFILTER TREATING TOLUENE AT HIGH LOADS

Antonio D. Dorado & Xavier Gamisans
Department of Mining Engineering and Natural Resources, Universitat Politècnica de Catalunya, Manresa, Spain

Juan A. Baeza, Javier Lafuente & David Gabriel
Department of Chemical Engineering, Universitat Autònoma de Barcelona, Barcelona, Spain

Biofiltration employs the metabolic activity of microorganisms immobilized on a packing material to degrade gas phase pollutants which are the energy source for microbial growth. However, biomass accumulation is one of the most critical parameters that need to be controlled along the operation of biofilters and biotrickling filters in order to achieve stable performance for long term operation. The excess of biomass inside the bioreactor significantly increases pressure drop, which ultimately may lead to wash part of the biomass from the system or to force the replacement of the support media. Additionally, excess biomass implies an increased operation cost (due to pressure drop increases) coupled to a reduced removal efficiency due to reduced specific surface areas and porosities. In addition, potential channeling because of non-uniform biomass growth might occur. Moreover, most of dynamic mathematical models in biofiltration rarely incorporate biomass accumulation processes meaning that predicted results could differ significantly from experimental data in the case of high loads of pollutant.

Several authors have addressed biomass growth and modeling from different approaches. Alonso et al. (1997)[1] concluded that the excessive accumulation of biomass in a reactor has a negative effect on contaminant removal efficiency. The backwashing technique was showed as a useful technique to avoid negative conditions of operation. Nevertheless, the corresponding model considers a homogeneous biomass for one type of microbial species. Deront et al. (1998)[2] evaluated the possibility of following the biomass growth by pressure drop measurements without monitoring other parameters like respiratory activity at the same time. Okkerse et al. (1999)[3] incorporated an active biomass fraction and inert biomass fraction to distinguish microbial components in the biofilm phase but not achieving biofilter clogging in the experimental operation.

In the present work, a lab-scale biofilter treating high loads of toluene was designed in order to quantify and understand biomass accumulation towards future modeling of the system. Pressure drop, reactor weight increase, carbon dioxide production, oxygen consumption, liquid pH, total volatile solids, biomass nitrogen content, optical density of the leachate and toluene gas phase abatement were monitored to characterize biomass growth. Parameters were monitored in the evolution of the biofilter from the inoculation to a forced clogging episode in five sampling ports along the height of the biofilter to determine the corresponding profiles during the period of operation. Biofilter performance was followed for 55 days of operation until pressure drop reached values above 30 cm $H_2O \cdot m^{-1}$. Subsequently, inlet toluene load was stopped and the development of the biofilter was observed in order to quantify the detachment of the biomass from the support media under starvation conditions. After 20 days without inlet feed, the biofilter was restarted to determine the competence of biomass for recovering the capacity to abate pollutant for a period of 45 days.

The biofilter, with a total volume of 16 liters, was inoculated with activated sludge from a waste water treatment plant. After 72 hours of operation, with an inlet load of 200 $g \cdot m^{-3} \cdot h^{-1}$ and an empty bed residence time of 35 seconds, the removal efficiency was around 97% in the half of the total height of the reactor (Figure 1). Afterwards, the inlet load was increased up to 600 $g \cdot m^{-3} \cdot h^{-1}$ and the metabolic activity was detected in the whole biofilter achieving a removal efficiency close to 99.5% in the outlet stream after 2 days. Microorganism activity was practically kept at the same elimination capacity for 45 days (Figure 1) until the clogging episode was detected (Figure 2). Some modules of the reactor showed an elimination capacity superior to 1000 $g \cdot m^{-3} \cdot h^{-1}$ in specific periods of time. Previous works achieving such high elimination capacities have not been found in the literature.

During the interruption of feed, watering and collecting of leachate was kept. Biomass detachment from the packing material was related to the decrease in the pressure drop (until 5 cm $H_2O \cdot m^{-1}$) and the increase in the nitrogen content in the leachate (data not shown). In this period, although the pollutant inlet was closed, carbon dioxide production reached a value of 70 $g \cdot m^{-3} \cdot h^{-1}$, which indicates that the biomass was not completely inactive after 20 days of starvation. It is supposed that the organic matter of packing material, the immobilized biomass and the nutrients incorporated with the irrigation water were enough to maintain partially active the biomass. The rapid decrease of pressure drop in this period demonstrates watering is an effective technique to wash the excess of biomass accumulated in the biofilter.

In the latter period, from day 75, the biofilter was progressively fed with toluene until an inlet load of 390 $g \cdot m^{-3} \cdot h^{-1}$. Immediately, after few hours, the removal efficiency was once again close to 99%, reflecting that biomass

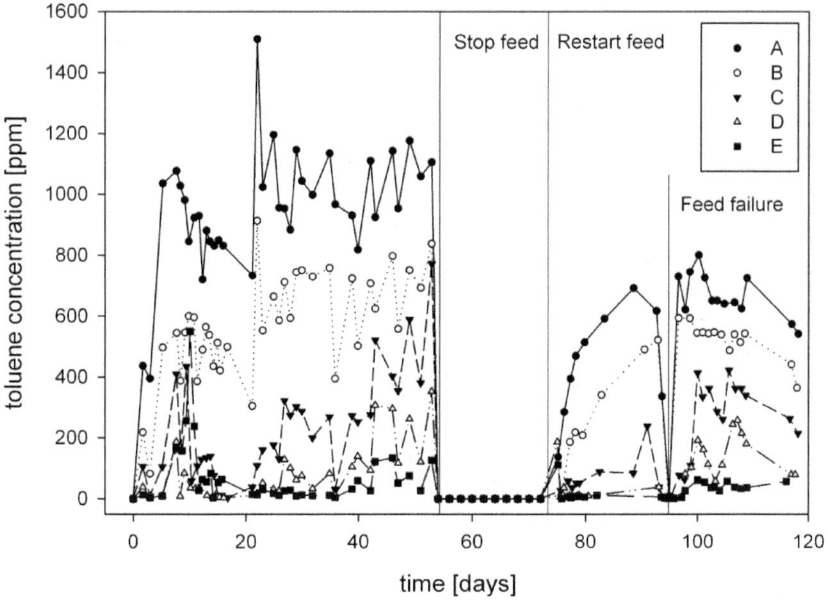

Figure 1 Evolution of toluene concentrations in the operation of the biofilter at the reactor inlet (A), first module (B), second module (C), third module (D) and reactor outlet (E).

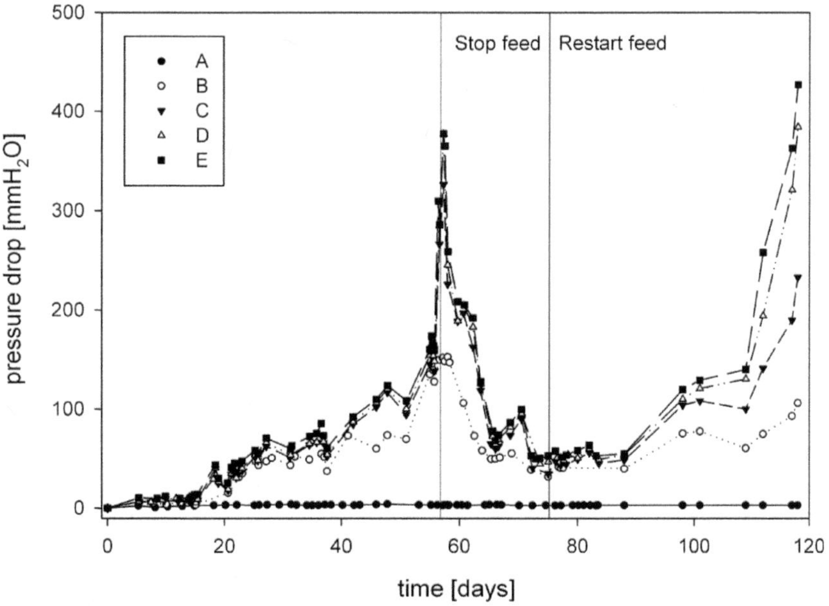

Figure 2 Evolution of differential pressures in the operation of the biofilter at the reactor inlet (A), first module (B), second module (C), third module (D) and reactor outlet (E).

was not completely inactive and a rapid acclimatization of microorganisms after the long starvation period. On day 95 a feed failure was registered but the operation was rapidly recuperated. Thus, biofilter was able to cope with intermittent loads without being substantially affected in the operation. In this period, after 45 days of operation, pressure drop increased up to 40 $cmH_2O \cdot m^{-1}$. The short time to achieve this maximum acceptable value of head loss is partially explained by the decrease of frequency in the water irrigation (1 L each 3 days). Apart from pressure drop, other parameters related to biomass accumulation as carbon dioxide production and oxygen consumption show a consistent evolution with present results (data not shown). Otherwise, the total weight of biomass increased up to 5500 g in the first period, subsequently decreased 1000 g in the starving period and finally achieved a value of 6000 g after the 120 days of operation.

In conclusion, results demonstrate that high loads of toluene previously not reported can be eliminated by biofiltration. However, it is necessary a strategy to control excess biomass accumulation avoiding reactor clogging to ensure a long time operation. Moreover, present results demonstrated that biofilter is able to recover rapidly the activity after several weeks of starving. Finally, experimental data collected during the operation time permitted to observe significant variations in the main parameters related to biomass accumulation which should permit to calibrate and validate a mathematical model which incorporates the biomass growth in the expressions describing the processes involved in a biofilter system.

REFERENCES

1 Alonso, C., Suidan, M.T., Sorial, G.A., Smith, F.L., Biswas, P., Smith, P.J. and Brenner, R.C. Gas treatment in trickle-bed biofilters: biomass, how much is enough? *Biotechnology and Bioengineering*. **54**(6): 583–594 (1997).
2 Deront, M., Samb, F.M., Adler, N. and Péringer, P. Biomass growth monitoring using pressure drop in a concurrent biofilter. *Biotechnology and Bioengineering*. **60**(1): 97–107 (1998).
3 Okkerse, W.J.H., Ottengraf, S.P.P., Osinga-Kupiers, B. and Okkerse, M. Biomass accumulation and clogging in biotrickling filters for waste gas treatment. Evaluation of a dynamic model using dichloromethane as a model pollutant. *Biotechnology and Bioengineering*. **63**(4): 418–430 (1999).

Chapter 38

EVALUATION OF INNOVATIVE PACKING MATERIALS FOR THE BIODEGRADATION OF H_2S: A comparative study

Eric Dumont & Yves Andrès
GEPEA, Ecole des Mines de Nantes, Nantes Cedex, France

INTRODUCTION

H_2S is an undesirable contaminant of various odorous effluents which could be, in case of low concentrations and high flow rates, treated by biofiltration. Such biological technique needs a good support media which must have a high surface area for biomass attachment, a good moisture retention capacity, a low pressure drops characteristics, be mechanically resistant, and can provide a source of nutrients for biomass.[1–2] The ideal packing material is then always prospected. To address several operational problems (low pH, lack of nutrients and dissolved carbonate), a new synthetic material has been developed in our laboratory (called UP20). It has been demonstrated[3] that this material offers real advantages by providing both nutrients for the biomass and buffering effect. Nonetheless, UP20 material must solely be considered as an additive material improving the degradation of the pollutant, i.e., it must be used in association with another economic packing material. In this way, the aim of this work was to test and to compare innovative configurations of packing materials in order to treat airstream polluted with H_2S. Critical comparison with fibrous peat (regarded as a "reference" packing material) was also carried out.

EXPERIMENTAL

PACKING MATERIALS

The packing materials (Table 1) used for comparison were: (1) the nutritional synthetic material UP20 containing CH_4N_2O, H_3PO_4, $CaCO_3$ (C/N/P molar

Table 1 Some characteristics of materials used.

	UP20[3]	*Sapwood*	*Pine bark*	*Pozzolan*	*Fibrous peat*

Density (kg/m^3)	920	143	370	1500	270
Specific surface area (m^2/g)	<1	<1	<1	<1	705
Water holding capacity (% by weight)	47	395	67	21	250
pH	8.1	–	4.5	6.9	4.5

Table 2 Configuration of packing materials in biofilter (bed high: H = 1 m).

Biofilter	*Packing materials*	*Configuration*	*Mass (kg)*	*Bed porosity*
N°1	peat	100%	0.840	0.60
N°2	sapwood	100%	0.622	0.44
N°3	pine bark	100%	1.55	0.48
N°4	peat-UP20	80/20% v/v stratified in 2 layers	0.67 + 0.88	0.56
N°5	sapwood-UP20	80/20% v/v stratified in 2 layers	0.612 + 0.842	0.32
N°6	pozzolan-UP20	80/20% v/v stratified in 2 layers	5.06 + 1.02	0.46
N°7	peat-UP20	80/20% v/v mixed	0.67 + 0.84	0.60

ratio: 100/5/1) and an organic binder (20% in mass) from Elotex industry. The dimensions of granules are 7 mm in diameter and 15 mm in length; (2) a mechanically crushed sapwood (Hortifibre®) which is a cheap material composed of 100% of wood fibers from cultivated forests. Sapwood looks like fibrous wood stick of 1 cm long and 2 mm diameter; (3) pine bark, chosen for its good physical properties (low bed density) and very low price; (4) pozzolan, which is a volcanic siliceous rock, inert and relatively cheap. Pozzolan used was mainly composed with SiO_2 (44%), Al_2O_3 (15%), Fe_2O_3 (15%) and CaO (10%) and appeared as irregular spherical beads with diameters ranging from 5 to 10 mm; (5) a fibrous peat purchased from "S.A. La Florentaise Company". Elemental analysis of the peat was C: 50.2%; H: 6.0%; O: 33.0%; N: 0.8%.

Table 1 presents some characteristics of packing materials used. The specific surface area and the density were measured by mercury intrusion porosimetry (Micromeritics AutoPore IV 9500). Biofilters were filled (bed high H = 1 m) with different configurations of packing materials as summarized in Table 2. For configuration "in layers", biofilters were filled with packing materials

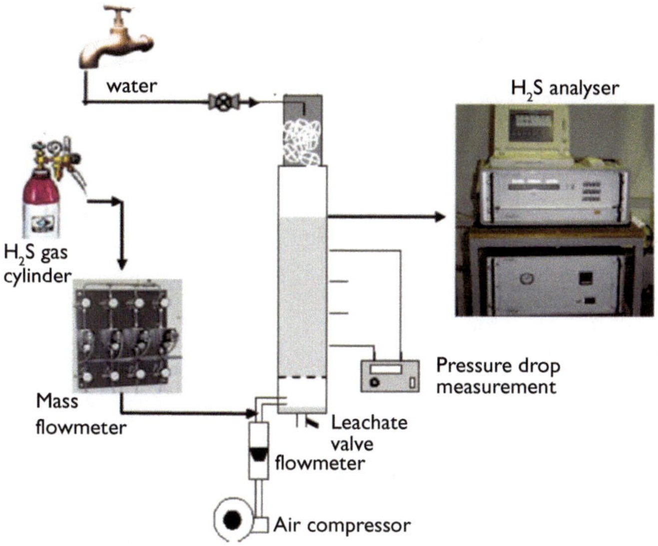

Figure 1 Schematic diagram of biofilter system for H_2S treatment.

(peat, sapwood or pozzolan; 0.8 m high) topped with a layer of UP20 (0.2 m high) corresponding to a 80/20 volume ratio.

EXPERIMENTAL SETUP

The treatment of a synthetic gas polluted with H_2S was carried out in laboratory scale biofilters in the same conditions (Figure 1). Biofilters were constructed with plastic cylinders (1.5 m in height and 0.1 m in diameter). The gas flow to be treated was obtained by mixing H_2S (99.7% purity from the gas cylinder) with the air stream. H_2S flow was controlled by a 5850S Brooks mass flow controller. Various concentrations of H_2S (up to 400 mg/m^3; measured using an Onyx 5220 device – Cosma France) were carried out to determine the optimum biofilter performances. The superficial velocity of the polluted gas on each biofilter was 65 m/h (0.018 m/s; gas flow rate 0.5 Nm^3/h) corresponding to an empty bed residence time (EBRT) of 57 s. To maintain the bed humidity, tap water was sprayed for 1 h on the top of the column (corresponding to roughly 33 liters), followed by 1 h stand-by. Both biofilters filled with pine bark and pozzolan-UP20 respectively were inoculated with washed activated sludge (1.35 g dry weight) per column twice in two weeks. The initial biomass came from the activated sludge of a wastewater treatment plant (city of Nantes, France) and was not acclimatized to treat H_2S. Other biofilters were neither inoculated with washed activated sludge nor acclimatized to treat H_2S. Apart from polluted air and UP20 material, no nutritive solution for feeding microorganisms was introduced into the biofilters.

RESULTS

Biofilters were operated continuously for at least 95 days. The elimination capacities ($EC = Q(C_{in}-C_{out})/V$; $g \cdot m^{-3} \cdot h^{-1}$) was calculated according to the loading rates ($LR = QC_{in}/V$; $g \cdot m^{-3} \cdot h^{-1}$) used in this study. Figure 2 summa-

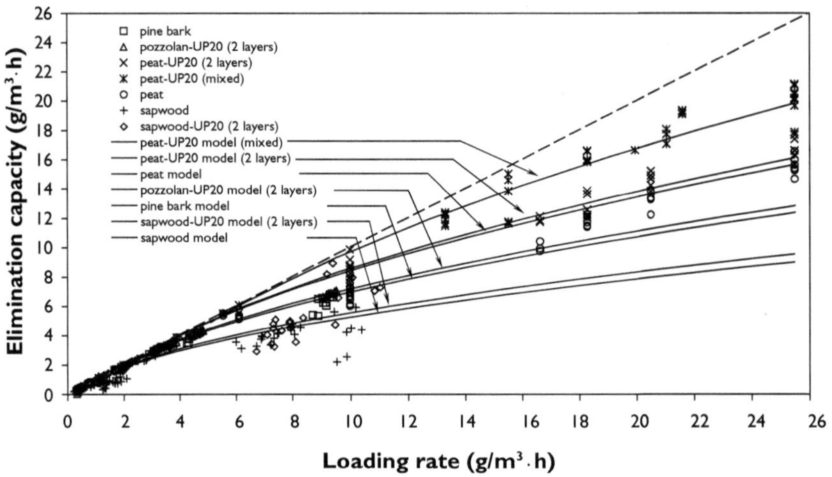

Figure 2 Elimination capacity versus loading rate for laboratory pilot-scale biofilters: EBRT = 57 s.

Table 3 Kinetic constants according to both models[3–4].

	Michaelis-Menten model			*Ottengraf model*
Packing materials	K_s $(mg \cdot m^{-3})$	V_m $(g_{H_2S} \cdot kg_{material}^{-1} \cdot day^{-1})$	V_m $(g_{H_2S} \cdot m^{-3} \cdot h^{-1})$	K_I $(g^{0.5}_{H_2S} \cdot m^{-1.5} \cdot h^{-1})$
N°1. 100% peat	113.5	4.75	53.4	15.3
N°2. 100% sapwood	17.5	1.39	8.3	7.9
N°3. 100% pine bark	35.9	1.07	16.5	11.4
N°4. peat-UP20 (2 layers)	111.8	2.74	–[a]	15.9
N°5. sapwood-UP20 (2 layers)	10.3	0.69	–[a]	8.4
N°6. pozzolan-UP20 (2 layers)	42.4	0.33	–[a]	11.9
N°7. peat-UP20 (mixed)	89.2	3.32	(55.3)[b]	21.3

[a] not calculated due to the difference of density between the "2 layers"; [b] calculated value using a pseudo density of the mixture).

rizes the evolution of the elimination capacities versus loading rates for all biofilters. In a purpose of modeling the biofilter performances, biokinetic constants were calculated using the Ottengraf model[4] and a modified Michaelis-Menten model[5] (Table 3).

From Figure 2 and Table 3, the main conclusions are:

- Peat is the best material to treat large H_2S concentrations. Moreover, peat performances can be significantly improved by addition of UP20 (in mixture).
- The association pozzolan-UP20 tested in this study (80/20% v/v in "2 layers") represents an interesting "packing material" to treat pollutant loading rates up to 5 $g \cdot m^{-3} \cdot h^{-1}$.
- For small H_2S concentrations, sapwood could be satisfactorily used instead of peat. This material has proven to efficiently degrade H_2S with a pollutant loading rates up to 4 $g \cdot m^{-3} \cdot h^{-1}$. However, the addition of UP20 (in a configuration "2 layers") does not contribute to give a real advantage in biofilter performances.

REFERENCES

1 Kennes C and Veiga MC, *Bioreactors for waste gas treatment*, Kluwer Academic Publishers, Dortrecht, The Netherlands (2001).

2 Dumont E, Andrès Y, Le Cloirec P and Gaudin F, Evaluation of a new packing material for H_2S removed by biofiltration. *Biochemical Engineering Journal* 42: 120–127 (2008).

3 Gaudin F, Andrès Y and Le Cloirec P, Packing material formulation for odorous emission biofiltration. *Chemosphere* 70: 958–966 (2008).

4 Ottengraf SPP and van den Oever AHC, Kinetics of organic compound removal from waste gas with a biological filter. *Biotechnol. Bioeng.* 25: 3089–3102 (1983).

5 Hirai M, Ohtake M and Shoda M, Removal kinetics of hydrogen sulfide, methanethiol and dimethyl sulfide by peat biofilters. *J. Fermentation Bioeng.* 70: 334–339 (1990).

Chapter 39

START-UP AND TUNING OF ORGANIC BIOFILTERS UNDER RESTRICTED MOISTURE CONDITIONS

A. Elias, A. Barona, G. Gallastegui, L. Gurtubay & N. Rojo
Department of Chemical and Environmental Engineering
University of the Basque Country, Bilbao, Spain

INTRODUCTION

The control of water or moisture content in conventional biofilters is a crucial factor for biofilter effectiveness because microorganisms require water to carry out their metabolic activities (Datta and Allen, 2005[1]). An inefficient water control or supply has been reported as responsible for 75% of biofilter failures` (Auria et al., 2000[2]). The drying-out effect can be partially solved by humidifying the gas stream before it enters the bioreactor. Nevertheless, additional moisture adjustments may also be required because bio-oxidation is an exothermic reaction.

MATERIALS AND METHODS

ACCLIMATION OPERATION

A sludge sample containing original biomass was collected in a wastewater treatment plant (Muskiz, Bizkaia, Spain) and a preliminary characterization and acclimation to p-xylene were carried out.

BIOFILTER OPERATION

A biofiltration system was assembled under downflow mode consisting of three interchangeable modules with a total volume of 4.5 l, an air-compressor, a flow meter, a humidification chamber, a vessel containing p-xylene and an active carbon chamber.

Sampling and measurement ports were located along the three PVC modules of the bioreactor. The packing material selected was made up of composted pig manure and sawdust, and has already been used in previous works (Elías et al., 2002[3]).

Throughout experimentation, the gas flow was introduced at a pseudo-constant rate of 1.0–1.5 l min^{-1}, corresponding to an EBRT varying between 270 and 180 s. The inlet and outlet concentrations of p-xylene at each module were measured by gas chromatography.

Biofiltration of an air stream containing p-xylene was conducted for 450 days at two different moisture conditions or periods. The first period (for the first 259 days of operation) comprised a "dry period" when the biofilter operated without any additional irrigation or moistening, and water was supplied solely by the saturated inlet gas flow. The second period ran from day 260 to day 450, involving a "wet period" when 400 ml of a standard nutrient solution was supplied every 25 days. Temperature and moisture content in the carrier material were measured after each irrigation of nutrient solution.

RESULTS AND DISCUSSION

The response of the biofilter during "dry" and "wet" operation is shown in Figure 1. During "dry" operation, p-xylene removal efficiency was lower than 70% for particularly reduced pollutant inlet loads (<10 $g\ m^3\ h^{-1}$). Thus, p-xylene was hardly degraded during this period, as the active biomass had another readily available carbon source provided by the support material itself. The lack of moisture caused a rapid growth of attached biomass, which could be visually detected and might be associated with fungal colonization. It has been reported that a fungal population is able to tolerate drier conditions than bacteria (Cox et al., 1996[4]; Auria et al., 2000[2]).

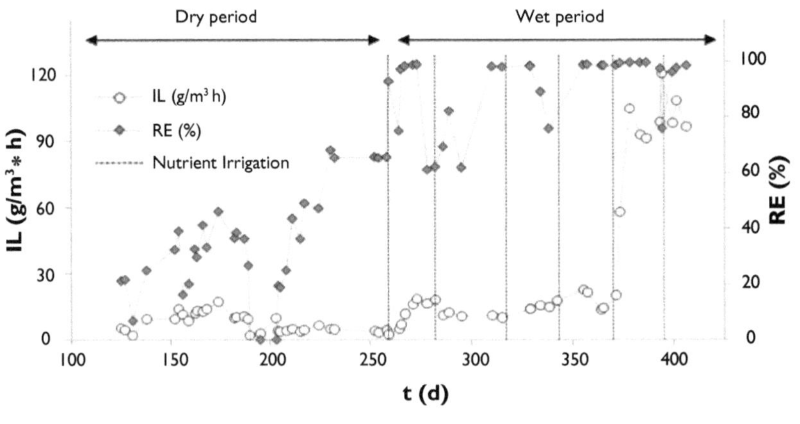

Figure 1 Inlet load (IL) and removal efficiency (RE) during dry and wet operation.

From the 260th day of operation onwards, and as a result of the regular addition of the nutrient solution (wet period), removal efficiency increased to 95% even after a sudden rise in the pollutant inlet load (from 20 to 90 g m^3 h^{-1} at the end of operation). It was concluded that the biofilter biomass was not seriously damaged by the dry period.

After each irrigation or nutrient supply, temperature and removal efficiency were monitored at different heights in the bioreactor (0, 10, 20, 30, 40 and 50 cm) during the following days. Figures 2 and 3 show the response of the biofilter after the fifth irrigation (similar results were obtained for the other nutrient additions).

In the first days after irrigation, the highest temperatures were recorded at the upper module due to the presence of more favourable biomass growth conditions and higher pollutant concentrations to be degraded. Hence, the high removal efficiency was basically attributed to the activity in the first module (Fig. 3). In the subsequent and final days, the middle module was the one mainly responsible for pollutant degradation (>50%), as shown in Figure 3. By the end of this period, the temperature in the third module rose due to the increasing removal efficiency in the lower module. As a conclusion, degradation activity started at the inlet of the bed and then spread to the

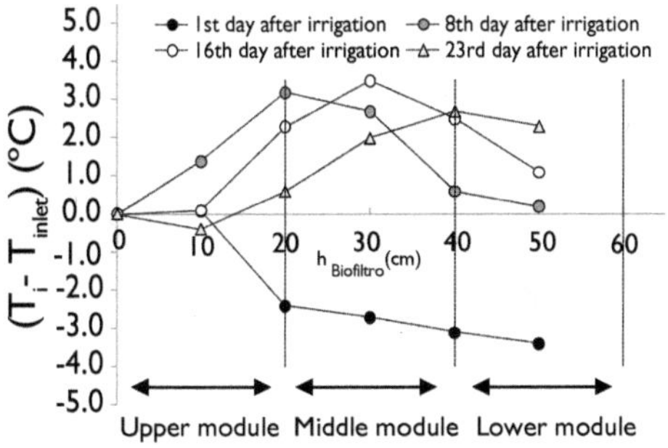

Figure 2 Temperature gradient along the filter bed height after the fifth irrigation.

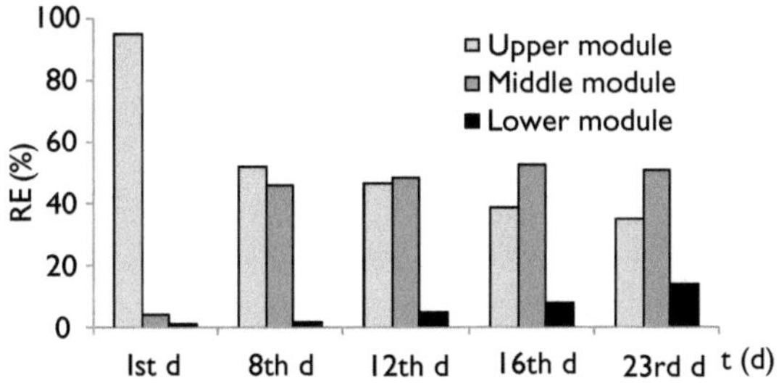

Figure 3 Removal efficiency in the modules over time after the fifth irrigation.

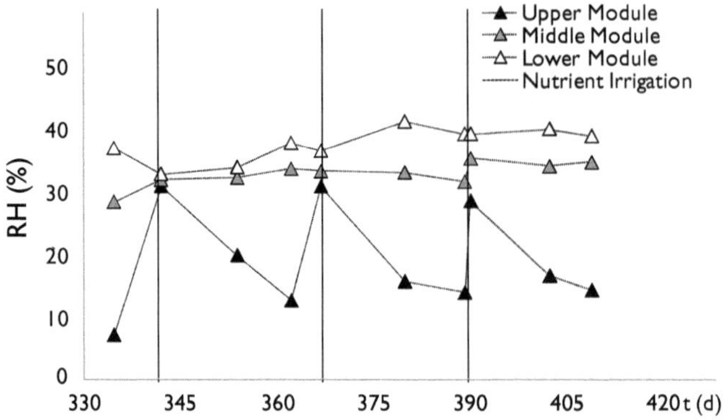

Figure 4 Relative humidity (RH) profiles in the three modules after three irrigations.

middle and lower sections as the upper section degradation rate decreased. The contribution of each module to the overall reactor performance therefore varied considerably over time.

The relative humidity profile in the three modules after three irrigations is shown in Figure 4. The highest water consumption was found in the upper module, and the highest moisture content was logically found in the lower module.

ACKNOWLEDGEMENTS

The authors acknowledge the Spanish Ministry of Science and Innovation (CTM2006-02460 financed jointly with ERDF funding) and the University of the Basque Country (Research group GIU05/12UPV and GIU08/10UPV) for the financial support for the project. The research work of G. Gallastegi was financially supported by the University of the Basque Country (PIFB001/2007/02 training research grant).

REFERENCES

1 Datta I and Allen DG, Biofilter Technology. Eds: Z. Shareefdeen and A. Singh, In *Biotechnology for Odor and Air Pollution Control*. Publisher Springer (2005).

2 Auria R, Frere G, Morales M, Acuña ME and Revah S, Influence of mixing and water addition on the removal rate of toluene vapors in a biofilter. *Biotechnology and Bioengineering* **68**: 448–455 (2000).

3 Elias A, Barona A, Arreguy A, Rios J, Aranguiz I and Peñas J, Evaluation of a packing material for the biodegradation of H_2S and product analysis. *Process Biochemistry* **37**: 813–820 (2002).

4 Cox HHJ, Magielsen FJ, Doddema HJ and Harder W, Influence of the water content and water activity on styrene degradation by Exophiala jeanselmei in biofilters. *Applied Microbiology and Biotechnology*. **45**: 851–856 (1996).

Chapter 40

OXIDATION OF BIOLOGICALLY-PRODUCED ELEMENTAL SULFUR AT NEUTROPHILIC CONDITIONS

Marc Fortuny, Carles Casas, Javier Lafuente & David Gabriel
Department of Chemical Engineering, Universitat Autònoma de Barcelona, Barcelona, Spain

Xavier Gamisans
Department of Mining Engineering and Natural Resources, Universitat Politècnica de Catalunya, Manresa, Spain

INTRODUCTION

Removal of Reduced Sulfur Compounds (RSC) such as hydrogen sulfide (H_2S) from waste or energy-rich gases has been traditionally carried out by means of physico-chemical systems.[1] However, their high operating costs and the increasing interest in energy recovery from such gaseous effluents have speed up the development of alternative biological systems.

The removal of RSC by sulfur-oxidizing bacteria (SOB) occurs by oxidation to S^0 or sulfate depending on oxygen availability (Eqs. 1, 2):[2]

$$HS^- + 0.5O_{2\,(aq)} \rightarrow S^0 + OH^- \quad \text{(biological reaction)} \tag{1}$$

$$HS^- + 2O_{2\,(aq)} \rightarrow SO_4^{2-} + H^+ \quad \text{(biological reaction)} \tag{2}$$

Although bioreactors for the treatment of low concentrations of H_2S are nowadays commonly used in industry,[3] the treatment of high H_2S concentrations is greatly constricted by a limited capability of oxygen supply for the complete biological oxidation of H_2S to sulfate, which leads to S^0 formation.[4] Since it is well known that SOB not only grow on sulfide oxidation but can also use S^0 according to Eq. 3,[2] the aim of this study was to investigate the biological oxidation of S^0 as a recovery strategy for a S^0-clogged biotrickling filter.

$$S^0 + 3/2O_2 + H_2O \rightarrow SO_4^{2-} + 2\,H^+ \quad \text{(biological reaction)} \tag{3}$$

MATERIALS AND METHODS

BIOTRICKLING FILTER REACTOR EXPERIMENT

In-situ S^0 oxidation was investigated in a lab-scale biotrickling filter reactor (BTFR) as described elsewhere.[5] The experiments where carried out after more than a year operating at an inlet H_2S concentration of 2,000 ppm_v (55.6 g H_2S $m^{-3}h^{-1}$), an EBRT of 180 s, a controlled pH range of 6–6.5 and excess oxygen supply. Prior to S^0 oxidation, the reactor was forced to produce S^0 for a period of 6 weeks with a maximum applied loading rate (LR) of 204.5 g H_2S m^{-3} h^{-1}. Subsequently, sulfide supply was stopped and the air flow increased. Previous research on the BTFR had shown that a high trickling liquid velocity (TLV) favored sulfate production and did not have a significant effect on accumulated S^0 wash down.[5] Therefore, two different TLV were also applied during the experiment, 3.8 and 17 m h^{-1}.

DISCONTINUOUS STIRRED TANK REACTOR EXPERIMENTS

Powdered biological S^0 oxidation was also studied into a 1.5 L discontinuous stirred tank reactor (DSTR) with biomass obtained from the same BTFR. A series of different runs (Table 1) were carried out either with powdered S^0 collected from the BTFR or H_2S ($H_2S_{(g)}$ 99%) supplied via a digital mass-flow controller (Bronkhorst, NL). All runs were performed under excess dissolved oxygen and carbonate concentrations (>5 mg O_2 L^{-1} and 3 g C L^{-1}).

Daily measures of sulfate, thiosulfate, biomass and S^0 liquid concentrations were performed as described elsewhere.[5]

RESULTS AND DISCUSSION

BTFR S^0 OXIDATION

A sulfur mass balance was carried out from the ionic sulfur species measurement and the sulfide LR and removal efficiency (RE) (Fig. 1). S^0 production

Table 1 Overview of the initial conditions for the different DSTR runs.

Run number	*1*	*2*	*3*	*4*
Substrate	S^0	S^0	S^0	H_2S
Load (mg S-H_2S L^{-1} h^{-1})	–	–	–	39.3
S^0 pulse (mg L^{-1})	298	1222	1128	–
Biomass (mg N L^{-1})	20.3	18.5	16.2	21.2
Inoculum origin	BTFR	Run 1	Run 2	Run 3

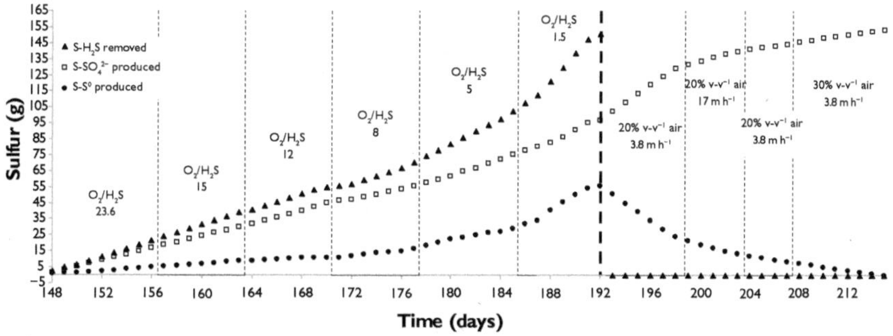

Figure 1 Sulfur mass balance in the BTFR. The thick dashed line separates the S^0 accumulation period (left) from the S^0 oxidation period (right). The thin dashed lines separate periods with different O_2/H_2S supplied ratios during S^0 accumulation (days 148 to 192) and different O_2 supply or trickling liquid velocity periods during S^0 oxidation (days 192 to 215).

was calculated by subtraction as previously reported.[6] Upon sulfide supply shutdown and air flow increase, the sulfate production rate increased up to 223 mg S-SO_4^{2-} h^{-1} (3.48 mmols SO_4^{2-} h^{-1} L^{-1} packed bed) which is a very similar rate to those reported for stirred reactors (3.5 mmols SO_4^{2-} h^{-1} L^{-1}).[6] Increasing the TLV on day 199 only caused a decreased sulfate production rate due to an important solids wash out. Therefore, it is speculated that the decrease in bioavailable S^0 and biomass were the main cause for the sudden sulfate production rate slowdown from day 199 onwards. Moreover, according to the sulfur balance, only 6 days after starting the S^0 oxidation period, 57% of the accumulated S^0 had been oxidized, which proves that even if S^0 accumulates inside a packed bed reactor, it can be rapidly biologically oxidized to sulfate if sulfide is not present.

DSTR S^0 OXIDATION

An overview of the results for the different reactor runs is given in Table 2. Run 4 data belong to the oxidation of the freshly, *in-situ* produced S^0 after sulfide supply shutdown.

When a pulse of dried, powdered S^0 was fed into the stirred reactor, an initial low S^0 oxidation rate was observed followed by a fast sulfate production phase and a final slowdown till complete stop (data not shown). Runs 2 and 3 from the DSTR reactor showed about three times smaller maximum volumetric sulfate production rates than that observed in the BTFR or that reported by Janssen *et al.*[6] Both the experiment performed by Janssen *et al.*[6] and the BTFR experiment studied the biological oxidation of freshly produced S^0. However, runs 1 to 3 were performed with externally added, dried and powdered S^0 as substrate. Therefore, it was speculated that the slower S^0 oxidation rates

Table 2 Overview of the results for the different DSTR runs.

Run number		*1*	*2*	*3*	*4*
Maximum volumetric sulfate production rate	mg S-SO_4^{2-} h^{-1} L^{-1}	7.20	34.09	35.31	108.69
	mmol SO_4^{2-} h^{-1} L^{-1}	0.23	1.07	1.10	3.40
Specific maximum sulfate production rate	mg S-SO_4^{2-} h^{-1} mg $N_{aver.}^{-1}$	0.48	1.69	1.77	2.87
Biomass growth rate	mg N h^{-1}	0.11	na	0.66	0.55

n.a: not available

observed in runs 2 and 3 were caused by a low substrate bioavailability, due to the lack of a proper sulfur-particles surface colonization and particles size distribution.

This assumption was afterwards confirmed upon biological oxidation of fresh, *in-situ* produced S^0 during run 4, after 6 days of *in-situ* S^0 formation inside the DSTR (Table 2). One day after sulfide supply shutdown, biomass and sulfate concentrations started increasing while the recently formed S^0 and thiosulfate concentrations decreased. Table 2 shows as the maximum volumetric sulfate production rate from run 4 was much higher than those for run 2 or 3 and actually was very similar to the value reported by Janssen *et al.*[6] or the value obtained from the BTFR experiment. Hence, it was hypothesized that the fresh, *in-situ* produced S^0 properties (particle size and surface characteristics) facilitated S^0 bioavailability and oxidation compared to dried sulfur properties.

CONCLUSIONS

Accumulation of S^0 turns out to be the main operational problem in a biotrickling filter treating high loads of H_2S if not properly operated. However, the results showed that using the SOB capacity to oxidize accumulated S^0 under sulfide starvation can be considered as a mechanism for reactor unclogging. Also, it was shown as the main rate limiting factor on S^0 biological oxidation is S^0 bioavailability.

REFERENCES

1 Woodcock KE, Gottlieb M, Natural gas, in *Kirk-Othmer Encyclopedia of Chemical Technology*. Wiley, **12:** 377–386 (2004).

2 Kuenen JG, Colourless sulphur bacteria and their role in the sulphur cycle, *Plant Soil* **43:** 49–76 (1975).

3 Syed M, Soreanu G, Falleta P, Béland M, Removal of hydrogen sulphide from gas streams using biological processes – A review, *Can Biosyst Eng* **48:** 2.1–2.14 (2006).
4 Fortuny M, Baeza JA, Deshusses MA, Gamisans X, Casas C, Lafuente J, Gabriel D, Biological sweetening of energy gases mimics in biotrickling filters. *Chemosphere* **71:** 10–17 (2008).
5 Fortuny M, Gonzalez-Sánchez A, Deshusses MA, Gamisans X, Casas C, Gabriel D, Lafuente J, Operational aspects of the biological sweetening of energy gases mimics in a lab-scale biotrickling filter, in *Proceedings of the USC-UAM Conference on Biofiltration for Air Pollution Control*, Long Beach, California (2008).
6 Janssen AJH, Sleyster R, van der Kaa C, Jochemsen A, Bontsema J, Lettinga G, Biological Sulphide Oxidation in a Fed-Batch Reactor. *Biotechnol Bioeng* **47:** 327–333 (1995).

Chapter 41

AIR PHYTOREMEDIATION – REMOVAL OF PARTICULAR MATTER, PAHs AND METALS BY HIGHER PLANTS

S.W. Gawronski, J. Graczyk, R. Popek & M. Korzeniewski
Laboratory of Basic Research in Horticulture, Warsaw University of Life Science, Warsaw, Poland

A. Sæbo
Bioforsk—Norwegian Institute for Agriculture and Environmental Research, Norway

H. Gawronska
Laboratory of Basic Research in Horticulture, Warsaw University of Life Science, Warsaw, Poland

ABSTRACT

In urban area inhabitants are permanently exposed to several pollutants, level of which depends on traffic intensity or vicinity of industrial emitters. At a certain moment the strength of the negative impacts of human beings became so great, that only the most resistant species are able to survive, very often on the cost of limited growth and development. Present state of knowledge and social awareness of negative anthropological impact on the environment leads to search for ways to decrease it and, as much as possible, to compensate for the damage already in place. One of the most difficult to protect components of the environment is air. In urban areas air pollution became one of the biggest challenges for environmental protection, both because of the ease of pollutants' transport in the atmosphere and the difficulty of their removal. It is worth stressing that plants are the only higher organisms, which have successfully inhabited sites polluted with, for example, heavy metals and organic compounds. As they remain in one place for a long time they had to elaborate protective mechanisms enabling them to survive in extreme environmental conditions. Plants affect chemical, physical and biological processes in the environment and steer them in such a way as to repair the environment and make it as close as possible to the optimum.

Observations of such processes led to a discovery of new possibilities of plants utilization, called phytoremediation, which gives a chance of 'reparation' of the degraded environment. The name 'phytoremediation' derives from a Greek word 'phyto', meaning 'plant', and Latin word 'remedium', which means 'a tool against bad'. The idea of using plants to decrease the amount of pollutants in the environment has been known for quite a long time and its origins are uncertain. Phenomenal discoveries of recent years, especially within the area of physiological and molecular bases of mechanisms of harmful substances uptake from the environment, together with knowledge on cultivation methods, enabled elaboration of the first technologies for cleaning the environment. Advantage of plants is their ability to take up many different pollutants at a time: metals and organic compounds from soil and together with these also gaseous pollutants (NO_x, CO_2, CO i O_3) and suspended particles from the air. Industrial technologies remove usually only one group of pollutants, which make phytoremediation advantageous. Plant biomass obtained from contaminated sites contains pollutants, but since it is biological material it can be utilized. Preferred method is combustion in special furnaces (in incinerators, combined heat and power stations, cement plants), where organic compounds are degraded to CO_2 and heavy metals remain in ashes, from which they are recovered or which are stored in old mines.

Phytoremediation in urban areas is applied on two types of sites. The first type is degraded post-industrial sites, where phytoremediation serves as a tool for recovery of these areas to a safe state, so that they can be further used as recreational grounds, business or even housing areas. The second, prospective, type of area for phytoremediation application is sites surrounding traffic routes, where actually phytoremediation should become a continuous process in order to remove pollutants on regular basis. In the midst of transportation arteries contaminated is air, soil and also water flowing from lanes and pavements. This section of phytoremediation is based on perennial plants, mainly trees, because of large area for air pollutants accumulation. At present, main sources of pollutants in urban areas are means of transport, which emit a number of dangerous toxins such as: heavy metals, polycyclic aromatic hydrocarbons (PAHs), NO_x, O_3 and particulate matter, which were recognized as one of the most dangerous in recent years. In smaller towns main polluting sources are private fire-places and heating systems based on coal and other solid fuels. The latter significantly increase emissions during winter. In the past few years role of industry as a great environmental polluter was radically diminished, still, however, combined heat and power plants are one of the main contamination sources in urban areas.

European Environment Agency acknowledged suspended particulate matter as the most dangerous air pollutant and estimates that they are reason for 348 000 fatal cases among EU citizens due to diseases caused by exposure to this pollutant. On average, suspended particles shorten EU citizens' life by 8, 6 months and Polish citizens' by 10, 7 months. The greatest life shortening in noted in Belgium (15 months), The Netherlands (13, 5 months) and the smallest in Finland (2, 5 months) and Estonia (3, 8 months). Data above are averages for each country, but unfortunately among three most polluted areas, beside of Northern Italy and Belgian-Dutch border and Silesia in Poland, were inhabitants' life is shortened by 2 years due to particulate matter pollution (1).

Depending on the origin, there are primary particulates, which are formed during processes in nature (fires, crunching of rocks), human activity, such as:

combustion (vehicle fuel combustion, heating, waste combustion), metallurgical industry, mining, construction processes, rubbing of vehicle tires, brakes, road surface, and bioparticles: spores and hyphae of fungi, bacteria and viruses. One of the most toxic are particles emitted during fuel combustion, both for heating and for propelling vehicles. Among the latter especially dangerous are Diesel engines emitting between 20 and 100 times more suspended particles than petrol engines.

In plant life air plays also an important role and they have developed various mechanisms in order to eliminate pollutants contained in the air. Suspended particles PM_{10} and $PM_{2,5}$ are deposited largely in waxes of leaves, young shoots and bark of trees and shrubs. Because waxes are washed off from leaves by rain and torn off by wind in autumn together with leaves we are able to collect only part of the particles. Even though the washed off particles get to the soil and increase its contamination, they are removed from the air. Similarly, the torn off waxes saturated with particles, as heavier are deposited on the ground surface. Apart from waxes, a great role in accumulating larger particles is played by hairs present on the above ground parts of many plant species. Amount of wax, its structure and amount of hairs play crucial role in accumulating particles by plants. Suspended particles which reach soil get through different processes, for example decomposition (organic compounds) or inhibition in sorption complex (heavy metals). This means that the role of plants in phytoremediation of air pollutants is much greater than what we can estimate just from the amount of pollutants in leaves. The possibility to collect leaves in the autumn and shoots during the spring cutting of trees and shrubs (which have been accumulating pollutants also throughout winter) is a great opportunity to remove the contaminants outside the threatened area and to utilize them. The most appropriate is to combust both leaves and shoots, in order to eliminate PAHs, and remaining ashes containing heavy metals should be properly taken care of. If transportation costs are high, in order to decrease the biomass, hot composting is often used.

Plant species characterised by good phytoremediation potential, planted in contaminated areas or close to pollution sources, not only have great esthetical value but can also greatly improve human health. It is not to be doubted that there are cities, in which the protective role of plants against pollutants is of top importance. Positive aspects of cultivating plants in urban areas have been known for long, but discovering great phytoremediation potential of some of the species should bring more attention to them. Significant differences are observed between species and also cultivars in their suitability to phytoremediation, depending on the levels and types of dominant pollutants.

A cooperative Project on air phytoremediation is run in our Labs, aiming at the evaluation of the capacity of various plant species, recommended for cultivation in urban area, for accumulation of PMs and to analyze 17 of PAHs and 17 of HMs using HPLC-MS ICP-MS methods respectively. Amount of PM_{10}, $PM_{2.5}$ and $PM_{0.2}$ deposited on leaf surface (rinsed with water) and captured by waxes (immersing in chlorophorm) were measured. The water rinsed fraction comprises the PM deposited on the surface and not very deeply into the wax, refers to PMs that in nature are removed with rain and probably also by wind. While the part removed by chloroform (removing all wax) refers to the remaining part of PM, which are too big to penetrate the plant cell, except small amount that can enter through open stomata.

In the case of *Canna x generalis*, perennial ornamental species often cultivated in cities, the amount of PM_{10} and of $PM_{2,5}$ rinsed with water was 12.4 μg cm^{-2} and of 0.86 μg cm^{-2} of leaves respectively and these values for PMs extracted with chloroform were 24.3 μg cm^{-2} and 2.57 μg cm^{-2}.Thus, about half of the particulate matter was deposited on the surface of waxes and hairs, meaning that they can be easily removed by water (and probably by heavy rain), while the other half penetrates deeper into waxes and can be removed from the plants only together with waxes.Capability for accumulation of PM of 44 plant taxons of trees and shrubs, recommended for cultivation in urban area, but grown under the same conditions in a nursery was evaluated in leaf samples collected at the end of the vegetation seasons of 2007 and 2008. From coniferous species, out of which the majority does not tolerate pollutants and very often gets damaged in the contaminated urban environment, to the very tolerant: *Platanus x hispanicus, Ginko biloba, Populus sp., Fracsinus sp., Acer sp.* or *Amopha fruticosa* and *Cornus alba*. One of the highest amount of suspended particles was accumulated by silver birch (*Betula pendula*), which has also the thickest wax cover.

Plants having leaves with a thick layer of wax or more hairs accumulate more PM. And species with the leaves covered with bigger amounts of waxes accumulate smaller PM (<2.5 μm) while those with hair accumulate bigger PM (10–2.5 μm). The amount of PM deposited on leaves of trees ranged, in average from 9.2 to 29.2 μg cm^{-2}, depending on species Shrubs accumulated similar amount of PM and collected from 9, 1 to 22.4 μg cm^{-2}. Analysis of 17 PAHs and 17 HMs are in progress and will be presented.

Generally cultivation of plants of high phytoremediation potential on contaminated sites and in their surrounding not only brings esthetical value, acts as a source of oxygen, increases air humidity and reduces noise, but most important can play a role of 'green liver' in urban environment.

ACKNOWLEDGEMENTS

Authors wish to express to their thanks to Polish-Norwegian Research Fund for financing research within: grant # PNRF – 193 – AI – 1/07.

REFERENCE

1 Air pollution in Europe 1990–2004. European Environment Agency Report No 2/2007. Copenhagen. Denmark.

Chapter 42

MODELLING THE BIOFILTRATION OF STYRENE-ACETONE-AIR MIXTURES

A.M. Gerrard
Teesside University, Middlesbrough, England

J. Skalicky & J. Paca
Institute of Chemical Technology, Prague, Czech Republic

INTRODUCTION

Biofiltration is a useful process for cleaning high flow-rates of gases containing low concentrations of volatile organic compounds. An experimental program at the ICT, Prague has produced data on the performance of a pilot scale rig removing pure styrene and styrene-acetone mixtures from air. The *pure* component data was initially modelled using the well known Ottengraf[1] equations. Often first order kinetics were found to adequately represent the concentration profiles found in the laboratory experiments.

EXPERIMENTAL

In the styrene-acetone-air mixture experiments, a laboratory-scale biofilter was used to degrade the styrene and acetone. The system had an internal diameter of 150 mm, the bed was 1 m in height and the styrene and acetone concentrations were measured at the inlet, middle and outlet of the bed. Pall rings were used as the packing. Measurements started 128 days after inoculation and 55 sets of data were collected over 67 days.

A constant air flow-rate (20 L/min) was used. The inlet styrene concentration was kept constant during the whole experiment (at around 50 mg/m^3) but the inlet acetone concentration was gradually increasing from 47 mg/m^3 to 1058 mg/m^3.

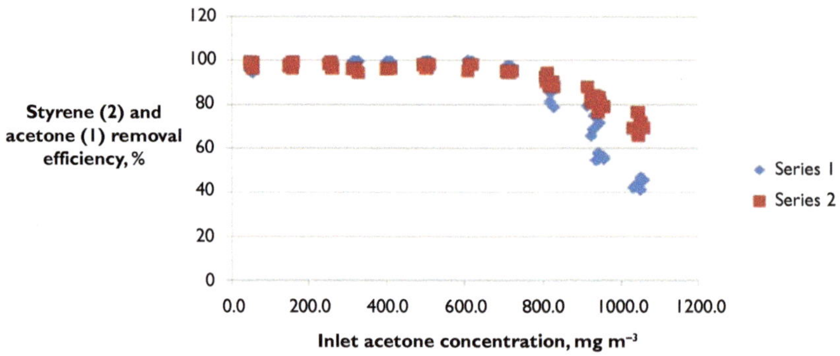

Figure 1 The effect of acetone concentration on removal efficiencies.

RESULTS

The effect of the increasing inlet concentrations of acetone was to reduce the efficiency of removal of both the styrene and the acetone, see figure 1.

MODELLING

This mixture was modelled using an analogous approach to an earlier study (Gerrard[2]) reported in Coruna in 2007 on gasoline mixtures.

First order models are suggested which are extended to include the inhibitory effect of the acetone. Some analytic and numerical techniques are used in the solution of the resultant equations.

In more detail, if we assume the acetone follows first order kinetics with some self-inhibition then:

$$\frac{dCa}{dh} = -B_1 \cdot Ca \cdot (1 - p \cdot Ca) \tag{1}$$

where p is the inhibition constant.

Thus:

$$\int_{Ca_{in}}^{Ca} \left(\frac{1}{Ca} + \frac{p}{1 - p \cdot Ca} \right) \cdot dCa = -B_1.h \tag{2}$$

This can be solved to give:

$$Ca = \frac{Ca_{in} \cdot \exp(-B_1 \cdot h)}{[1 - p \cdot Ca_{in} \cdot (1 - \exp(-B_1 \cdot h))]} \tag{3}$$

The digestion of the styrene is also assumed to be first order and to be inhibited by the presence of the more easily digested acetone (which is preferentially consumed by the micro-organisms.)

$$\frac{dCs}{dh} = -B_2 \cdot Cs \cdot (1 - p' \cdot Ca) \tag{4}$$

Combining equations (3) and (4):

$$\frac{dCs}{dh} = -B_2 \cdot Cs \cdot \left(1 - \left[\frac{p' \cdot Ca_{in} \cdot \exp(-B_1 \cdot h)}{1 - p \cdot Ca_{in} \cdot (1 - \exp(-B_1 \cdot h))}\right]\right) \tag{5}$$

Hence:

$$Cs = Cs_{in} \cdot \exp\left(-B_2 \cdot \int_0^h \left[1 - \frac{p' \cdot Ca_{in} \cdot \exp(-B_1 \cdot h)}{[1 - p \cdot Ca_{in} \cdot (1 - \exp(-B_1 \cdot h))]}\right] \cdot dh\right) \tag{6}$$

This integral in equation (6) has been solved analytically but it is more convenient to solve it numerically using the Trapezium rule.)

Then the parameters B_1, B_2, p, p' were altered to give the best fit to the measured concentration profiles of acetone and styrene down the bed's length.

The best values were found to be:

Table 1 Values of fitted parameters.

Parameter	
B_1	3.6
p	0.00083
B_2	3.85
p′	0.00093

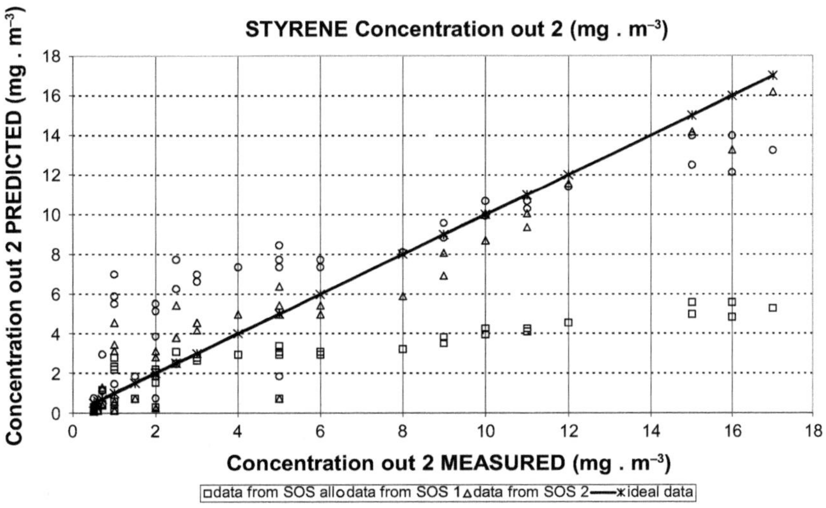

Figure 2 Values of calculated (predicted) outlet concentrations of styrene in styrene-acetone mixture biofilter obtained by using different B_1, B_2, p and p' parameters.

Figure 2 is a scatter diagram to show the adequate fit of this model to the experimental data.

CONCLUSIONS

The experimental data showed that the acetone inlet concentration affected its own removal rate and that of styrene.

A new model has been proposed to describe this behaviour.

ACKNOWLEDGEMENT

This study was financially supported by EU Socrates scheme which funded Mr Skalicky's visit to the University of Teesside and, the Ministry of Education of the Czech Republic, Project MSM 6046137305.

REFERENCES

1 Ottengraf SPP, *Biofilters in Biotechnology*, VCH, Verlagsgeselltschaft, Weinheim, pp. 436–446, (1986).
2 Gerrard AM, Halecky M and Paca J, Gasoline biofiltration: an analytic model.
3 Paper given to the International Congress on Biotechniques for Air Pollution Control, La Coruna, 4–5 October (2007).

NOMENCLATURE

B_1, B_2 kinetic parameters
C gas concentration, g · m^{-3}
h filter bed height, m
p, p' inhibition parameters

Subscripts

$_a$ acetone
$_{in}$ inlet condition
$_s$ styrene

Chapter 43

ALKALINE BIOFILTRATION OF VOLATILE SULFUR COMPOUND ODORS

Armando González-Sánchez, Guillermo Baquerizo & Sergio Revah
Departamento de Procesos y Tecnología, Universidad Autónoma Metropolitana-Cuajimalpa, Mexico City, Mexico

Luis Arellano-García & Sergio Hernández-Jiménez
Departamento de Procesos e Hidráulica, Universidad Autónoma Metropolitana-Iztapalapa, Mexico City, Mexico

ABSTRACT

BACKGROUND. An alkaline mineral medium was used previously to enhance the absorption rate of hydrogen sulfide in a odor biofiltration process. In this work a high operating pH (≈10) was used to perform the biotrickling filtration of carbon disulfide and ethanethiol odors, here a mathematical model was developed and validated under dynamic operation.

RESULTS. The solubility of carbon disulfide and ethanethiol was increased around 50% and 80% respectively by using an alkaline mineral medium instead of pure water. The alkaline biotrickling filter reached maximum elimination capacities for carbon disulfide of 2.3 g m^{-3} h^{-1} and 3.5 g m^{-3} h^{-1} for ethanethiol case, both with removal efficiencies around 50%.

CONCLUSION. The absorption of carbon disulfide and ethanethiol under alkaline condition is enhanced, nonetheless, the elimination capacity of the alkaline biotrickling filter was limited by the bacterial consortium activity, which was corroborated by the mathematical model.

Keywords: Carbon disulfide, ethanethiol, alkaline biotrickling filter, odors, biofiltration modeling

INTRODUCTION

As far as volatile sulfur compounds (VSC) are concerned, odor contamination is usually related with hydrogen sulfide (H_2S), carbon disulfide (CS_2), thiols (R-SH) and organic sulfides (R-S-R).

In a previous work,[1] an alkaline mineral medium and an alkaliphilic sulfooxidizing bacterial consortium (ASBC) were succesfully deployed in a biotrickling filter (BTF) at pH 10.0 to remove H_2S odors. Authors said that the alkalinity reduces H_2S gas liquid mass transfer resistances. Thus, the alkaline condition could result attractive for the biofiltration of other VSC. The objective of this work was to show the feasibility of using an alkaline BTF to remove gaseous CS_2 and ethanethiol (ET).

MATERIALS AND METHODS

GAS-LIQUID PARTITION COEFFICIENT IN ALKALINE MEDIUM

The effect of an alkaline mineral medium[2] (AMM) with pH 10.0 on the VSC solubility was determined by the evaluation of the partition coefficient (m). Sealed glass serum bottles (125 mL) were used to contain 25 mL of AMM and a determined quantity of gaseous CS_2 or ET. After enough time to attain equilibrium, gas samples (200 μL) were extracted from the bottles and analyzed in a gas chromatograph equipped with a FPD detector.

RESPIROMETRY

The activity of the ASBC was evaluated by measuring the oxygen consumption rates in liquid stirred cultures as a function of initial CS_2 or ET concentrations (between 0.1 and 8.00 mmol L^{-1}). To represent the dependence of the specific oxygen respiration rate (qO_2) with substrate concentration, a double Monod model (sulfur substrate and dissolved oxygen) was proposed. The experimental data were fitted to double Monod using a direct random search method published elsewhere[3] in order to estimate the biokinetic parameter values.

BIOTRICKLING FILTER (BTF)

Rectangular pilot scale alkaline biotrickling filters (BTFs) with a 5.2 L bed volume were used to treat the CS_2 or ET containing gases. The packing material was open pore polyurethane foam cubes (EDT, Germany) with a specific area 600 m^2 m^{-3}, a density of 35 kg m^{-3}, and a porosity of 0.97. The AMM was used as trickling liquid with a flow rate of 12 L h^{-1}. Distilled water was added at a flow rate of 0.02 L h^{-1} to compensate the water evaporation loses. Air and trickling liquid flowed co-currently for the case of CS_2 biofiltration in order to avoid pollutant stripping while counter-current mode was applied for ET.

The BTFs were inoculated with ASBC consortium from a 0.5 L batch culture that was bubbled with gaseous CS_2 or ET for 30 days. The pH of the system was controlled at 10.0 by addition of 2 N NaOH. The system was maintained at 25°C in a temperature controlled room.

MODEL DEVELOPMENT

The process was modeled in a basis of general mass balances of ET and oxygen. Model assumptions can be found elsewhere[4] but assuming that the biofilm is fully wetted and the main resistance to the ET mass transfer is related to gas film since the BTF was operated under alkaline condition.[1]

RESULTS AND DISCUSSION

The solubility of CS_2 and ET increased around 50% and 80%, respectively when AMM was used as absorbent instead of deionized water; see partition coefficients in Table 1.

The experimental data of the respirometry experiments were adjusted to the double Monod kinetics. Values for maximum qO_2 induced by the oxidation of CS_2 and ET were 7.0 and 11.0 $\mu molO_2\ g_{prot}^{-1}\ min^{-1}$ respectively. Meanwhile the saturation constants (K_S) for CS_2 and ET oxidation were 460 and 2910 $\mu mol\ L^{-1}$ respectively. These data show a lower biological activity compared with other biological CS_2 oxidation reports, i.e. *Thiobacillus thioparus* TK-m[5] and *Thiomonas* sp WZW[6] for which a maximum qO_2 not less than 500 $\mu mol_{O2}\ g_{prot}^{-1}\ min^{-1}$ and a K_s around 20 mmol L^{-1} were reported. No previous reported results for ET biological oxidation exist.

Figures 1 and 2 show the loading rates and removal efficiencies (RE) during the BTFs operation. For CS_2 a maximum RE of 50% was attained for a loading rate of a 4.7 $g_{CS_2}\ m^{-3}\ h^{-1}$, meanwhile for ET the maximum RE was 50% when a loading rate of 7.2 $g_{ET}\ m^{-3}\ h^{-1}$ was applied. The low oxidizing activity of the ASBC determined lower elimination capacities compared with other reported no alkaline biotrickling filters. For example, CS_2 elimination capacities of 152 $g_{CS_2}\ m^{-3}\ h^{-1}$ (6) and 68 $g_{CS_2}\ m^{-3}\ h^{-1}$ (7) have been reported, both with RE close to 100%.

The Mathematical model examination reveals that the mass transfer process did not limit the elimination capacities of the BTF because the alkaline operation, then the biological reaction was corroborated to be the rate limiting step in the alkaline biofiltration process. The model showed that oxygen concentration was high enough to have no influence on the ET consumption rate.

Table 1 Experimental partition coefficients in deionized water and alkaline mineral medium.

Compound	*m (H_2O, pH 7.0)*	*m (MMA, pH 10.0)*
CS_2	0.480 ± 078	0.210 ± 0.066
ET	0.100 ± 0.033	0.020 ± 0.045

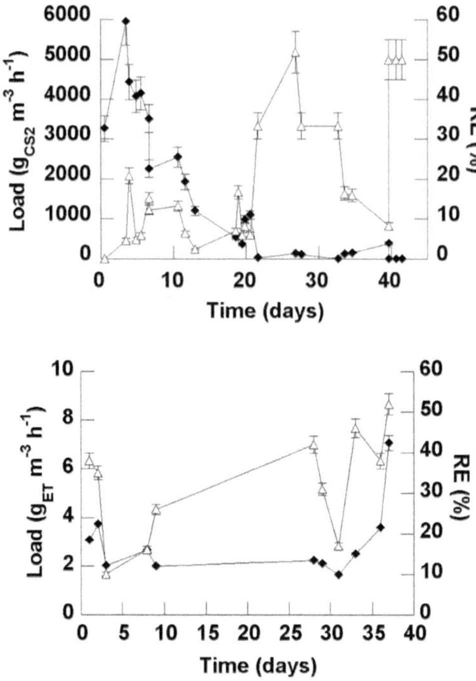

Figure 1 and 2 Performance of the ASBC inoculated BTF treating CS_2 (left) or ET (right). (♦) load, (Δ) removal efficiency.

REFERENCES

1 González-Sánchez A, Revah S and Deshusses M, Alkaline biofiltration of H_2S odors. *Environl Sci Technol* **42**(19):7398–7404 (2008).

2 Sorokin DY, Lysenko AM, Mityushina LL, Tourova TP, Jones BE, Rainey FA, Robertson LA and Kuenen GJ, *Thioalkalimicrobium aerophilum* gen. nov., sp. nov. and *Thioalkalimicrobium sibericum* sp. nov., and *Thioalkalivibrio versutus* gen. nov., sp. nov., *Thioalkalivibrio nitratis* sp. nov. and *Thioalkalivibrio denitrificans* sp. nov., novel obligately alkaliphilic and obligately chemolithoautotrophic sulfur-oxidizing bacteria from soda lakes. *Int. J Syst Evol Microbiol* **51**:565–580 (2001).

3 Luus R and Jakola THI, Optimization by direct search and systematic reduction of the size of search region, *AICHE J* **19**(4):760–766 (1977).

4 Kim S and Deshusses MA, Development and experimental validation of a conceptual model for biotrickling filtration of H_2S. *Environ Prog* **22**(2):119–128 (2003).

5 Smith NA and Kelly DP, Oxidation of carbon disulphide as the sole source of energy for the autotrophic growth of *Thiobacillus thioparus* strain TK-m. *J Gen Microbiol* **134**:3041–3048 (1988).

6 Pol A, van der Drift C and Op den Camp HJM, Isolation of a carbon disulfide utilizing *Thiomonas* sp. and its application in a biotrickling filter. *Appl Microbiol Biot* **74**:439–446 (2007).

7 Lobo R, Revah S and Viveros-García T, An analysis of a trickle-bed bioreactor: Carbon disulfide removal. *Biotechnol Bioeng* **63**: 98–109 (1999).

Chapter 44

SHORT TERM CHARACTERIZATION OF A H_2S BIOTRICKING FILTER PACKING USING A GASEOUS-LIQUID RESPIROMETER

Armando González-Sánchez, Javier Lafuente & David Gabriel
Department d'Enginyera Química, Universitat Autònoma de Barcelona, Barcelona, Spain

Xavier Gamisans
Department d'Enginyera Minera i Recursos Naturals, Universitat Politecnica de Catalunya, Manresa, Spain

Armando González-Sánchez & Guillermo Baquerizo
Departamento de Procesos y Tecnología, Universidad Autónoma Metropolitana-Cuajimalpa, Mexico City, Mexico

ABSTRACT

A lab scale biotrickling filter called gaseous-liquid respirometer (GLR) was used to characterize the main biofiltration properties of a representative sample of packing material from a H_2S operating biotrickling filter. Abiotic and biotic short term experiments besides a simplified mathematical model allowed estimating the external mass transport and bio-kinetic parameters.

INTRODUCTION

The biofilters and biotrickling filters have been widely used for the H_2S removal from energetic and waste gases streams.[1] The biological removal process consists in the gaseous H_2S transference to an aqueous scrubbing solution or an attached biofilm, where the pollutant is degraded by sulfide-oxidizing microorganisms to non-volatile species such as elemental sulfur and sulfate.[2] Several mathematical models for H_2S biotrickling filtration have reported that biokinetic parameters play a major role in the process performance,[3,4] however external mass transfer resistance could also influence the process when short empty bed residence times are applied.[5] In addition, most of the H_2S biofiltration models do not consider when the oxygen becomes a limiting substrate. Later situation can be reached when high H_2S loads are supplied to the biofilter. This work reports a lab scale biotrickling filter called gaseous-liquid respirometer (GLR) besides a simplified mathematical model, both were used to characterize the main phenomena occurring during the biotricking filtration of high H_2S loads.

MATERIALS AND METHOS

EXPERIMENTAL SETUP

The experimental system consisted in a cylindrical biotricking filter made on acrylic, with an internal diameter of 0.047 m and a overall height of 0.70 m, which 0.35 m was packed with a sample of a packing material namely polypropylene HD Q-PAC (Lantec Products Inc., CA, USA, with a 4 × 4 mm grid opening). The reactor was operated in a counter-current mode. The gaseous and liquid phases were continuously recycled by pumping (diaphragm vacuum pump MZ 2C, Vacuubrand, Germany and peristaltic pump 77200-12, Cole Parmer, USA).

The GLR was provided with an oxygen gas analyzer (SIDOR module OXOR-P, SICK, Germany) and a polarographic dissolved oxygen electrode (CellOx 325, WTW, Germany). The oxygen measurements were continuously recorded in a personal computer. The gas temperature was kept in 23°C.

GASEOUS LIQUID RESPIROMETER (GLR) CHARACTERIZATION

Abiotic and biotic experiments were performed in order to characterize the GLR. From abiotic experiments (i.e., using clean packing) the overall liquid mass transfer coefficient for oxygen (i.e., K_La) was determined as a function of different liquid and gaseous flowing rates. An additional N_2 or air flow (F_{gas}) was supplied to the GLR in order to induce the O_2 striping or O_2 saturation respectively (Figure 1).

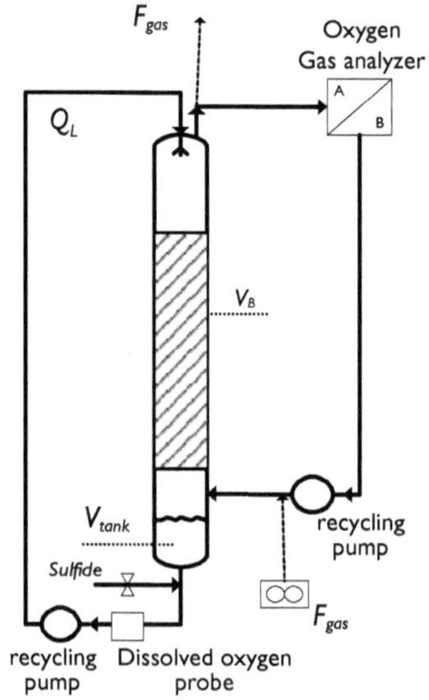

Figure 1 Experimental system: Gaseous-Liquid Respirometer (GLR).

For biotic tests, the GLR was repacked with a colonized packing material taken from a biotrickling filter continuously fed with H_2S.[6] The experiment started with the addition of 300 mL of mineral medium[6] with a 100 mmol L^{-1} of ionic strength phosphate buffer (pH around 7.0), which was continuously recycled and aerated for some hours in order to stabilize the biofilm. Afterwards, the GLR was closed and immediately 3 mL of dissolved sulfide pulse (1 mol L^{-1}) was injected to obtain a final concentration of 5 mmol L^{-1} in the liquid phase.

MODEL DEVELOPMENT

Mass balances for oxygen and H_2S in the biotrickling filter were stated. Due to the recycling operating mode of the GLR, an ideally mixed regime was assumed for both phases. The free volume gas and the liquid recycling tank were also considered in the balances. Model assumptions can be found elsewhere[5] but assuming that the biofilm is fully wetted.

Gas phase

$$\frac{\partial C_{g,i}}{\partial t} = \frac{Q_g}{V_g}(C_{g,i}^{free} - C_{g,i}) - \frac{V_L}{V_g} \cdot K_L a \left(\frac{C_{g,i}}{H_i} - C_{l,i} \right) \quad i = O_2, H_2S \tag{1}$$

$$\frac{\partial C_{g,i}^{free}}{\partial t} = \frac{Q_g}{V_g^{free}}(C_{g,i} - C_{g,i}^{free}) - \frac{V_{tank}}{V_g^{free}} \cdot K_L a \left(\frac{C_{g,i}^{free}}{H_i} - C_{l,i}^{tank} \right) \quad i = O_2, H_2S \tag{2}$$

Liquid phase

$$\frac{dC_{l,i}}{dt} = \frac{Q_L}{V_L}(C_{l,i}^{tank} - C_{l,i}) + K_L a \left(\frac{C_{g,i}}{H_i} - C_{l,i} \right) - \frac{D_i \cdot a \cdot V_B}{V_L} \left(\frac{C_{l,i} - C_{b,i}\big|_{x=0}}{\delta / N} \right) - r_{l,i} \quad i = O_2, H_2S \tag{3}$$

$$\frac{dC_{l,i}^{tank}}{dt} = \frac{Q_L}{V_{tank}}(C_{l,i} - C_{l,i}^{tank}) + K_L a \left(\frac{C_{g,i}^{free}}{H_i} - C_{l,i}^{tank} \right) - r_{l,i} \quad i = O_2, H_2S \tag{4}$$

Biofilm

$$\frac{dC_{b,i}}{dt} = D_i \frac{\partial^2 C_{b,i}}{\partial x^2} - r_{b,i} \quad i = O_2, H_2S \tag{5}$$

where $C_{g,i}$, $C_{l,i}$ and $C_{b,i}$ are the concentrations of component i in the bulk gas phase, bulk liquid phase and biofilm respectively (g m^{-3}); $C_{g,i}^{free}$, $C_{l,i}^{tank}$, are the concentrations of component i in the free gas volume and in the liquid recycling tank respectively (g m^{-3}); H_i is the gas/liquid partition coefficient of component i (dimensionless); a is specific surface area (m^2 m^{-3}); D_i is the diffusion coefficient of component i in the biofilm (m^2 h^{-1}), x is the thickness position in the biofilm (m); $r_{l,i}$, $r_{b,i}$, are the consumption rates of component i in the bulk liquid and in the biofilm respectively (g m^{-3} h^{-1}); δ is the biofilm thickness (m) and N is the total number of layers that thickness biofilm was divided for the numerical resolution of the mathematical model.

MICROBIAL KINETIC

The specific consumption for oxygen was described by a double Monod-type kinetic expression depending on sulfide and oxygen concentrations.

$$r_{b,i} = \frac{\mu_{max}}{Y_i} X \left(\frac{C_{b,O_2}}{C_{b,O_2} + K_{S,O_2}} \right) \left(\frac{C_{b,H_2S}}{C_{b,O_2} + K_{S,H_2S}} \right) \quad i = O_2, H_2S \tag{6}$$

where μ_{max} maximum specific growth rate (h^{-1}), Y_i biomass yield coefficient for component i (g biomass g^{-1} substrate), X biomass concentration (g m^{-3}) and $K_{S,i}$ half saturation constant for component i (g m^{-3}). The expression for $r_{l.i}$ can be easily deduced from equation (6).

RESULTS AND DISCUSSION

The performance of a biotic experiment is shown in Figure 2. These data were employed to calibrate the GLR model, using optimized values of kinetic parameter obtained from a separated set of biotic experiments.

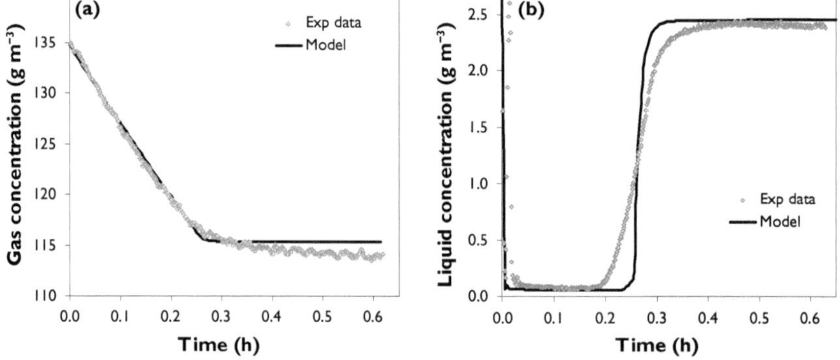

Figure 2 Simulated results and experimental results for biotic experiments: (a) gaseous oxygen concentration, (b) dissolved oxygen.

Table 1 Values of physicochemical and biological parameters used in GLR model.

Parameter	*Symbol*	*Value*	*Units*	*Reference*
Specific surface area	a	433	$m^2\ m^{-3}$	From provider
Liquid Holdup	V_L	$0.51 \cdot 10^{-3}$	m^3	Abiotic experiment
Overall liquid mass transfer coefficient	K_La	20	h^{-1}	Abiotic experiment
Maximum consumption rate for H_2S	$\frac{\mu_{max}}{Y_{H_2S}}$	0.87	$g\ gprot^{-1}\ h^{-1}$	Adjusted for simulation
Maximum consumption rate for DO	$\frac{\mu_{max}}{Y_{O_2}}$	0.38	$g\ gprot^{-1}\ h^{-1}$	Adjusted for simulation
Half saturation constant for H_2S	K_{s,H_2s}	0.03	$g\ m^{-3}$	Adjusted for simulation
Half saturation constant for O_2	K_{s,O_2}	1.0	$g\ m^{-3}$	Adjusted for simulation
Biomass concentration	X	24	$g\ prot\ L^{-1}$	Adjusted by simulation
Biofilm thickness	δ	130	μm	Adapted from (5)

The best fitting values together with mass transfer coefficient (from abiotic experiments) are shown in Table 1. Obtained values were in concordance with results reported previously[2] and showing that H_2S was mainly oxidized to elemental sulfur. The model predictions are in good agreement with experimental results indicating that transport and biological processes were satisfactorily included in the GLR model.

Model sensitivity analysis (results not shown) revealed that mass transport phenomena play a notable impact on model predictions. Also, a high influence of the free gas volume was found, suggesting to improve the design of the GLR.

The GLR showed to be able to characterize the basic biofiltration properties (i.e., transport and biological phenomena) using a small and representative piece of packing material from an operating biotrickling filter in a short term.

REFERENCES

1 Gabriel D and Deshusses MA, Retrofitting existing chemical scrubbers to biotrickling filters for H_2S emission control. *Proc. Natl. Acad. Sci. U.S.A.* 100(11):6308–6312 (2003).

2 Gonzalez-Sanchez A, Tomas M, Dorado AD, Gamisans X, Guisasola A, Lafuente J and Gabriel D, Development of a kinetic model for elemental sulfur and sulfate

formation from the autotrophic sulfide oxidation using respirometric techniques. *Water Sci. & Technol.* **59**(7):1323–1329 (2009).
3 Martin RW, Li HB, Mihelcic JR, Crittenden JC, Lueking DR, Hatch CR and Ball P, Optimization of biofiltration for odour control: Model calibration, validation, and applications. *Water Environ. Res.* **74**(1):17–27 (2002).
4 Oyarzun P, Arancibia F, Canales C and Aroca GE. Biofiltration of high concentration of hydrogen sulphide using *Thiobacillus thioparus. Process Biochemistry* **39**(2):165–170 (2003).
5 Kim S and Deshusses MA, Development and experimental validation of a conceptual model for biotrickling filtration of H_2S. *Environ Prog* **22**(2):119–128 (2003).
6 Fortuny M, Baeza JA, Gamisans X, Casas C, Lafuente J, Deshusses MA and Gabriel D. Biological sweetening of energy gases mimics in biotrickling filters. *Chemosphere* **71**(1):10–17 (2008).

Chapter 45

INFLUENCE OF BARK pH AND ELECTRICAL CONDUCTIVITY ON LICHEN VALUE DIVERSITY AROUND POWER PLANTS "KOSOVA A AND B"

B. Mustafa, A. Hajdari & Esat Hoxha
Department of Biology, Faculty of Mathematical and Natural Science, University of Prishtina, Prishtinë, Kosovo

INTRODUCTION

Presence of power plants, heavy traffic, high density of population etc., in Prishtina region, is the main sources of SO_2, NO_{2x} and dust. The power plant "Kosova A and B" have a total effective capacity 1355 MW. "Kosova B" discharges approximately *260 g/s* SO_2, *556 g/s* NO_x *and 376 g/s dust*. Only one unit of Kosova B with capacity 200MW emits about 25 tones of dust and ash per hour, which is 74 times more than the EU standards for dust pollution (WHO 2002). This pollution becomes more severe in regard of traffic pollution and mineral exploitation. All these pollutants cause the chemical and physical changes in tree bark, which is a natural habitat of epiphytic lichens.

Lichens are highly sensitive in air pollution and have been identified as indicators of environmental quality, particularly air quality, since as early (Nylander, 1866), latter many such investigations was done in different industrial areas (Pejćinović & Hoxha, 1986; Sommerfeldt & Volker, 2001). Lichens are considered to be the best plants to use as biomonitors of air pollution because, these species obtain nutrients from wet and dry deposition, they do not have real roots and cuticle is lacking or reduced.

The aim, of this study has been to investigate lichen value diversity, and its correlation with bark pH and electrical conductivity around power plants area.

MATERIAL AND METHODS

The study region expands 150 km^2 around power plants "Kosova A and B" located on the NW of Prishtina. Epiphytic lichens were investigated on the species of the genera: *Populus and Quercus*. The standardized EU lichen recording method was used to determinate the lichen value diversity (Asta et al., 2002). The sampling units were selected in different sites around power plants to investigate the effects of environmental stress on lichen. This sampling strategy was designed to record variation of epiphytic lichen distribution. Most of the sites are concentrated near the power plants, where most changes was expected and in SE, S and SW corresponding to the wind direction (Tab. 1). Within the study region 20 sampling sites were randomly selected. Sampling was done exclusively in trees out the woodlands. The size of sample unit was 0.5 × 0.5 km. All sites were marked using GPS (Garmine-gekos 201). Sampling units were selected based on presence of suitable trees and no road disturbance. In six sampling units LVD, pH and electrical conductivity were determined on the oak tree and in fourteen sampling units d on the popular species. Only trees of a single species were selected within a survey unit to determinate LVD.

In the same localities where were determined the LVD, also was collected the bark of trees. On each sampling site on 10 samples of bark were taken. Bark tree was grinded and used to determine the pH and electrical conductivity. 1.5 grams of finely milled bark was mixed with 15 ml of deionised water and left to stand for 24 hours. Conductivity and pH were measured with a combined pH and conductivity meter (Hana pH and CD meter).

RESULTS

To investigate the impact of substrate changes on lichen value diversity, bark pH and electrical conductivity was determined in area around power plants Kosovo A and B. Measurements were done in 20 localities. During the investigation for determination of the lichen value diversity in Prishtina region 23 epiphytic lichen species were registered. Most of them were foliose species (11), 7 were crustose, whereas 5 fruticose species.

In most cases lichen value diversity was lower near the pollution sources. This value was increased toward the increased distance from pollution sources. Higher lichen value diversity was recorded in the east and west of pollution sources, where sites have rich vegetation, whereas lower values were recorded near power plants and main roads which were characterized with higher level of air pollution and less vegetation.

The lichen value diversity has been considerable influenced by the nature of the substrate on which lichens was developed. The highest, values of pH were measured, in localities with the highest lichen value diversity and vice versa. This value, was also dependent from the trunk of tree species, hence the lowest value of the correlation between pH and LVD was measured on the trunks of genera *Populus* ($R = 0.1$), whereas, the highest value of correlation was measured on the trunks from genera *Quercus* ($R = 0.8$; $p > 0.05$). It is well-known that the pH of bark has an influence on epiphytic lichen. (Türk & Wirth 1974).

Regarding, to bark electrical conductivity the highest value was recorded in localities with higher level of air pollution and vice versa. The results, show that the highest negative correlation between electrical conductivity and lichen value diversity was in *Quercus* species (R = –0.4) whereas, the value of this correlation in *Populus* species was R = –0.3.The negative correlation between bark electrical conductivity and lichen value diversity was recorded from other authors too (Poikolainen, 2004; Lippo et al. 1995; Lötschert & Köhm, 1978). It is also recognized that rougher bark as in case of *Quercus* trees provided a better habitat for the development of a wider spectrum of lichen species (Brodekova et al., 2006).

Table 1 Main characteristics of the samples sites, each locality site is identified by number, name, coordinate (UTM, zone 33), altitude, number of lichen species, number of lichen species found per locality, pH, electrical conductivity, lichen values diversity.

No.	*Localities*	*Longitude*	*Latitude*	*Altitude*	*Orientation*	*Nr. of species*	*pH*	*Elect. condu. µs/cm*	*LVD*
1	Obiliq	505504	4725866	544	CENTER	3	5.8	857.0	41.6
2	Shkabaj	508754	4723802	524	E	5	6.8	696.3	49.4
3	Dragodan	512423	4724164	684	E	7	5.7	582.3	58.7
4	Taukbashqe	514925	4723669	556	E	5	7.0	632.0	64.7
5	Mramor	524571	4722231	749	E	8	6.3	354.1	96.2
6	Lagja e spitalit	513695	4721458	618	SE	5	6.1	1086.3	59.2
7	Graqanic	517397	4716609	593	SE	5	4.9	959.0	78.1
8	Slivovë	527992	4715081	785	SE	14	6.4	294.6	115.4
9	Prelez	514421	4700887	575	S	3	6.2	547.3	75.5
10	Ferizaj	515127	4690707	552	S	4	5.0	332.5	56.8
11	Lipjan	509178	4707023	562	SSW	5	6.6	473.3	70.0
12	Magure	501126	4710160	590	SWW	4	5.4	406.3	79.2
13	Kroi i mbaretit	495762	4716647	627	SW	6	9.5	712.6	97.4
14	Sllatin	500190	4715930	510	SWW	4	6.6	473.3	77.6
15	Lismir	505909	4724004	544	W	3	9.2	945.3	36.2
16	Grabovc	499305	4723078	570	W	8	5.0	564.0	64.6
17	Drenas	492142	4720304	596	W	9	8.5	469.4	93.1
18	Vushtrri	499684	4738598	541	NW	4	5.4	839.6	76.8
19	Prugovc	504977	4726742	543	N	2	10.3	568.0	72.0
20	Bardhosh	512264	4729313	628	NE	5	5.9	645.2	70.5

REFERENCES

1 Asta J, Erhardt W, Ferretti M, Fornasier F, Kirschbaum U and Nimis P, et al. Mapping lichen diversity as an indicator of environmental quality. NATO Science Series. IV. Vol. 7. 273–279 (2002).
2 Brodekova L, Gilmer A, Dowding P, Fox H and Guttova A. An assisment of epiphtic lichen diversity and environmental quality in Knocksink woodnature reserve. Irland. *Biology abd environment. Royal Irish Academy.* 106 B. 3. 215–223 (2006).
3 Lippo H, Poikolainen J and Kubin E. The use of moss lichen and pine bark in the nationwide monitoring of atmospheric heavy metal deposition in Finland. *Water. Air. and Soil Pollution* 85. 2241–2246 (1995).
4 Lötschert W and Köhm J. Characteristics of tree bark as an indicator in high immission areas II. Contents of heavy metals. *Oecologia* 37. 121–132 (1978).
5 Poikolainen J. Epiphytic lichens and tree bark as biomonitors for airpollutants – specifically for heavy metals in regional surveys. Oulu: Oulu University Press. (2004).
6 Sommerfeldt M and Volker J. Evaluation of a method for the Reassessment of air quality by Lichen mapping in the city of Izmir. Türkey. *Turk J. Bot.* 25. 45–55 (2001).
7 WHO & UNEP, Urban Air Pollution in Megacities of the World. Blackwell. London. (1992).

Chapter 46

MOSSES AND LICHENS AS INDICATOR OF AIR QUALITY AROUND POWER PLANTS KOSOVA

A. Hajdari, B. Mustafa & Esat Hoxha
Department of Biology, Faculty of Mathematical and Natural Science, University of Prishtina, Prishtinë, Kosovo

INTRODUCTION

The power plant "Kosova A and Kosova B" have a total effective capacity 1355 MW, only "Kosova B" discharge approximately *260 g/s* SO_2, *556 g/s* NO_x *and 376 g/s dust.* Chemical analysis of the dust particles which originated from power plant contains different chemical elements including radionucleotides Pb^{214}, Ti^{218}, Cs^{137}, U^{235} Adrović, et al. (1996) and toxic compounds as nitrogen oxides, SO_2, As, Pb, Cd, Ni, Cr, Co etc., which have negative impact on the living organisms.

Lichens are highly sensitive in air pollution and have been identified as indicators of environmental quality, particularly air quality, since as early (Nylander, 1866), latter many investigations was done (Pejćinović & Hoxha, 1986; Seaward M., 2003; Loppi S., 2004).

In other hand mosses are highly efficient accumulators of persistent pollutants, such as heavy metals. Mosses and lichens are considered to be the best plants for use as biomonitors of air pollutants (Rühling & Tyler 1968) because, these species do not have real roots, cuticle is lacking or reduced, their leaves are only a one cell-thick-layer, obtain nutrients as well as heavy metals mainly from wet and dry depositions.

The aim of this study was to determinate the lichen value diversity and its correlation with heavy metal deposition on the mosses around power plants area.

MATERIALS AND METHODS

The study region expands 150 km^2 around power plants "Kosova A and B" located on the NW of Prishtina. Epiphytic lichens were investigated on the tree species of the genera: *Populus and Quercus*. The standardized EU lichen recording method was used to determinate the lichen value diversity (Asta et al., 2002). The sampling units were selected in different sites around power plants to investigate the effects of environmental stress on lichen. This sampling strategy was designed to record variation of epiphytic lichen distribution. Most of the sites are concentrated near the power plants, where most changes was expected and in SE, S and SW corresponding to the wind direction (Table 1). Within the study region 20 sampling sites were randomly selected. Sampling was done exclusively in trees out the woodlands. The size of sample unit was 0.5 × 0.5 km. All sites were marked using GPS (Garmine-gekos 201). Sampling units were selected based on presence of suitable trees and no road disturbance. In six sampling units LVD were determinate on the oak tree and in fourteen sampling units on the popular species. Only trees of a single species were selected within a survey unit to determinate LVD.

Table 1 Main characteristics of the samples sites, name, coordinate (UTM, zone 33), altitude, lichen diversity values and concentration of heavy metals.

No.	*Localities*	*Longitude*	*Latitude*	*Altitude*	*LVD*	*Cd*	*Pb*	*Cu*	*Zn*	*Co*	*Ni*
1	Obiliq	505504	4725866	544	41.6	0.62	19.00	15.40	66.33	7.60	4.50
2	Shkabaj	508754	4723802	524	49.4	0.41	5.87	15.67	64.56	1.19	6.87
3	Dragodan	512423	4724164	684	58.7	0.35	21.73	19.73	69.89	8.61	4.12
4	Taukbashqe	514925	4723669	556	64.7	0.45	19.20	27.00	90.33	7.53	5.50
5	Mramor	524571	4722231	749	96.2	0.76	19.60	25.50	73.00	14.72	1.28
6	Lagja e spitalit	513695	4721458	618	59.2	0.34	17.90	17.87	80.22	6.73	3.71
7	Graqanic	517397	4716609	593	78.1	0.40	66.36	25.04	136.47	2.59	4.19
8	Slivovë	527992	4715081	785	115.4	0.86	13.40	16.20	81.17	1.20	1.94
9	Prelez	514421	4700887	575	75.5	0.34	20.50	18.00	84.22	3.55	3.11
10	Ferizaj	515127	4690707	552	56.8	0.34	14.00	14.80	65.33	8.77	1.55
11	Lipjan	509178	4707023	562	70.0	0.46	30.00	20.20	94.67	2.13	3.06
12	Magure	501126	4710160	590	79.2	1.69	18.00	10.10	89.00	9.36	10.46
13	Kroi i mbaretit	495762	4716647	627	97.4	0.49	19.20	12.33	76.67	14.07	5.89
14	Sllatin	500190	4715930	510	77.6	0.66	9.25	10.45	56.50	11.5	6.87
15	Lismir	505909	4724004	544	36.2	0.33	2.10	16.2	58.16	1.06	2.15
16	Grabovc	499305	4723078	570	64.6	0.63	21.12	26.96	69.67	0.60	3.29
17	Drenas	492142	4720304	596	93.1	0.55	44.20	17.10	84.33	8.53	3.92
18	Vushtrri	499684	4738598	541	76.8	1.23	42.93	17.60	99.67	4.32	5.37
19	Prugovc	504977	4726742	543	72.0	0.46	26.80	22.40	84.00	6.45	10.59
20	Bardhosh	512264	4729313	628	70.5	0.32	10.33	21.07	83.00	4.87	3.11

In the same localities where was determined the LVD, also was collected mosses (*Brachytecium rutabulum (Hedw.) Schimp*), to measure the heavy metals concentration. The moss samples were collected in sites which were located at least 300 m from the main roads, at least 100 m from any road or single house and 3 m away nearest tree. On each sampling site on average 3 subsamples were taken within an area of 50 m × 50 m. The samples were dried at temperature 40°C. One gram of mosses from each sample was burn in the temperature 450°C for three hours. Heavy metals concentration (Cd, Pb, Cu, Zn, Co and Ni) were determined with standard methods using AAS – atomic absorption spectroscopy (UNICAM 929 spectrometer). The concentrations of heavy metals were expressed in µg/g (micrograms per gram) of dry weight.

RESULTS

During the investigation for determination the lichen value diversity around power plants "Kosova A and B" were registered 23 epiphytic lichen species. Most of them were foliose species (11), 7 were crustose, whereas and 5 fruticose species. There was a different value of the LVD depending from the distance of source of air pollution, the highest levels of LVD were in the cleanest area, but this value is decreased on moving towards to the polluted areas (Table 1).

HEAVY METAL CONCENTRATION

Concentrations of hevy metals (Zn, Pb, Cu, Co, Ni, Cd) measured in the different moss samples varied depending from distance of sites from pollution sources. The maximal concentrations were found at the sites surrounding power plants and mines whereas, in these sites the lichen value diversity had minimal value. From heavy metal higher concentration had Zn 56.5–136.6 $\mu g \cdot g^{-1}$ and this value had decease as follow: Pb: 2.10–66.3 $\mu g \cdot g^{-1}$; Cu 10.10–27.00 $\mu g \cdot g^{-1}$; Co 0.60–14.72 $\mu g \cdot g^{-1}$; Ni: 1.29–10.59 $\mu g \cdot g^{-1}$; Cd 0.32–1.69 $\mu g \cdot g^{-1}$; the higher concentration of heavy metals was found in localities near pollution sources (localities Graçanic near Hajvalia mine where is exploited Pb, Zn, Ag, Au etc., Vushtrri near Stanterg mine Pb, Zn, Ag, Au etc., Drenas near ferronickel factory Table 1).

CORRELATIONS OF THE LVD WITH HEAVY METAL CONCENTRATION AND COORRELATION BETWEEN HEAVY METALS IN MOSSES

In the Table 2 is given the correlation between lichen value diversity and heavy metals, and correlation between heavy metals. The LDV had highest positive

Table 2 Correlations of the LVD with heavy metal concentration in mosses.

	LVD					
Cd	0.362	Cd				
Pb	0.296	0.100	Pb			
Cu	−0.017	−0.326	0.355	Cu		
Zn	0.177	−0.459(*)	0.682(**)	0.584(**)	Zn	
Co	0.300	0.165	−0.069	−0.249	−0.292	Co
Ni	−0.015	0.422	0.046	−0.283	−0.329	0.182

*Correlation is significant at the 0.05 level (2-tailed).
**Correlation is significant at the 0.01 level (2-tailed).

correlation with cadmium, cobalt and lead whereas, the negative correlation had with cupper and nickel but neither hadn't significant correlation. Correlation between heavy metal concentrations was between zinc-lead and zinc-cupper which have significance at the level ($p > 0.01$) whereas, zinc-cadmium had negative correlation ($p > 0.05$).

REFERENCES

1 Adrović F, Popović R and Ninković M, The gama doze rates of radiation in the air and closer and further surrounding of the Kosovian Coal Power Plants. *Univ. Thought. Nat. sci.* 1996:87–90 (1996).
2 Asta J, Erhardt W, Ferretti M, Fornasier F, Kirschbaum U and Nimis P, et al. Mapping lichen diversity as an indicator of environmental quality. NATO Science Series, IV, 273–279 (2002).
3 Loppi S, Mapping the effects of air pollution, nitrogen deposition, agriculture and dust by the diversity of epipytic lichens in central Italy. *English Nature* nr. 525, 37–41 (2004).
4 Nylander W, Les lichenes du Jardin du Lixemburg. *Bull Soc. Bot Tidskr* Fr.13: 364–372 (1866).
5 Pejćinović D and Hoxha E, Liśajevi kao indikatori stepena zagađenosti živote sredine. Prishtinë: Natyra e Kosovës, Enti Krahinor për Mbrojtjen e Natyrës. (1986).
6 Rühling A, Atmospheric heavy metal deposition in Europe – estimations based on moss analyses. *NORD*, 9, 1–53 (1994).
7 Seaward R and Coppins J, Lichens and hypertrophication. *Bibliotheca Lichenologica* 88, 561–572 (2004).

Chapter 47

THERMODYNAMIC MODELING FOR HYDROGEN PRODUCTION FROM BIOMASS AND EVALUATION OF BIOMASS ENERGY TECHNOLOGIES

Sh. Hemmati, Y. Saboohi & N. Hashemi
Department of Energy Engineering, Sharif University of Technology, Tehran, Iran

M. Vossoughi & G.R. Pazuki
Department of Chemical and Petroleum Engineering, Sharif University of Technology, Tehran, Iran

ABSTRACT

Compared with fossil fuel, biomass is a clean energy with zero CO_2 emission, because CO_2 is fixed by photosynthesis during biomass growth and released again during utilization. Due to its low energy density, direct use of biomass is not convenient. Thus, it is necessary to convert biomass to fuel gas, such as hydrogen, which can be used cleanly and highly efficiently in fuel cell. Thermo-chemical gasification is likely to be the most cost-effective conversion process and it is promising technology for renewable hydrogen production by utilizing biomass. Biomass gasification produces a mixture of gases (mainly consisting of H_2, CO, CO_2, CH_4 and higher hydrocarbons), solids (char) and liquids (aromatic hydrocarbons) known as 'tars'. The relative proportion of each constituent depends on the operating conditions such as temperature, pressure, type of gasifying medium, biomass type; biomass feed rate, heating rate, flow rate of gasifying medium, physical characteristics of biomass such as particle size, shape, surface area to volume ratio and the gasifier design.

So far various experimental investigations into gasification of biomass carried out, but the work on thermodynamic analysis is relatively limited. Thermodynamic analysis is very helpful in providing theoretical guidance for optimization of design and operation of biomass gasification system. So, in this work a thermodynamic equilibrium model was used to predict the chemical composition of the products of biomass gasification. Gibbs energy minimization approach was used to determine the product gas composition. The code of MATLAB software was used. Gasifier, the most critical component of any biomass gasification system, was modeled as an equilibrium reactor. Equilibrium calculations were compared with experimental data from literature. Finally investigation of evaluation of biomass energy technologies carried out.

Keywords: Biomass, Hydrogen, Energy, Thermodynamic, Modeling

INTRODUCTION

Dependence on fossil fuels as the main energy sources has led to serious energy crisis and environmental problems, i.e. fossil fuel depletion and pollutant emission. The increasing energy demands will speed up the exhaustion of the finite fossil fuel. Moreover, combustion of fossil fuel produces substantial greenhouse and toxic gases, such as CO_2, SO_2, NO_x and other pollutants, causing global warming and acid rain [1].

Compared with fossil fuel, biomass is a clean energy with zero CO_2 emission. Hydrogen production plays a very important role in the development of hydrogen economy [2].

BIOMASS

Biomass is the fourth largest source of energy in the world, accounting for about 15% of the world's primary energy consumption and about 38% of the primary energy consumption in developing countries. Biomasses consist of all the living material present on earth. Biomass may vary significantly in its physical and chemical properties due to different variety and diversified origin. Usual composition of biomass comprises cellulose, hemi-cellulose, lignin, extractives and inert ash [3].

HYDROGEN PRODUCTION FROM BIOMASS

A large number of different processes are being developed for the production of hydrogen from biomass. Figure 1 shows pathways from biomass to hydrogen storable intermediates [4, 5].

THERMODYNAMIC MODELING AND ANALYSIS OF BIOMASS GASIFICATION FOR HYDROGEN PRODUCTION IN SUPERCRITICAL WATER

Thermo-chemical gasification is likely to be the most cost-effective conversion process. However, a large portion of biomass is wet, and this causes high drying costs in classical thermo-chemical gasification process. Biomass gasification in supercritical water (SCW) is a promising technology for the utilization of wet biomass [6]. Figure 2 shows the simulated system for biomass gasification in SCW.

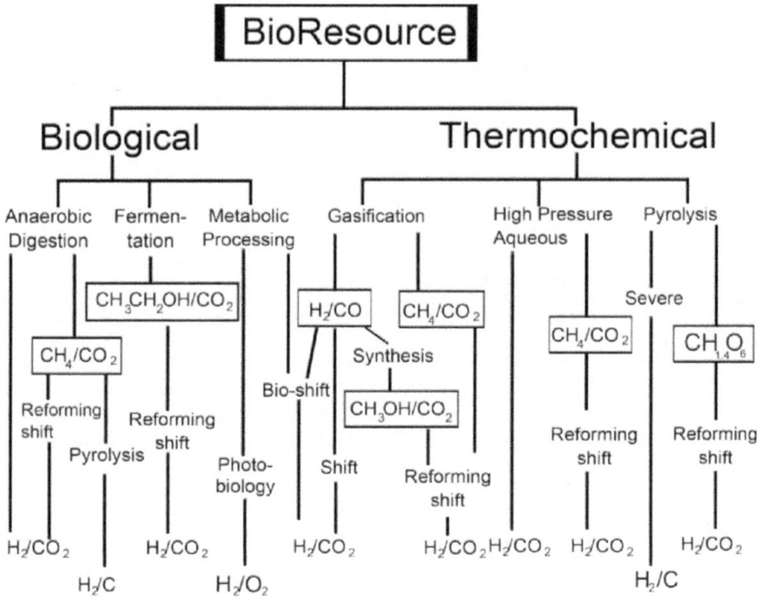

Figure 1 Pathways from biomass to hydrogen storable intermediates.

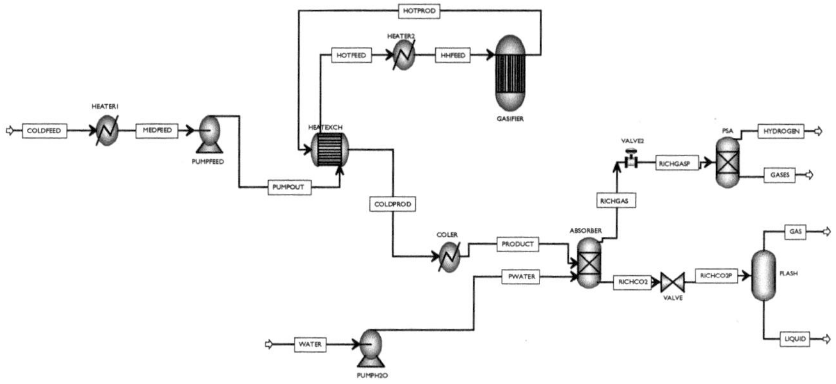

Figure 2 The schematic of simulated system for biomass gasification in SCW.

Thermodynamic models for chemical equilibrium in the reactor were conducted. Gibbs free energy of a system, with fixed T and P, with PR equation of state used [7].

RESULTS OF THERMODYNAMIC MODELING

The product gas was considered to include mainly H_2, CH_4, CO and CO_2 in dry basis. Figure 3 shows the mol number of these four components from glucose gasification in SCW. We compared our model predictions to the work of Boukis et al [8].

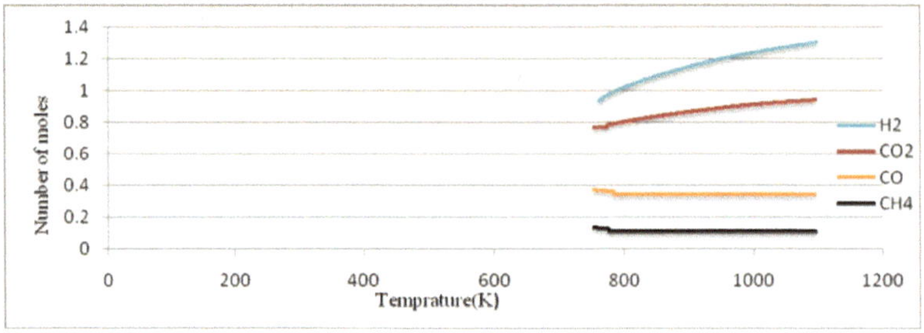

Figure 3 The mol number of four components from glucose gasification in SCW.

EVALUATION OF BIOMASS ENERGY TECHNOLOGIES

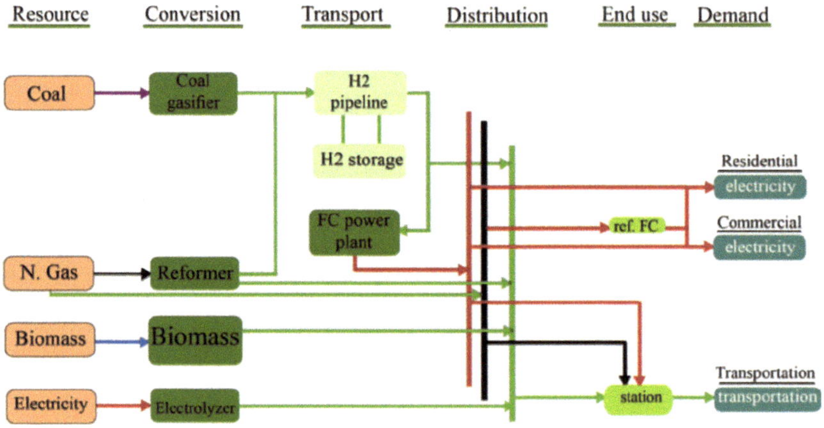

Figure 4 Hydrogen supply flow diagram [9].

ACKNOWLEDGMENTS

This work is supported by the Iranian Fuel Conservation Organization.

REFERENCES

1 Meng Ni, Dennis Y.C. Leung, Michael K.H. Leung, and K. Sumathy. An overview of hydrogen production from biomass, *Fuel Processing Technology* 87 (2006) 461–472.
2 Youjun Lu, Liejin Guo, Ximin Zhang, and Qiuhui Yan. Thermodynamic modeling and analysis of biomass gasification for hydrogen production in supercritical water, *Chemical Engineering Journal* 131 (2007) 233–244.
3 Donald L. Klass, *Biomass for Renewable Energy and Fuels*. Entech International, Inc. Barrington, Illinois, United States.

4 Thomas A. Milne, Carolyn C. Elam and Robert J. Evans. Hydrogen from Biomass State of the Art and Research Challenges, National Renewable Energy Laboratory Golden, CO, USA. 2001.
5 Hemmes K, Degroot A, and Demuil H. Bio H_2, application potential of biomass related hydrogen production technologies to the Dutch energy infrastructure of 2020–2050.
6 Calzavara Y, Joussot-Dubien C, Boissonnet G and Sarrade S. Evaluation of biomass gasification in supercritical water process for hydrogen production, *Energy Conversion and Management* 46 (2005) 615–631.
7 Huiqing Tang and Kuniyuki Kitagawa. Supercritical water gasification of biomass: thermodynamic analysis with direct Gibbs free energy minimization, *Chemical Engineering Journal* 106 (2005) 261–267.
8 Boukis N, Diem V, Habicut W and Dinjus E. Methanol reforming insupercritical water, *Ind. Eng. Chem. Res.* 42 (2003) 728–735.
9 Meysam Qadrdan, Yadollah Saboohi and Jalal Shayegan. A model for investigation of optimal hydrogen pathway, and evaluation of environmental impacts of hydrogen supply system, *International Journal of Hydrogen Energy* 33 (2008) 7314–7325.

Chapter 48

IMPROVED ODOUR CONTROL WITH FULL-SCALE MULTI-LAYERED BIOLOGICAL TREATMENT METHODS

J. Cesca
CH2MHill, Chadswood NSW, Australia

N.J.R. Kraakman
BIOWAY, Manly East NSW, Australia

INTRODUCTION

Melbourne Water installed the largest biotrickling filter system in the southern hemisphere (at the time) at the Western Treatment Plant in 2006, treating 126,000 m^3/h of odorous air from the Western Trunk Sewer which conveys over half of Melbourne's sewage to the wastewater treatment plant.

Extensive Process Proving Trials conducted on-site found that the multi-stage biotrickling filter system (BTF) exceeded its design performance requirements. In more recent times hydrogen sulphide concentration has increased by about 50% from original design requirements due to increased sewage concentration and travel time resulting from water conservation activities. Despite this increase in hydrogen sulphide loading, the installation still meets or exceeds its original performance requirements. The odour control facility has proved to be robust and has coped well with altered process conditions and requires minimal maintenance.

The biological technology was determined to be the most appropriate for this site due to the high H_2S load as well as a number of other economic, social and environmental considerations. The system with a multi-stage tower design, including structured inert media, is guaranteed for 20 years, leading to much lower operating costs in comparison to conventional biofilters. It also has a significantly smaller footprint. Although the capital cost of BTF is about double that of a comparable chemical scrubber installation, they did not require expensive chemicals for treatment. Based on the estimated H_2S load, the payback period for the extra capital cost was less than three years. The BTF is also a more sustainable alternative as they only require water and a small amount of nutrient for effective treatment.

METHODS

The odour treatment facility comprises 10 biotowers (PURSPRING™) designed by BIOWAY and constructed by a joint venture between United Group Infrastructure and Bioway Technologies Australia Pty Ltd in 2006. The biological odour treatment system uses six different layers to provide the different environmental conditions for the mix of micro-organisms. The primary control system philosophy of the odour treatment facility is to provide the ability to operate the plant unmanned and automatically under normal operation, with supervisory control provided from the Main Operator Control Room.

The Process Proving Trial was undertaken during a four week period. The air sampling took place over a four (4) week period, with odour samples taken at six (6) day intervals to capture different daily trends. Five (5) inlet and outlet odour samples were taken on each day resulting in a total of ten (10) inlet and ten (10) outlet odour samples.

Continuous inlet and outlet H_2S measurements were also taken throughout this period as well as number of other operational parameters as outlined below.

An online H_2S transmitter (Drager, type Polytron 7000 H_2S detector) and an H_2S datalogger (Odalog, type 0–200 ppm_v) at the inlet sampling ports measured inlet foul air H_2S concentrations to the OCF.

Outlet treated air H_2S concentrations from the system were measured by an online H_2S transmitter (Honeywell, type MDA Model SPM "Chemcassette" analyser), and an H_2S datalogger (Odalog, type 0–2 ppm_v) installed at the outlet sampling ports.

Odour concentrations were measured by collecting bag samples and analysing them in the NATA registered laboratory EML in accordance with

Table 1 Design specification and results of Process Proving Trial.

Parameter	*Design specifications*	*Process Proving Trial Results*
Inlet Odour Concentration (OU)	6,000–76,000	29,950 18,000–60,000
Outlet Odour Concentration (OU) – average and range	<2000*	460 170–1,900
Odour Removal Efficiency (%)	96%	98.5%
Inlet H_2S Concentration, (ppm_v) average and range	6–45.9	0–41.3
Outlet H_2S Concentration, (ppm_v) – average and range	<0.5**	0.010 0–0.080
H_2S Removal Efficiency (%)	99.5%	>99.6%

* Readings shall be measured in accordance with the procedures as specified in the Australian Standard.
** To apply to at least 99.5% of all H_2S readings, measured as the average of any ten consecutive readings taken at two minute intervals.

the requirements of the AS/NZS method 4323.3.2001. This standard is the Australian equivalent of the international standard method EN13725. Samples from the inlet and outlet airstream were taken simultaneously.

RESULTS AND DISCUSSION

A summary of the Odour Treatment Facility performance in comparison to the original design specifications are presented in Table 1 below.

No major failures or process upsets were recorded during the Process Proving Trial, with stable operation recorded. Minor process upsets experienced

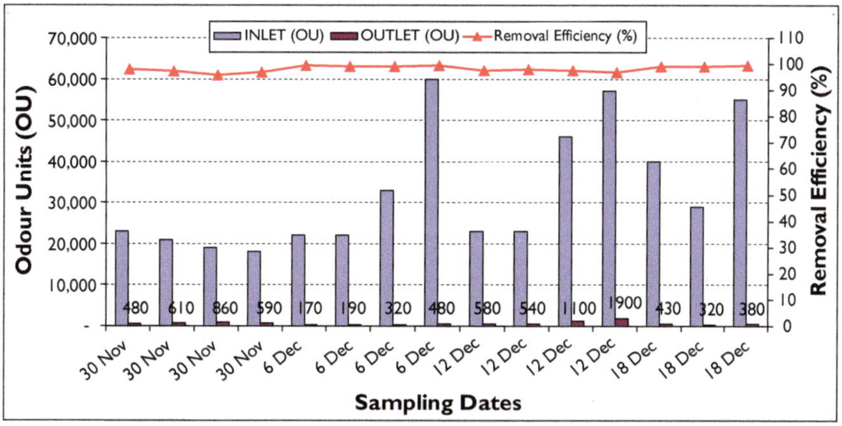

Figure 1 Odour sampling data.

Figure 2 Biological odour control system with ten (10) Bioway™ multi-layered biotowers.

prior to or during the Process Proving Trial itself did not influence the operating efficiency of the odour treatment system.

CONCLUSION

After more than 3 years of operation the Odour Control Facility still meets or exceeds its performance requirements. This is despite the increase in sewer hydrogen sulphide levels since its design. The odour control facility has proved to be robust and has coped well with altered process conditions, and requires minimal operator attention and maintenance.

Chapter 49

THE MBR FOR BIOLOGICAL DESULFURIZATION OF BIOGAS:
Miniaturizing the BioReactor with the DMT BioSulfurex® HCP process

R. Lems & E.H.M. Dirkse
DMT, Joure, The Netherlands

ABSTRACT

DMT has developed a process for biological desulfurization at low pH for the treatment of biogas. This process (BioSulfurex®) is pH, temperature, nutrient and oxygen controlled. During an intensive research period the elimination capacity (EC) and removal efficiencies (RE) at different process conditions were determined. The effect of variations in H_2S loading, cleaning procedures and operational stops are demonstrated with field data. The next step in the development of biological desulphurisation of biogas is the introduction of the DMT-HS packing, introducing EC's of 500 $gH_2S/m^3/h$ or more. This system is self-cleaning, produces mainly solid elemental Sulphur, has no chemical need and no or minimum supply water requirement.

Keywords: Biological desulfurization, H_2S oxidation, DMT-BioSulfurex®, H_2S, biogas, EC, Elimination Capacity

INTRODUCTION

Hydrogen Sulfide (H_2S) has been a problematic gas compound for centuries. In biogas the acid gas can cause corrosion of f.e. piping and engines, preventing gas utilization without pre-treatment. Until recent the removal of H_2S is usually done by a chemical system due to the ultra high concentrations of H_2S (up to 5%v/v). This leads to acidification and/or clogging of bio- (trickling) filters. Chemical systems are also thought to be much more reliable and they occupy a much smaller footprint. But with the development of sophisticated plastic materials as bacterial carrier material for bioreactors it is now also feasible to biologically treat polluted (bio)gas stream with high concentrations of H_2S.[1] In the near future biological treatment may even surpass chemical treatment of H_2S as most favorable process.

THEORETICAL BACK GROUND

BIO-REACTOR FOR BIOGAS DESULFURIZATION

DMT developed the Bio-Sulfurex® for biological removal of H_2S from biogas. This is a bio-trickling filter where water circulates over an inert packing material (Pall Rings) with specific micro-organisms. The packing material has a large surface area at which the micro-organisms can attach but which is also coarse enough to prevent clogging. In figure 1, a process flow diagram is shown of the Bio-Sulfurex®. The polluted gas is mixed with small amounts of air containing the oxygen for the biological oxidation of the H_2S. The H_2S is degraded by the bacteria *Acidothiobacillus Thiooxidans*. These bacteria convert H_2S into sulphuric acid at sufficient oxygen concentrations and even at very low pH values (equations 1 and 2).

$$H_2S + \tfrac{1}{2}O_2 \rightarrow S + 2H_2O \qquad (1)$$

$$S + 1\tfrac{1}{2}O_2 + H_2O \rightarrow H_2SO_4 \qquad (2)$$

At low oxygen concentration the H_2S is mainly converted to elemental sulphur[2] (equation 1) which is washed out with the circulation water. A minimum of fresh water is added to assure a very low pH which favours only the growth of *Acido Thiobacillus*. The micro-organisms are kept at the optimal process condition[3,4] by temperature regulation and if necessary, nutrients are added with the supply water.

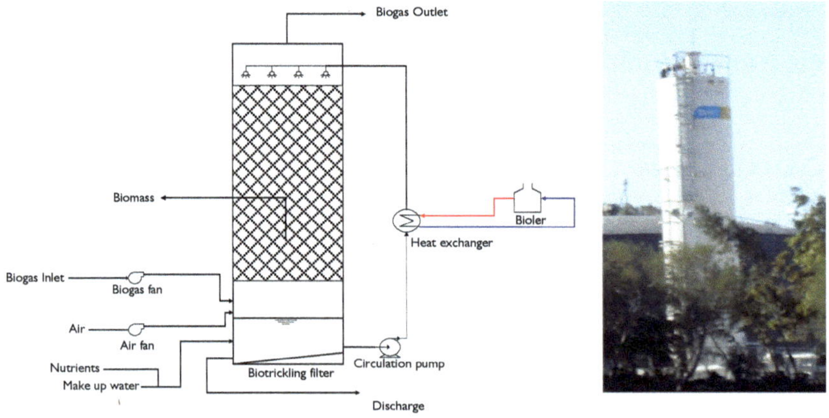

Figure 1 Left; PFD, Right; photo of the BioSulfurex.

The first bio-reactors for biogas were built with an EC of 25–40 gH_2S/m^3*h. DMT has increased the bacteriological activity by a strict control and optimization of the process parameters. The newest desulphurization systems are highly sophisticated biological scrubbers where pH, temperature, nutrient and oxygen are controlled. In this way elimination capacities of 80–100 gH_2S/m^3*h can be reached with efficiencies of >97–99%. By control of the oxygen concentration, the ratio between conversion of H_2S to sulfuric acid and elemental sulfur can be predefined. Nutrient dosing is calculated and monitored to reach the maximum EC without inducing excessive bacterial growth.

PRACTICAL DATA

MAXIMIZING THE EC

Two installations (located in Israel and Greece) have been extensively tested and monitored.[5,6] These reactors are about 4 meter in diameter and 14 meter in total height (see figure 1; right). They are designed to treat up to 2000 m^3/hr biogas with 2000–3500 ppm of H_2S. The results for the start-up of the reactor in Greece are displace in figure 2 (left). The different optimizing periods are clearly visible. The first start-up (1) was at low temperatures with no nutrients which resulted in a low EC (10 $gH_2S/m^3/h$). The second period (2) was at higher temperatures increasing the EC to approximately 15 $gH_2S/m^3/h$. In the third (3) period the EC dropped slightly due to maintenance of the gas engines. The bio-reactor was only in operation for 1 to 2 hours per day. In the fourth (4) period, nutrients (ammonium-phosphate) were added to the system. This resulted in an substantial increase in the EC (>45 $gH_2S/m^3/h$) even though the plant was still running 1 to 2 hours per day during the first days of this period. During the fourth period the removal efficiency reached >97% but the H_2S load decreased. This is also why the EC decreased. The maximum EC observed was 55 $gH_2S/m^3/h$.

The plant in Israel has been started with optimal process conditions from the start and was inoculated with sludge from a WWTP (See Figure 2; Right).

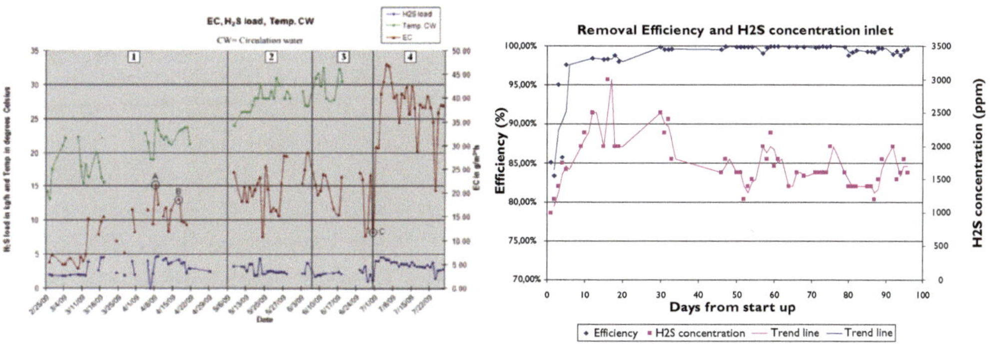

Figure 2 Start-up of the BioSulfurex®, Left; Greece, Right; Israel.

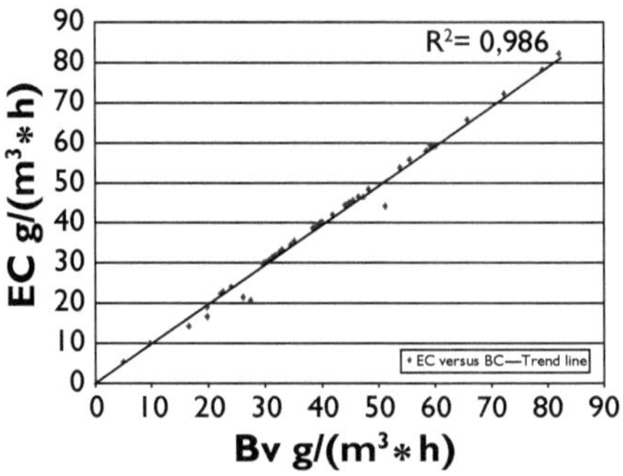

Figure 3 EC/Bv graph.

The temperatures was about 30–35°C, the pH ranged from 1.2 till 2.5, the oxygen concentration was 2–2.5% and there were more than enough nutrients. After 1 month almost 100% removal efficiency has been reached. Additional testing with variation in pH (0.5–3.5) and temperature (28–41°C) showed no significant change in removal efficiency and EC. The oxygen concentration (1–3%v/v) also had a minimum effect of the EC but did decrease the required water supply. This is due to extra formation of elemental sulfur and less sulphuric acid.

The plant in Israel has been running at different H_2S loadings (Bv in g/m³/h) due to variations in flow and H_2S concentration which resulted in the EC/Bv plot (figure 3). The straight line indicates that at the maximum observed EC (observed as 81.5 gH_2S/m³/h.) still >99% RE is possible and that this is not maximum possible EC. At higher loading also higher ECs are expected.

PRACTICAL SYSTEM OPERATION

VARIATIONS AND STOPS

In general the bioreactor is though to be very sensitive to variations in H_2S load (flow × concentration) and (maintenance) stops. Practical data however has proven the opposite for the BioSulfurex® system (see figure 4). After an overnight stops of 16 hrs the bioreactor was back at full operation within 5 minutes (H_2S <20 ppm). After a 3 day stop it took only 30 minutes to reach complete removal efficiency again (>99%). Also variations in H_2S loading have only a small effect on the total removal efficiency. Both an 65% increase in load within 30 h and an increase of almost 300% in about 1 week did not effect the efficiency significant. The H_2S outlet stayed below 50 ppm. (H_2S inlet ±1500–2000 ppm)

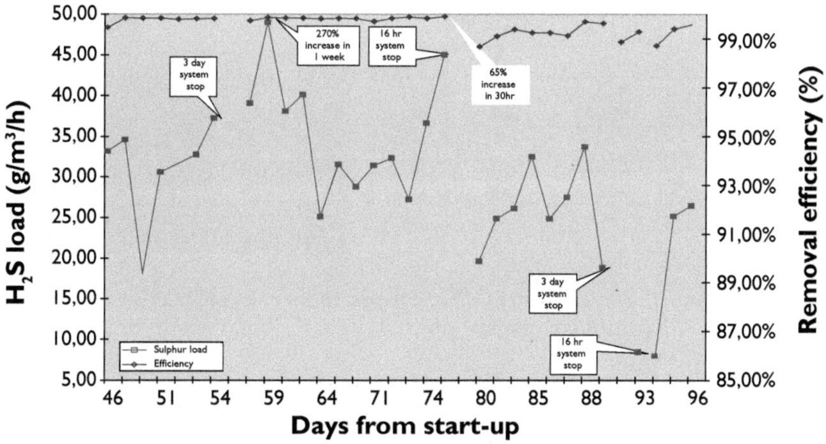

Figure 4 Effect of system stops and H2S loading on the BioSulfurex® performance.

SLUDGE GROWTH (FOULING) AND CLEANING

The second disadvantage of a bioreactor is supposed to be the frequent problems with excessive sludge growth and resulting difficult cleaning actions. However these problems usually arise by insufficient process control which leads to an increase in a large amount of different fast growing bacteria populations while *Acidothiobacillus Th.* is a very slow growing bacteria as most autotrophic bacteria. By control over the pH, nutrients and the oxygen concentration in combination with the high H_2S concentrations a culture of >95% pure *Thiooxidans* can be maintained in the bioreactor which results in minimal sludge formation. To further minimize the growth of *Thiooxidans* it is important to keep the temperature just above the optimal temperature at 35–38°C and add just enough or below the optimum amount of nutrients (±200:10:1/S:N:P). Excess sludge (fouling) can also be caused by solid sulphur. By a precise control over the oxygen concentration the optimum ratio between sulfuric acid (needs water supply for dilution) and solid sulphur (creates sludge) can be retained. The sludge which is formed on the contact media is about 85% organic and 15% sulphur and it easily removed from the contact media (see below for DMT-SSR cleaning procedure) and plastic/metallic surfaces, when the above conditions are applied.

By strict control over these parameters the sludge growth within the BioSulfurex® is reduced to a minimal speed. The bioreactor will experience a small fouling rate in time which is observed by the pressure loss over the bioreactor. Starting from 1–3 mbar it will increase to 20–50 mbar in 1 or 2 years (depending on process control and sulphur load). The fouling can be removed by a simple maintenance procedure developed by DMT. The DMT-SSR (Selective Sludge Removal) cleaning procedure which takes about 6–8 hours after

	7-May	8-May	
pH	1.67	2.58	
System Temperature	27	26	°C
Biogas flow	900	857	m^3/h
Pressure loss	55	6	mbar
H_2S BioSulfurex® in	±2500	±2500	ppm
H_2S half way	238	1100	ppm
H_2S BioSulfurex® out	4.3	24	ppm

Table 1 & Figure 5 Results with DMT-SSR procedure. Israel (table) and random bioreactor photo's.

which the biological performance will be effected only slightly. E.g., in Israel the bioreactor was cleaned over night from 7th to the 8th of May (see table 1). After which the pressure loss was back from 55 to 6 mbar again. The removal efficiency remained at >99% even though a lot of excess sludge was removed. The EC of the first (bottom)half of the reactor decreased slightly but this was compensated by an increase in EC at the second (top)half.

In figure 5 (down left) it is visible what happens when there is no control over the bioreactor (excessive nutrients, no oxygen, temperature nor pH control). Excessive bacterial growth resulted in complete clogging of the reactor. Even this bioreactor could be cleaned by the DMT-SSR cleaning procedure (figure 5; top right).

THE MINIATURE BIOREACTOR (MBR) FOR BIOLOGICAL DESULPHURIZATION

BIOSULFUREX® HSP

The current BioSulfurex® resembles a chemical system already quite closely due to the stable operation and high efficiencies. To further improve the biological system compared to a chemical treatment the EC has to be further improved to reduce footprint.

The future processes which DMT are already testing are using highly structured packing materials (e.g. DMT-HS packing) with a large specific surface. Due to the structure of the packing the material is self-cleaning and excessive sludge growth is impossible. Results from a similar pilot with the DMT-HS packing, which treated a H_2S polluted air stream, demonstrated that the EC for DMT-HS is approximately 6–8 times higher than for Pall-Rings. This would result in ECs of 500–800 $gH_2S/m^3 * h$ for biogas. Also research from Marc Fortuny et al.[7] demonstrate that EC's of minimal 250 $gH_2S/m^3 * h$ for (bio) gas are possible with highly structured packing at conditions which were still limited in carbon source, oxygen, pH and temperature. Marc Fortuny

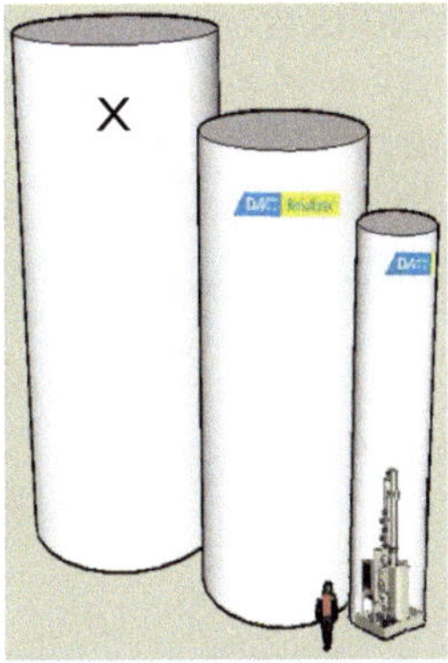

Figure 6 Scale model for an Conventional bioreactor (left), the DMT-BioSulfurex® (middle), The BioSulfurex® HSP (right) and a chemical scrubber (front) for the treatment of 2000 m^3/hr biogas with 2000 ppm H_2S.

et al also demonstrated that at oxygen limitations about 94–95% of the H_2S is converted to elemental sulfur instead of sulfuric acid.

The BioSulfurex® HSP has several advantages due to the use of the DMT-HS packing material. Both the elimination capacity and gas load can be tremendously increased. This results in a very small bio-reactor with a minimum foot print (See figure 6) which is almost comparable to a chemical system. At the same time the system is self-cleaning and there is no effect of clogging from the formed sludge and/or elemental sulfur. This makes it possible to add a minimal amount of oxygen, converting all H_2S into elemental sulfur. This ensures a maximum caloric value of the treated gas and at the same time the sulfur can be collected and re-used in various processes (e.g., fertilizer). There will be minimal to no water consumption because a minimum amount of H_2SO_4 is formed and the operation at low pH. In most cases the condensation from the biogas would be sufficient as clean water supply. No chemicals are needed and only minimal waste stream of acid water is produced which can be directly discharged in a WWTP. The smaller design also reduces power consumption and makes operation and maintenance more easy.

CONCLUSION

Biological desulfurization of biogas can be used for H_2S removal at high concentration (>3500 ppm) with high efficiency (>99%) resulting in very low outgoing H_2S concentration (<50 ppm). Within the practical operational window, the EC is highly dependent on temperature and available nutrients but less to pH and O_2. An EC of 80–100 gH_2S/m^3/h and more can easily be reached with conventional packing materials and a good process control. A good process control also results in minimal sludge growth and easy cleaning procedures. Process stability and removal efficiencies of the BioSulfurex® system are equal to or even surpass a chemical system. Future DMT bioreactors will be designed by the MBR principle. The Miniature BioReactor will have a self-cleaning packing with a high population density of bacteria operating at EC's of 500–800 $gH_2S/m^3/h$ or more reducing the footprint comparable to a chemical scrubber. The H_2S will be converted to elemental Sulfur for absolute minimum to no water consumption.

REFERENCES

1 Dirkse, E.H.M. Biogas Desulphurisation using the DMT multiple stage Sulfurex® process, 11th European Biosolids Conference, Manchester (2006).
2 van der Zee, F.P. et al, Sulfide removal by moderate oxygenation of anaerobic sludge environments, *Bioresource Technology*, 98:518–524 (2007).
3 Silver, M. Dinardo, O. Factors affecting oxidation of thiosalts by thiobacilli, *Applied and environmental microbiology*, Vol. 41, No. 6: 1301–1309 (1981).
4 Germán Aroca et al, Comparison on the removal of hydrogen sulfide in biotrickling filters inoculated with thiobacillus thioparus and acidithiobacillus thiooxidans, *Electronic journal of biotechnology*, Vol. 10, No. 4 (2007).
5 Bruné, E. Biological biogas desulfurization at Emek Hefer, internal graduation report, 2007.
6 van der Linden, P. Optimization of the BioSulfurex®, internal graduation report, (2009).
7 Fortuny M. et al, Biological sweetening of energy gases mimics in biotrickling filters, *Chemosphere*, 71, 10–17 (2008).

Chapter 50

MULTIPLE STAGE DMT ODOUREX®: The next generation Biological Odour control systems

R. Lems, E.H.M. Dirkse & J. Edens
DMT, Joure, The Netherlands

ABSTRACT

DMT Environmental Technology B.V. has developed in the period medio 2006 until medio 2008 an improved technique to remove H_2S, mercaptanes and other VOC's in a biological way. The Odourex® system is a 3-stage bio-trickling filter with an acid and a neutral stage followed by a polishing step. During a demonstration project the Odourex® systems has show to perform 3 to 5 times better than the conventional filters (lava-rock/pall rings). The elimination capacity for H_2S has proven to be at least 100 $gH_2S/m^3 * h$ at optimum conditions. The filter proofed to be self-cleaning with minimum air pressure drop during the year of testing. The footprint of the Odourex® system is only 20% compared to a conventional lava rock filter.

Keywords: Bio-trickling filter, bio filter, DMT-HS, H_2S, mercaptanes, Multistage, odour control

INTRODUCTION

Odour is usually caused by the presence of a great variety of components like: hydrogen sulphide, mercaptanes, ammonia and different kinds of VOC's. An odour control system therefore will have to treat not only hydrogen sulphide, but also other pollutants to reduce the total mal odour. In biological processes every micro-organism prefers it's own process conditions; especially pH, temperature and substrate are playing important roles. So placement of only one biological filter often appears not to be sufficient. For that reason a combination of processes/filter steps is necessary.

THEORETICAL BACK GROUND

DEVELOPMENT OF PACKING/FILTER MEDIA FOR BIOLOGICAL FILTERS

The first bio-filters were large beds containing soil and/or compost and could only remove small concentrations of a large variety of (organic) compounds. Different filter materials have been developed to improve these bio-filters. Pure material or mixtures of compost, manure, bark, wood and coconut fibres have been used to increase the capacity of the bio-filters and decrease the footprint.[1] A second step in the development of biological treatment, which was more focussed on H_2S rich air streams, was the introduction of lava rock bio-trickling filters (BTFs). The high specific surface of the lava rock increased the population density of the (H_2S degrading) bacteria reducing the total required filter volume. Due to the water circulation it was easier to control process parameters like humidity, temperature and pH, which made it possible to operate near optimum conditions.[2] At the same time higher filters were possible due to the decrease in pressure drop which reduced the footprint significant. This resulted in higher removal efficiencies and EC's up to 20–60 $gH_2S/m^3 * h$[3,4] at an air load of 300–350 m^3/m^2. Also the first multi stage biological treatment systems were introduced, which consist of two lava rock in series, if necessary followed by a conventional bio-filter. In this way different environmental systems (e.g., pH) can be created to degrade different pollutants by specialized bacteria. However the lava rock is difficult to handle, has a large weight and is susceptible to clogging especially at high H_2S concentrations and can therefore only be used for small to medium air flows with maximum H_2S concentrations of 100–500 ppm.

HIGHLY STRUCTURED MEDIA

The last years plastic and foam packing materials for bio trickling filters have been developed to increase the biofilm surface and at the same time minimize resistance and weight.[5] This resulted in either high EC's but increased the chance of clogging (foam), or reduced the chance of clogging but did not increase the EC significant (Pall Rings). The plastic filter materials do result in an excellent process control (pH, T, pressure loss etc.) due to higher water circulation capacities. DMT is now introducing a new plastic packing material for BTF's; the DMT-HS filter media. Due to its highly structured, open media, clogging problems and declining removal efficiencies are being solved. The specific round spikes stimulate the growth of the micro organisms which increases the total bio film surface (increase in spike diameter) to approximately 1000 m^2/m^3. When the bio film becomes to thick/heavy it falls of the vertical positioned pins due to the shear of the water circulation. This makes the media self cleaning.

PRACTICAL DATA

A demonstration project was started in March 2007 at the WWTP of Eerbeek (the Netherlands) to test and compare the DMT-HS packing material in a practical situation.[6] Three types of media in BTF's filters were tested: Lavarock, Pall Rings and DMT-HS (See Figure 1). The main objectives of this field research was to compare removal efficiencies, elimination capacities and fouling rates. The second objective was to find the optimal process parameters for different components.

OBTAINED MEASUREMENTS

In the next graph (Figure 2, top) the H_2S load is displayed against the elimination capacities. The diagonal line presents a 100% removal efficiency. The graph demonstrates a EC for the DMT-HS of >100 $g/m^3/hr$ with still no reasonable decline in removal efficiency (>95%). For the pall rings the EC is maximum near 50 $g/m^3/hr$ with decreasing removal efficiencies. The lava rock has a maximum EC of about 15–18 $g/m^3/hr$. At >95% removal the maximum is 22 $g/m^3/hr$ for Pall and 9 $g/m^3/hr$ for lava rock.

Measurements concerning temperature showed that below 15°C, removal efficiency dropped quickly to an average 20–30% at 8°C (all systems, see Figure 1, middle). The pH had a negligible effect on the systems between 0.5–3 (all systems, see Figure 1, bottom). Nutrient addition was very important for all systems but had the most effect on the plastic filter materials. A mixture of ammonium and phosphate salts performed as good as effluent and even as good as a special nutrient mixtures with trace elements. A ratio of 100:30:1 (DMT-HS system, S:N:P) was found optimal. GCMS analysis showed that 50% of total VOC and 90% of the alcohols were removed with DMT-HS.

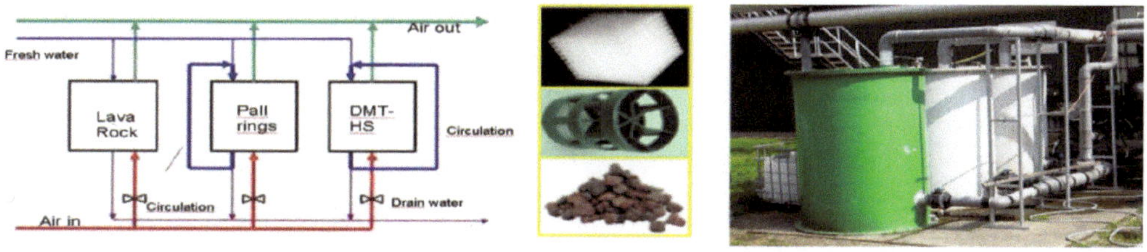

Figure 1 Left; PFD, middle; filter materials, right; photo of the demonstration plant.

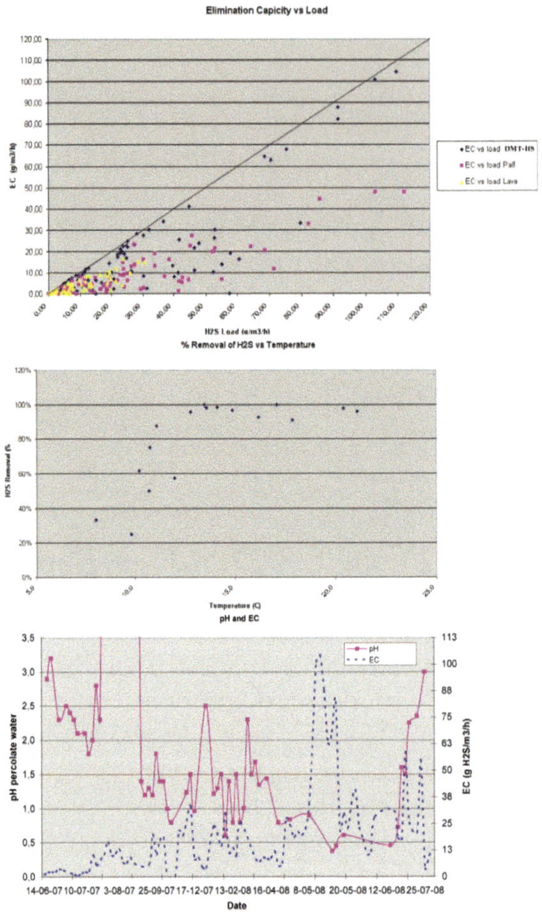

Figure 2 Top; EC different filter media, middle; relation temperature and EC; bottom; relation pH and EC.

During testing, measurements concerning pressure loss showed that DMT-HS caused a pressure drop between 20 and 50 Pascal (versus 350–450 Pa for Lava Rock) without clogging even at over-loading and dehydration conditions. After artificial clogging, the system was self cleaned within 1 day. The lava rocks clogged frequently (two monthly). See Table 1 for an overview of all test results.

DMT ODOUREX®

The results from the demonstration plant in Eerbeek were combined with field data of multiple BTF's for treatment of various pollutants which have

Table 1 Overview test results.

Parameter	*DMT-HS*	*2"Pall*	*Lava*
EC (g $H_2S/m_3/h$)	35–100	5–50	5–20
EBCT (s)	10–20	10–30	30–70
P. drop (Pa/m)	10–20	20–100	150–200
H_2S %	80–100	50–80	70–100
Mercaptanes %	>70	30–50	40–50
VOC %	50–90		

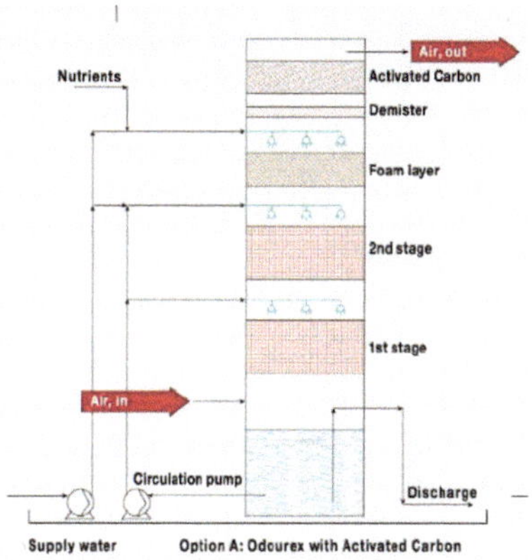

Figure 3 Odourex® system.

been extrapolated to create a multi-stage system for complete odour removal (see Figure 3). This single tower, three stage purification plant uses the DMT-HS in two stages and a foam filter as a polishing step as a third stage. All odour components can be degraded in a biological way at their own optimal conditions like T and pH. H_2S, NH_3 and alcohols are removed in the first stage at a acidic pH (0.5–1.5). Mercaptanes and other VOC's will be degraded at the second stage at a pH of 6–9. The last step is a polishing stage with foam, creating a very dense population of bacteria which are directly fed with fresh, nutrient rich water to accomplish maximum removal efficiency on all left trace components.

Compared with 2 lava rock filters in serie, followed by activated carbon as polishing step this results in a 4 times smaller volume and even 5 times smaller footprint (see Table 2 and Figure 3).

Still the pressure drop is less, the load can be higher and it is self cleaning. With these parameters the biological odour removal is even better for the Odourex® system (99.8%) than for the lava rock system (98.3%).

	Odourex®	Lava
A (m^2)	5.7	28.2
Vt (m^3)	20	80
Dp (Pa)	200 Pa	2100
RE (5)	99.8	98.3

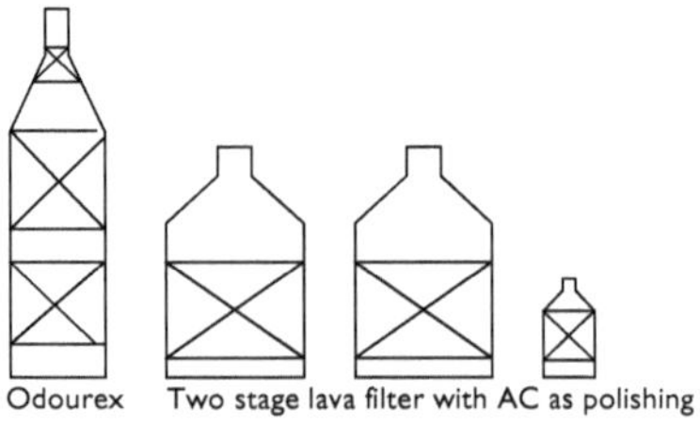

Table 2 and Figure 4 Overview of an odor control system design for a DMT-HS and a lava rock for system for 4400 m^3/hr and 50–75 ppm H_2S, 5–7.5 ppm Mercaptanes and 10–30 ppm VOC.

CONCLUSION

The development of bio filters for odour abatement has been driven by problems with clogging, footprint and removal efficiencies. This has resulted in the multi stage DMT Odourex® which treats different odour components like H_2S, mercaptanes and VOC's in different stages at optimal process parameters. The elimination capacity of the DMT-HS packing material is more than 3 to 5 times higher than lava rock and 2" pall-rings. This results in removal efficiencies on odour of >99% on a very small footprint with a self cleaning clogging free packing material.

REFERENCES

1 G.R. Moosavi et al, Biotechnology advances in treatment of air streams containing H_2S, *Journal of Biological Sciences*, 5 (2):170–175 (2005).
2 Smet, E. (1996). *Biofiltration of Volatile Organic Sulfur Compounds*. Gent University, Gent.
3 Evaluation of Biological air treatment systems at WWTPs (Dutch), 2003.
4 Various internal inspection reports.
5 Deshusses, M.A. and H.H.J. Cox. (2000). Biotrickling filters for air pollution control, In: G. Bitton (Editor-in- chief), *The Encyclopedia for environmental microbiology*, New York.
6 Dirkse, E.H.M. (2007). Odour control on Waste Water Treatment Plants and Pumping Stations using DMT Biotrickling Filtration, DMT environmental Technology. 12th European Biosolids Conference, Manchester.

Chapter 51

THE EMISSION OF MICROBIAL AEROSOLS FROM WASTEWATER TREATMENT PLANT WITH ORBAL OXIDATION DITCH PROCESS

Lin Li, Min Gao & Junxin Liu
Research Center for Eco-Environmental Sciences, Chinese Academy of Sciences, Beijing, PR China

ABSTRACT

In this report, conventional sampling and detection method, as well as PCR method were applied to estimate airborne microorganisms present in a municipal wastewater treatment plant (WWTP) environments in Beijing. Results showed that the concentration varied over a wide range for cultivable airborne bacteria and fungi. High concentrations of airborne bacteria and fungi were obtained in sludge dewatering house and fine grit chamber, respectively. The highest concentration of cultivable airborne bacteria was detected in the pore sizes of 2.1–3.3 μm. More than 67% airborne fungal were located in the 1.1–3.3 μm size range. The PCR method, based on DNA amplification techniques, can be applied to enhance the monitoring or verification the presence of airborne bacteria.

Keywords: wastewater treatment plant (WWTP), airborne microorganisms, community structure, particle characteristics

INTRODUCTION

Odors control has been of a concern and research topic in waste and wastewater treatment processes for many years. However, bioaerosols, another air emission from the wastewater, gained little attention by now. They are a significant subgroup of atmospheric aerosols and their transportation has been proved to be related to spread of human, animal and plant disease epidemics.[1]

Their concentrations and species differed from place to place depending on the kind of water treated, process selected and site location. Many studies have focused on determining airborne microorganisms concentration at WWTP[2–4] and traditional detection methods were introduced for bioaerosol exposure assessment. The traditional detection methods may be underestimation of total airborne microorganism concentration because of some microorganisms are uncultivable. Molecular biology techniques can provide promising tools to detecting almost total bacteria, which has been applied to analyze air samples for the presence of airborne microorganisms associated with adverse health effects.[5] To date, there are no reports of PCR in evaluation the airborne microorganisms emitted from WWTP attached to Beijing.

The main objective of the present study was to analyze the emission and distribution characteristics of airborne microorganisms in a WWTP. Conventional sampling and detection method, combined with PCR method were applied to describe the concentrations, size distributions and biodiversity of airborne microorganisms.

MATERIALS AND METHODS

Samples were collected from coarse screen (CS), fine grit chamber (FGC), near the rotating brush of the Orbal oxidation ditch (N-RB), far from the rotating brush and near the outlet of the oxidation ditch (F-RB), sludge thickening tank (STT) and sludge dewatering house (SDH).

A Six-stage Andersen sampler (FA-1, China) was used at a constant flow of 28.3 L min^{-1}, with sampling time 2 minutes, place at 1.5 m above ground. The airborne bacteria and Fungi were incubated in Nutrient agar at 30°C for 2 days and Rose Bengal Medium (BR, Aoboxing Biotech, Co. China) at 30°C for 7 days, respectively. For PCR analysis of total airborne bacteria (TB), samples were transferred in sterile water by drawing the gas for 20 min with a flow rate of 500 ml min^{-1}. Colonies of different stages incubated in Nutrient agar were washed and collected by sterile water for analysis of cultivable bacteria (CB). DNA of cells was extracted according to the instruction on the test kit (omega, USA). Primers of 357f and 518r were used to amplify the segment of 16S rDNA. The GC clamp was added to the forward primers to facilitate the DGGE. 30 μl of the PCR product was separated on an 8% acrylamide gel with a linear denaturant gradient increasing from 30% to 60% using the Bio-Rad D-GENE System. The resulting gel was visualized with Gel Red (Biotium USA). The images were analyzed with Quantity One Software (Bio-Rad, USA).

All the experimental data were analyzed using SPSS Versuib 10.0 (SPSS, Standard Version) for one way analysis of variance (ANOVA).The bacterial diversity was estimated from the intensity and number of bands by using the Shannon index (*H*).

RESULTS AND DISCUSSION

CONCENTRATION OF AIRBORNE BACTERIA AND FUNGI

Results showed that the highest concentrations of cultivable airborne bacteria (5565 CFU m^{-3}) were obtained in the SDH and lowest (459 CFU m^{-3}) obtained from the F-RB. The highest concentrations of cultivable airborne fungi (1590 CFU m^{-3}) were achieved in the FGC. The mean concentrations of the heterotrophic bacteria were 14CFU m^{-3} in the indoor control, which were approximately 400 times lower than that in the SDH. At the same time, the mean concentrations of fungi were 693CFU m^{-3} in the indoor control, which were over 56% lower than that in the FGC (Fig. 1).

The mechanical agitation of wastewater will accelerate the generation of airborne microorganisms.[9] The high concentration of bacteria as well as fungi(N-RB) could be due to the aeration of the rotating brush installed at the oxidation ditch enhanced significantly the emission of bioaerosols. The fact of that more airborne bacteria and fungi were found at CS than that at STT may also due to the rotation of grille sludge removal machine.

PARTICLE SIZE DISTRIBUTIONS OF AIRBORNE CULTIVABLE BACTERIA AND FUNGI

Figure 2. reveals that highest concentration of cultivable airborne bacteria (5521 CFU m^{-3}) were detected in stage 4 with pore sizes of 2.1–3.3 μm, and the lowest (1316 CFU m^{-3}) were found in stage 6 with pore sizes of 0.65–1.1 μm. More than 67% fungal were located in the 1.1–3.3 μm size range only 5.0% were below 1.1 μm. The distributions of bioaerosols particle size were quite different from place to place.

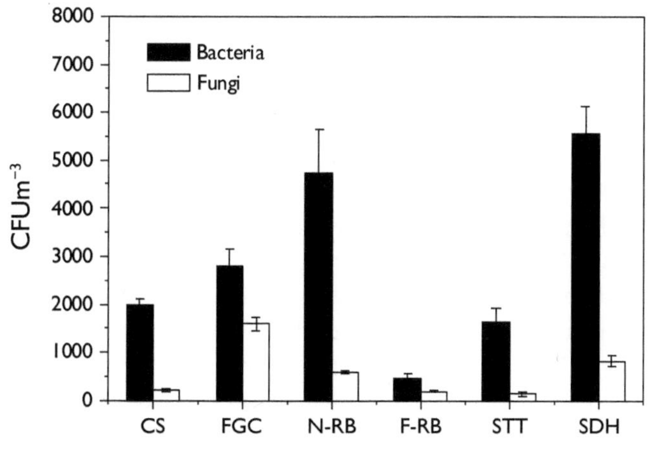

Figure 1 Total concentration of airborne bacteria and fungi from each sampling point.

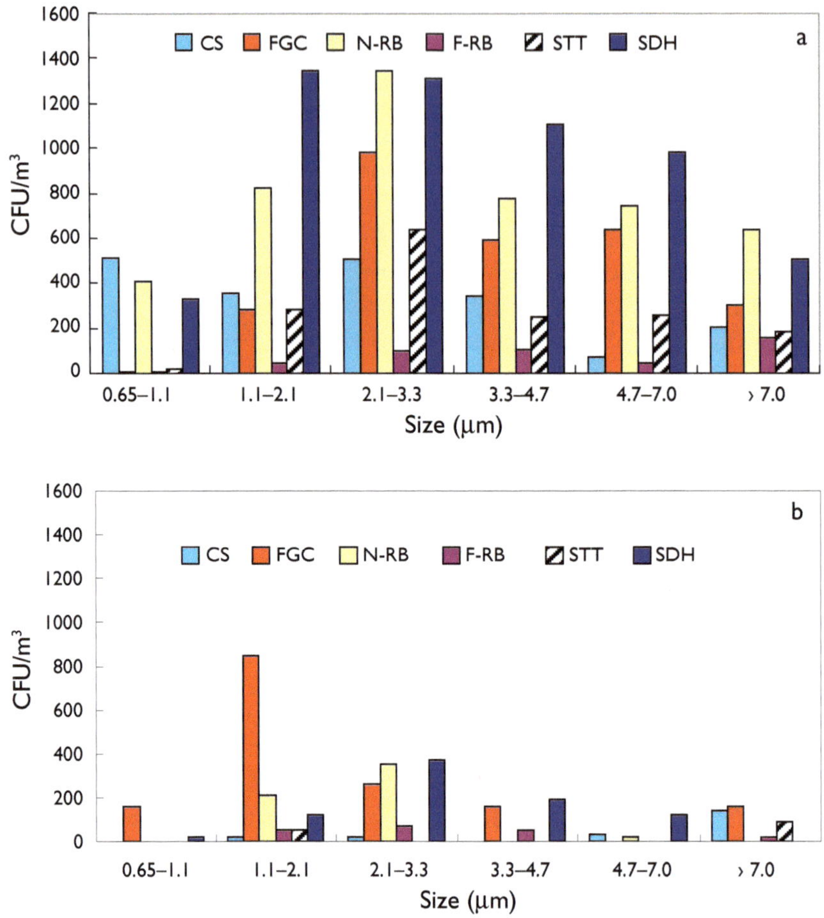

Figure 2 Particle size distributions of airborne cultivable bacteria and fungi (CFU m^{-3}). (a) bacteria; (b) fungi.

PCR AND DGGE

The profile of DGGE bands, which demonstrates that both CB and TB presented high diversity of the bacterial community. The number of TB bands in different process was greater than that of the CB. The distant genetic evolutionary relationship of different lane was represented in Fig. 3(b). The lanes of TB sampling from different stages had close evolutionary relationship, so did CB. However, on the same sampling point, there was no evident relationship between CB and TB. The Shannon index of each ATB was higher than the corresponding one of CB, which indicated that the population diversity of ATB was greater than that of CB.

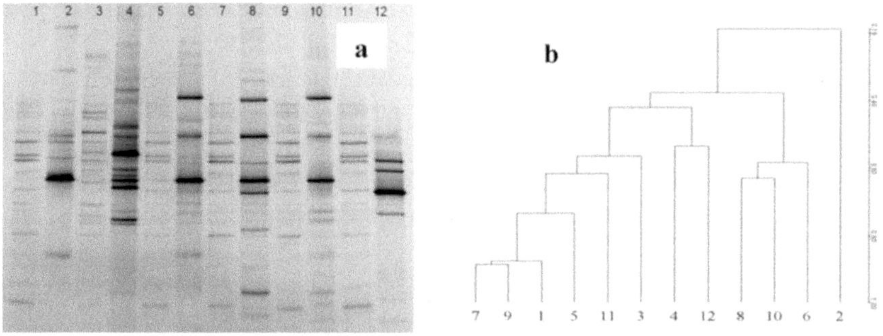

Figure 3 (a) DGGE profiles of CB and TB in each sampling ports. (b) Phylogenetic tree of CB and TB from three independent experiments. (1) CB of CS, (2) TB of CS, (3) CB of FGC, (4) TB of FGC, (5) CB of N-RB, (6) TB of N-RB, (7) CB of F-RB, (8) TB of F-RB, (9) CB of STT, (10) TB of STT, (11) CB of SDH, (12) TB of SDH.

CONCLUSION

The concentration recorded in the WWTP environments in Beijing varied over a wide range for cultivable bacteria and fungi. High concentrations of cultivable airborne bacteria and fungi obtained at the SDH and the FGC, respectively. The highest concentration of cultivable airborne bacteria was detected in the pore sizes of 2.1–3.3 μm, and the lowest were found within 0.65–1.1 μm. The mechanical agitation of rotating brush will accelerate airborne microorganisms emission. The PCR method, based on DNA amplification techniques, can provide promising tools to monitor the presence of airborne microorganisms.

ACKNOWLEDGEMENTS

We appreciate Dr. Jianwei Liu and Dr. Xin Zhou for their kind help in samples collection. The work was financially supported by the National Natural Science Foundation of China (No. 50621804 and No. 50678171).

REFERENCES

1 Sánchez-Monedero MA and Stentiford EI. Generation and dispersion of airborne microorganisms from composting facilities. Transactions of the Institution of Chemical Engineers. Part B, *Process Safety and Environmental Protection* 81: 166–170 (2003).
2 Ranalli G, Principi P and Sorlini C. Bacterial aerosol emission from wastewater treatment plants: Culture methods and bio-molecular tools. *Aerobiologia* 16: 39–46 (2000).

3 Luca GD, Zanetti F, Perari AC and Stampi S. Airborne coagulase negative staphylococci produced by a sewage treatment plant. *International Journal of Hygiene and Environmental Health* 204: 231–238 (2001).
4 Bauer H, Fuerhacker M, Zibuschka F, Schmid H and Puxbaum H. Bacteria and fungi in aerosols generated by two different types of wastewater treatment plants. *Water Research* 36: 3965–3970 (2002).
5 Cruz-Perez P, Buttner MP and Stetzenbach LD. Detection and quantitation of Aspergillus fumigatus in pure culture using polymerase chain reaction. *Molecular and Cellular Probes* 15: 81–88 (2001).

Chapter 52

PERFORMANCE OF A MONOLITH BIOREACTOR FOR THE REMOVAL OF STYRENE FROM POLLUTED AIR

Eldon R. Rene, M. Estefania López, María C. Veiga & Christian Kennes
Faculty of Sciences, Chemical Engineering Laboratory, University of La Coruña, La Coruña, Spain

ABSTRACT

The biological removal of gas – phase styrene using the fungus *Sporothrix variecibatus* was evaluated in a novel monolith bioreactor. Continuous experiments were carried out for 92 d at different inlet styrene concentrations, ranging between 0.07–2.5 g m^{-3}, and at two different flow rates corresponding to empty bed residence times (EBRTs) of 77 and 19 s respectively. A maximum elimination capacity of 67.4 g m^{-3} h^{-1} was observed at an inlet styrene load of 73.5 g m^{-3} h^{-1}. To ascertain the instantaneous response of the attached fungus to withstand fluctuations in loading conditions, short – term shock load was conducted. Though, the performance of the monolith reactor was moderately high and satisfactory for gas – phase styrene, there are certain critical factors that have to be addressed and satisfied before implementing monolith bioreactors for full – scale applications.

Keywords: Monolith reactor, *Sporothrix variecibatus*, styrene, transient – state operation

INTRODUCTION

Biofiltration is one of the most cost – effective and efficient means to remove volatile organic compounds (VOCs) and other hazardous air pollutants (HAPs) emanating from chemical process vent streams, prior to their discharge into the environment. The removal of styrene vapours, amongst others, emitted from various chemical process industries such as plastic moulding, fibre – glass, pipes and automobile component manufacture industries, is very important to meet pollution standards. Styrene is an important organic chemical classified as a VOC, that has an odour threshold of 0.43 mg m^{-3} and a saturation constant of 25.6 g cm^{-3} in air at 20°C.[1] Biological treatment systems such as biofilters (BF) and biotrickling filters (BTF) have proven to be efficient in handling styrene vapours, at different inlet loading rates (ILR), in lab – scale conditions.[2] Recently, the pursuance among researchers to examine different packing materials as support matrix for biomass growth in bioreactors applied to waste gas treatment has increased exponentially. Under this line of progressive research, our laboratory has undertaken a few initiatives for the development of new reactor systems to treat hydrophilic and hydrophobic VOCs using a novel monolith reactor.[3,4] This paper reports the performance of such a monolith bioreactor under steady – state and transient – state operating conditions.

MATERIALS AND METHODS

The ceramic monolith was supplied by Rauschert Verfahrenstechnik GmbH (Germany). The experimental setup and its operational mechanism have been described elsewhere.[3]

Sporothrix variecibatus (Fig. 1), isolated from a previously operated biofilter[5] treating styrene vapours was used for inoculating the monolith reactor. Before inoculation in the monolith reactor, the enrichment culture was grown in 1 L Erlenmeyer flasks (working volume – 250 mL) on styrene, having an initial concentration of about 50 mg L^{-1} at 30°C and at constant shaking at 175 rpm. After 48–56 h of growth, having reached an optical density (OD) of 0.9 at 600 nm, the cells were harvested by centrifugation at 5000 rpm for 20 min and re-suspended in 2 L fresh nutrient medium[5] to be used as inocu-

Figure 1 SEM photograph of *Sporothrix sp.*

lum for the continuous monolith bioreactor. After homogenizing, these cultures were stored in a reservoir tank, so as to re-circulate it over the top of the monolith reactor.

Gas samples collected from sampling ports, before and after the monolith packing, were subjected to chromatographic analysis, on a HP 5890 gas chromatograph, using a 50 m TRACER column and a FID.

RESULTS AND DISCUSSION

After acclimation (22 d), continuous experiments were carried out in two phases, at an EBRT of 77 and 19 s and at concentrations ranging from 0.06–2.5 g m^{-3}. The degree of pollutant degrading capacity of a bioreactor is commonly quantified as a function of pollutant loading by determining the elimination capacity values. Also, EC is frequently used as a performance and design criterion for gas – phase biofilters without further verification of the microbial quantity and activity. At both EBRT, the styrene elimination rate per unit packing volume of monolith increased linearly with an increase in the styrene inlet loading rate. At an EBRT of 77 s, for ILR < 80 g m^{-3} h^{-1}, nearly 90% of gas – phase styrene was removed by the attached *Sporothrix sp.* and converted to CO_2. Conversely, at an EBRT of 19 s, the critical ILR to reach more than 90% RE was just about 40 g m^{-3} h^{-1}. The maximum EC was lower at 19 s than at 77 s (Fig. 2). The results are not surprising, as the critical load in biological waste gas treatment systems has shown to decrease with an increase in the gas flow rate, i.e., decreasing EBRTs.[5,6]

In practically operated bioreactors, some decrease in the performance of fungal systems should be expected in response to continuous fluctuating loads.

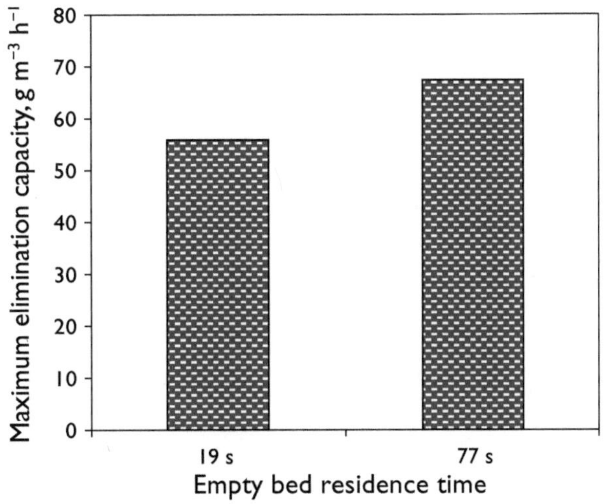

Figure 2 Maximum EC as a function of the residence time.

Therefore our objective was to simulate the expected industrial type loading fluctuations and observe the response in the laboratory scale monolith bioreactor. For this purpose, a 6 – h short – term study at an EBRT of 77 s, giving rise to medium shock loads was planned. The results indicated that the styrene laden monolith bioreactor was sensitive to fluctuations in loading rates. A sudden increase in the ILR from 10 to 85 g m^{-3} h^{-1} decreased the RE from nearly 100% to 61% during medium shock loads.

Though, the performance of the monolith reactor was moderately high and satisfactory for gas – phase styrene, the bed pressure drop increased to about 7 cm H_2O/m bed height. This was due to excess fungal growth at the different channel openings of the monolith that eventually led to channelling and by passing of the contaminated air. The absence of a proper liquid distribution system at the top could have also led to reduced reactor performance. One possibility would be to modify the distributor by properly placing a bed of alumina or glass spheres at the top of the monolith surface. This simple construction would allow the incoming liquid to break up into thin rivulets and spread across the cross – section of the alumina bed, and deliver in a free falling fashion on the monolith channels underneath. Besides, the distance separating the distributor and top surface has to be optimized, but this factor should be significant enough, so that the entire top surface of the monolith receives the liquid.

CONCLUSIONS

The elimination capacity profiles, at constant gas – flow rates, increased with increase in styrene concentration, up to a maximum and then decreased when the concentration was increased further. The results from this study provide sufficient steady state and non – steady state information on fungal – monolith bioreactors and its potential for use in industrial purposes.

ACKNOWLEDGEMENTS

The present research was funded through project CTM2007-62700/TECNO of the Spanish Ministry of Science and Innovation and through European FEDER funds.

REFERENCES

1 James DH and Castor WM. In: *Ullmann's Encyclopedia of Industrial Chemistry*, V: **A25**, p: 329 (2000).
2 Kennes C, Rene ER and Veiga MC. Bioprocesses for air pollution control. *Journal of Chemical Technology and Biotechnology*. In Press, DOI: 10.1002/jctb.2216, (2009).

3 Jin Y, Veiga MC and Kennes C. Performance optimization of the fungal biodegradation of α – pinene in gas–phase biofilter. *Process Biochemistry* **41**; 1722–1728 (2006).
4 Jin Y, Veiga MC and Kennes C. Removal of methanol from air in a low-pH trickling monolith bioreactor. *Process Biochemistry* **43**; 925–931 (2008).
5 Rene ER, Veiga MC and Kennes C. Performance of a biofilter for the removal of high concentrations of styrene under steady and non – steady state conditions. *Journal of Hazardous Materials* **168**; 282–290 (2009).
6 Kennes C and Veiga MC. *Bioreactors for Waste Gas Treatment*. Kluwer Academic Publishers: Dordrecht; Boston (2001).

Chapter 53

REMOVAL OF α-PINENE FROM WASTE GASES THROUGH AIR DIFFUSION INTO ONE- AND TWO- LIQUID-PHASE SUSPENDED-GROWTH BIOREACTORS

María Montes, María C. Veiga & Christian Kennes
Faculty of Sciences, Chemical Engineering Laboratory, University of La Coruña, La Coruña, Spain

ABSTRACT

Over the past decades, research on the use of an organic liquid phase for the treatment of polluted air has significantly increased. The objective of the present study was to study the removal of α-pinene from a waste air stream in a continuous stirred tank bioreactor (CSTB), using a single-liquid aqueous phase or a combination of an aqueous-organic liquid phase. Continuous experiments were carried out at different inlet α-pinene concentrations, ranging between 0.28 g m^{-3}–25.1 g m^{-3} and at three different flow rates corresponding to empty bed residence times (EBRTs) of 60 s, 36 s and 26 s, respectively. The results confirm that the reactor's performance improved when adding the organic phase since the one-liquid-phase CSTB did not allow reaching maximum elimination capacities as high as the two-liquid-phase bioreactor.

Keywords: α-pinene, continuous stirred tank bioreactor, silicone oil, *Rhodococcus erythropolis*, waste gases

INTRODUCTION

α-pinene is a hydrophobic and recalcitrant volatile organic compound with a boiling point of 156°C.[1,2] Its degradation is considered complex.[3] It is typically found in waste gases from the pulp and paper industry and the forest products industry. An emerging technology for the treatment of VOCs is the two-liquid-phase suspended-growth bioreactor, also known as biphasic aqueous/organic reactor or partitioning bioreactor. Some recent studies have shown that the addition of a second liquid, organic phase, in suspended growth bioreactors is mainly useful for the removal of poorly water soluble pollutants, although the elimination efficiency of less hydrophobic pollutants may also significantly be increased in such systems.[4]

Two-liquid-phase systems consist of an aqueous phase and an organic liquid phase. Generally, in presence of an organic solvent, volatile pollutants do not reach high concentrations in the suspended-growth water phase, so microorganisms in the aqueous phase will not be inhibited. From a chemical engineering and thermodynamic point of view, silicone oil is still the most popular and most commonly used organic phase because of its interesting characteristics.[5]

The main purpose of the present research was to determine the potential of a two-liquid phase-bioreactor for the treatment of α-pinene. The removal of α-pinene from polluted air was studied in a continuous stirred tank bioreactor (CSTB) in two steps: firstly, using a single-liquid aqueous phase and secondly, using a combination of an aqueous –organic liquid phase. In this step, different concentrations of the organic liquid phase were used: 2% and 5% (vol:vol).

MATERIALS AND METHODS

The experimental set-up has been described elsewhere.[4] The reactor was set to operate at a temperature of 25°C, an agitation speed of 400 rpm and a total working volume of 1.5 L. The pH of the culture was kept constant, at 7.0. The total reactor volume was maintained constant. The polluted gas phase was generated by mixing two different air streams. A small stream of air was bubbled through a vial containing pure α-pinene which was mixed afterwards with a large stream of compressed air in a mixing chamber. The flow rates of both gas streams were measured and regulated with flowmeters.

The influence of silicone oil on α-pinene biodegradation was evaluated in the stirred tank bioreactor. The reactor was filled with different concentrations of silicone oil: 2% (vol:vol) (1.47 L aqueous mineral medium, 30 mL of silicone oil) and 5% (vol:vol) (1.42 L aqueous mineral medium, 75 mL of silicone oil), and inoculated with a strain of *Rhodococcus erythropolis*. The reactor was operated under constant experimental conditions, as described above.

RESULTS AND DISCUSSIONS

Experiments were carried out by varying the flow rate (0.084 $m^3 h^{-1}$ – 0.21 $m^3 h^{-1}$) and α-pinene's concentration (0.28 g m^{-3} – 25.1 g m^{-3}). The removal of α–pinene was investigated in different phases corresponding to decreasing empty bed residence times (EBRTs) of 60 s, 36 s and 26 s.

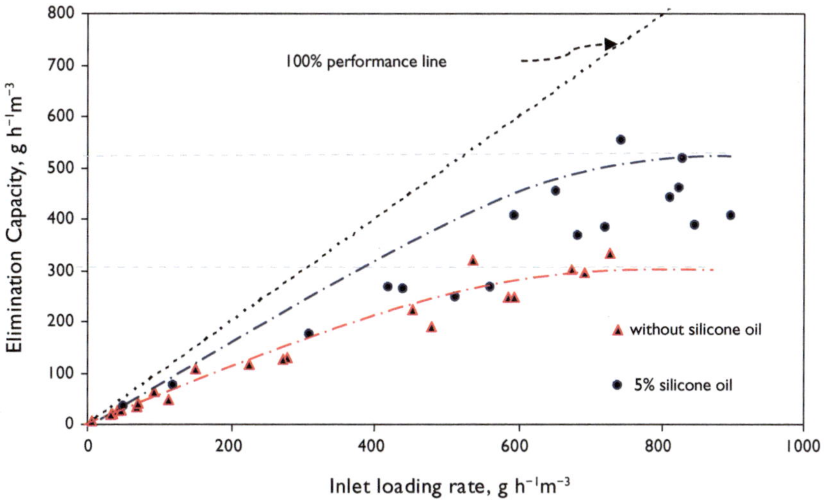

Figure 1 Effect of the addition of 5% silicone oil on the elimination capacity of the continuous stirred tank bioreactor at an EBRT of 60 s.

The presence of the organic phase, *i.e.*, silicone oil, at a volume ratio of 2% and 5% of the liquid phase, increased the maximum elimination capacity (EC). When adding 5% silicone oil to the CSTB (Fig. 1) the maximum EC increased by 45% reaching 550 g m^{-3} h^{-1} with a removal efficiency of 60% to be compared to 300 g m^{-3} h^{-1} as maximum EC in the one liquid phase system, with a removal efficiency of 50%. The present data suggest that the one-liquid-phase CSTB does not allow reaching maximum elimination capacities as high as two-liquid-phase reactor.

The results from this study showed that the introduction of an organic phase such as silicone oil is highly favourable for an optimal and fast mass transfer of the pollutant from the organic phase to the aqueous phase. Moreover, regarding CO_2 data, generally higher carbon dioxide concentrations were generated in the reactor than expected from the α-pinene load and stoichiometric equation. This was sometimes already observed and reported previously for a fungal biofilter treating α-pinene polluted air as well.[1] It can be attributed to different phenomena as, among others, endogenous respiration, but confirms that the polluted was basically completely converted into end products.

CONCLUSIONS

The results show that the one-liquid-phase CSTB does not allow reaching maximum elimination capacities as high as two-liquid-phase bioreactor. The presence of an organic phase has shown to be favorable as it reduces inhibition.

In the case of a single aqueous phase bioreactor, the system becomes inhibited at high loads.

ACKNOWLEDGEMENTS

The present research and the Ph.D. contract of M. Montes were financed through a project of the Spanish Ministry of Science and Innovation (CTM2007-62700/TECNO) and through European FEDER funds.

REFERENCES

1 Jin Y, Veiga MC and Kennes C. Performance optimization of the fungal biodegradation of α – pinene in gas–phase biofilter. *Process Biochemistry* **41**: 1722–1728 (2006).
2 Jin Y, Guo L, Veiga MC and Kennes C. Fungal biofiltration of α – pinene: effects of temperature, relative humidity and transient loads. *Biotechnol Bioeng* **96**: 433–443 (2007).
3 Kleinheinz GT, Bagley ST, St. John WP and Rughani JR. McGinnis GD, Characterisation of alpha-pinene degrading microorganisms and application to a bench-scale biofiltration system for VOC degradation. *Archives of Environmental Contamination and Toxicology* **37**: 151–157 (1999).
4 Bailón L, Nikolausz M, Kästner M, Veiga MC and Kennes C. Removal of dichloromethane from waste gases in one- and two- liquid-phase stirred tank bioreactors and biotrickling filters. *Water Res* **43**: 11–20 (2009).
5 Kennes C, Rene ER and Veiga MC. Bioprocesses for air pollution control. *Journal of Chemical Technology and Biotechnology*. In Press, DOI: 10.1002/jctb.2216 (2009).

Chapter 54

APPLICATION OF PLANTS FOR CONTROLLING AIR POLLUTION WITH CHLORINE-CONTAINING COMPOUNDS IN YEREVAN, ARMENIA

Gayane S. Nersisyan & Hasmik A. Hovhannisyan
The Center for Ecological-Noosphere Studies of the National Academy of Sciences of the Republic of Armenia, Yerevan, Armenia

ABSTRACT

One of pollutants of Yerevan's atmosphere is chlorine (Cl) and its compounds, which basic sources are chemical plants operating on the city's territory. The goal of this research was to reveal arboreous species with good indication and phyto-melioration properties in respect to chlorine-containing compounds so as to use them in controlling and cleaning up the air of Yerevan. The research was performed between 2005 and 2007. The studied objects were the soils and leaves of 5 arboreous species. The soils were low in chlorine, whereas the plants displayed high concentrations varying 0,50 to 1,77%. The produced schematic maps of chlorine distribution allover the city's territory allowed indication of the most polluted zones located mostly in the south and south-western parts of the city as well as on some sites in the direction of the rose wind. To control ambient air pollution with chlorine the best testing systems are found out to be the leaves of Robinia pseudoacacia L. and as the most efficient phytofilters Robinia pseudoacacia L. and Fraxinus excelsior L. can be used.

The research outcomes were included in the Functional Tree Planting Program for the city of Yerevan.

Keywords: chlorine, plants, accumulation, phytoindication, air control

INTRODUCTION

Air pollution with diverse toxicants on Armenia's urbanized sites has been intensifying as a result of industry development and the increasing amount of vehicles. Yerevan -the capital of Armenia – homes the major part of industrial enterprises; the density of urban population is high, and peculiarities of the relief and climatic conditions favor far reaching dispersion of gaseous toxicants and their accumulation by environmental compartments. The soil and vegetation of different districts of the city is high in lead and other heavy metals, nitrogen and chlorine compounds. Besides, the area of pollution zones has increased for recent years. Accumulation of diverse toxicants has also been established in farming crops cultivated on the territory of Yerevan and its vicinities.[1,2]

One of atmospheric pollutants is known to be chlorine and its compounds, originating predominantly as a result of operation of chemical plants and incineration of domestic refuse. Studying environmental pollution with this biogenic and at the same time extremely toxic element is essential to Yerevan as on its territory a scientific and industrial enterprise (SIE) "Nairit" operates producing chloroprene-synthetic rubber and other chlorine-containing products.

The bioindication method based on test system plants is widely applied in different countries and is the best and cost-effective one to control ambient air pollution with diverse toxicants.[3,4] We used the method to indicate heavy metal pollution of Armenia's urbanized sites.[5–7]

The goal of this research was to find out tree species with good indication and phytomelioration properties in respect to chlorine-containing compounds to apply them in controlling and cleaning up the air of Yerevan. To achieve the goal, the following tasks were to be managed:

- studying chlorine absorption and accumulation by the soils and plants of Yerevan,
- revealing species – chlorine accumulators and their ecological tolerance under urban conditions,
- assessing and mapping chlorine pollution of the city's territory using the revealed indicator plants.

EXPERIMENTAL

The research was performed in period 2005 to 2007. The studied objects were the soils and leaves of 5 arboreous species: Robinia pseudoacacia L., Fraxinus excelsior L., Populus alba L., Morus alba L., Vitis vinifera L. Samples were collected from 35 different points throughout the city in the middle of vegetation period falling on July (Fig. 1).

Chlorine contents were determined in water soil extraction following ISO 11464 standard, and in plants – through the method of titration with $Ag(NO_3)$.[8,9] The control contents were determined in soil and vegetation samples collected at a 50 km distance northeastward of the city.

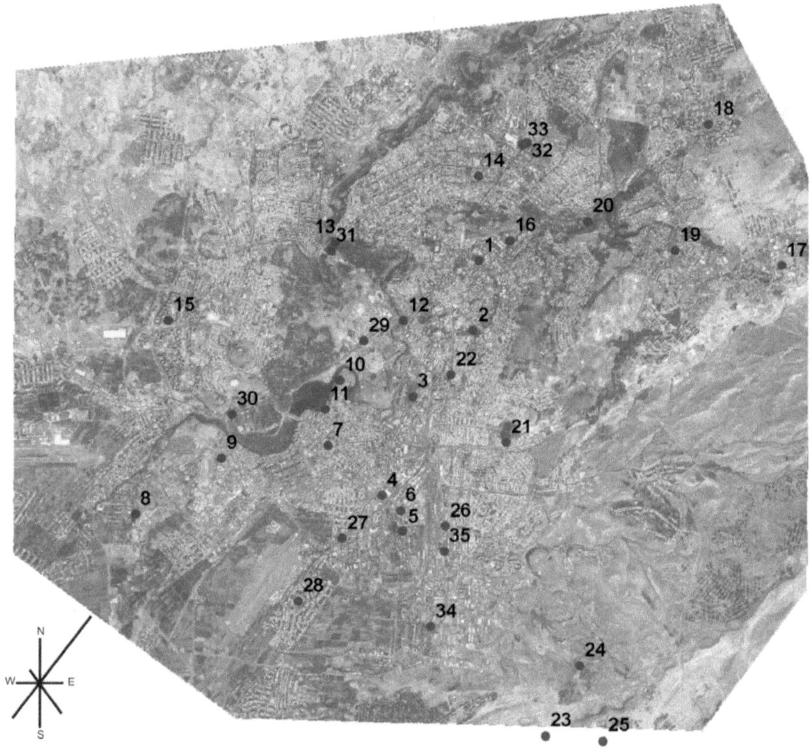

Figure 1 The sampling points on the satellite image of the city.

Based on a compiled database and using ArcGIS software, schematic maps were produced of Yerevan's soil and vegetation pollution with chlorine.

RESULTS AND DISCUSSION

The obtained research results indicated, that wholly the level of chlorine concentrations in the city's soils were not excessive vs. the control (0.02%). However, in the southwest part of the city, in the vicinities of SIE "Nairit" and the Water Cleanup Station local pollution zones were found out where chlorine concentrations in soils were 1.05–1.52 time excessive vs. the control (Fig. 2).

The major part of the city lies in the zone of mean concentrations of chlorine (0,009–0,019%) corresponding to the control. The overall area of local zones of increased concentrations of chlorine (0,019–0,03%) in the soils made 1,12% (2,49 km^2) of the entire territory of the city.

The studied soils were low in chlorine, however we indicated its accumulation by plants, this evidencing the presence of the element in the ambient

Figure 2 A schematic map of chlorine concentration distribution in Yerevan's soils.

Table 1 The dynamics of chlorine contents in the leaves of diverse arboreous species in Yerevan (%).

	Years								
	2005			*2006*			*2007*		
Species	*min*	*aver.*	*max*	*min*	*aver.*	*max*	*min*	*aver.*	*max*
Robinia pseudoacacia L.	0.59	1.06	1.64	0.61	0.92	1.35	0.50	0.81	1.77
Fraxinus excelsior L.	0.47	0.95	1.77	0.59	0.75	0.99	0.43	0.61	1.09
Populus alba L.	0.51	0.84	1.21	0.55	0.80	1.16	0.46	0.70	1.12
Morus alba L.	0.62	0.91	1.48	0.63	0.83	1.00	0.65	0.88	1.22
Vitis vinifera L.	0.50	0.81	1.39	0.52	0.77	1.35	0.49	0.78	1.34

air of Yerevan. Chlorine concentrations varied between 0,5 and 1,77% in all the 5 studied species, and maximal values exceeded the control (0,5%) by 3,6 times (Tab. 1).

A three-year long study of the dynamics of chlorine in plants enabled us to establish an inclination to some decrease of mean concentrations in 2007; however, maximal concentrations of the element remained high especially in variant with Robinia pseudoacacia L. No drastic changes in chlorine concentration both by years and sampling points were observed in Morus alba L. and Vitis vinifera L. All the said as well as the fact that the latter are fruit bearing species and are met with on the whole territory of the city, supports a conclusion that they are unsuitable for bioindication and biomonitoring purposes.

Subsequently, for approbation of the phyto-indication method we selected 3 decorative species widely spread in Yerevan vis. Robinia pseudoacacia L., Fraxinus excelsior L. and Populus alba L. It should be noted that such a selection was fully justified by our previous research outcomes which proved their being efficient test systems to monitor heavy metal pollution of the city.[5,7]

Fig. 3 gives mean concentrations of chlorine in the leaves of the species for all the studied points. As indicated, all the 3 species absorbed and accumulated chlorine. Most studied points displayed excessive concentrations vs. the control. Maximal concentrations of chlorine in the leaves of Robinia pseudoacacia L. exceeded the control by 1,5–3,6 times, and in the leaves of Fraxinus excelsior L. and Populus alba L. – by 1,45–2,2 times. The study indicated that of all the three species, it was Robinia pseudoacacia L., which had the best absorption properties (Fig. 3).

The schematic maps we produced on a basis of the obtained data indicate that the major part of the city lies within the zone of mean concentrations of chlorine (0,64–0,81%).

High concentration zones were established in the south and southwest parts of Yerevan, where the major sources of chlorine are situated: SIE "Nairit", Water Cleanup Station and the municipal refuse disposal site. We also found out some local pollution sites in the direction of the wind rose. Especially clear the said is seen in the variant with Robinia pseudoacacia L. (Fig. 4).

Phenological surveys allowed indication of high ecological tolerance of Robinia pseudoacacia L. and Fraxinus excelsior L.: even in strong pollution zones they kept good look and behaved as efficient phytofilters. All this evidence that these species can be widely used to control air pollution with chlorine-containing compounds and for phytomelioration purposes.

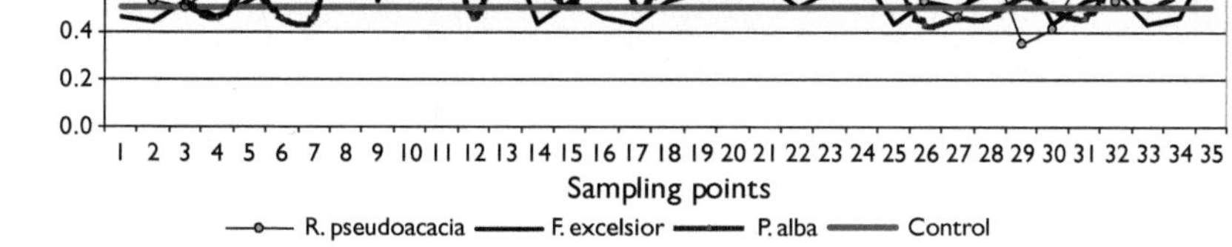

Figure 3 Mean concentrations of chlorine in plants.

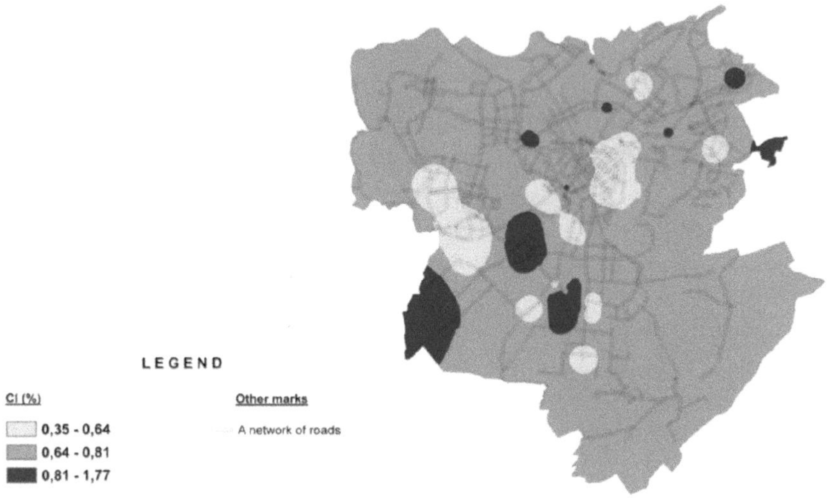

Figure 4 A schematic map of chlorine distribution in Yerevan. (*Robinia pseudoacacia L.* database).

CONCLUSIONS

Thus, the performed studies enabled us to conclude that

- chlorine accumulates and absorbs by the plants growing in the city and the major portion of chlorine enters the plants from the atmosphere,
- distribution of high concentrations of the element throughout the city depends on nearness of pollution sources, rose wind and vegetation species,
- the best chlorine absorption and phytomelioration property as well as ecological tolerance under conditions of chlorine pollution is typical of Robinia pseudoacacia L. and Fraxinus excelsior L.,
- it is advisable using the leaves of Robinia pseudoiacacia L. as a test system for bioindication and biomonitoring of chlorine pollution of the air basin of Yerevan.

All this evidences that the phytoindication method can be used to control ambient air pollution with chlorine-containing compounds, and chlorine-tolerant tree species such as Robinia pseudoacacia L. and Fraxinus excelsior L. – as efficient phytofilters for cleaning up the air of Yerevan.

The obtained research results were transferred to the Municipality and then included in the Program of Functional Tree Planting in the city of Yerevan.

REFERENCES

1 Hovhannisyan HA, Ecological monitoring of vegetables grown in urban sites. *Electronic J. of Natural Sciences*, National Academy of Sciences RA, 2(3): pp. 10–13 (2004), http://search.epnet.com/login.aspx?direct=true&AuthType=cookie,ip,url,uid&db=aph&an=15491672.

2 Hovhannisyan HA and Nersisyan GS, Studing concentration of some trace elements in urban agrosystems of Armenia. In Book of abstracts "*The X Symposium on Trace elements in the environment-ecological and methodical problems*", Polland, Koszalin-Mielno, 11–14 May, pp. 128–129 (2008).

3 Manning WJ and Feder WA, Biomonitoring Air Pollutants with plants. *Appl. Sci. Publ.* LTD, London, 140 p. (1980).

4 *Bioindication of pollution of terrestrial ecosystems,* ed. Schubert R. Mir, Moscow, 350 p. (1988) (in Russian).

5 Arevshatyan SH, Hovhannisyan HA, Muradyan AH, Avetisyan MH and Navasardyan IA, Biogeochemical peculiarities of Fraxinus excelsior under conditions of pollution of Yerevan with heavy metals. Sci. Reports of Yerevan State Univ., 1, pp. 114–120 (2000) (in Russian).

6 Hovhannisyan HA, Avetisyan MH and Nersisyan GS, *Phytoindication of lead pollution for city of Yerevan.* Vestnik IAELPS, Saint-Petersburg, Russia, V. 13, No. 4, issue 2, pp. 54–56 (2008) (in Russian).

7 Hovhannisyan HA and Nersisyan GS, A study of lead pollution of vegetation inYerevan (Armenia). In Book of proc. of 5th Int. Conf. "*Environmental Accounting-Sustainable Development Indicators*", Czech Republic, Prague, April, 23–24, 7 pp. (CD) (2009), http://ea-sdi.ujep.cz/en/conference-2009/full-paper-submissions.html, Session C: Environmental Modeling.

8 Fomin GS and Fomin AG, *Soils* (Controlling the quality and ecological safety by international standards), Reference book. Protector, Moscow, 304 p. (pp. 108–109) (2001) (in Russian).

9 Pochinok KhN, The methods of biochemic analysis of plants. *Naukova dumka, Kiev*, pp. 27–28 (1976) (in Russian).

Chapter 55

PHYSICOCHEMICAL CHARACTERIZATION OF A BIOFILTER PACKING FOR THE MODELING OF DISPERSION, AND SORPTION OF AMMONIA

Francesc Xavier Prenafeta-Boldú, Albert Magrí & Guillermo Baquerizo
GIRO Technological Centre, Barcelona, Spain

Guillermo Baquerizo
Departamento de Procesos y Tecnología, Universidad Autónoma Metropolitana-Cuajimalpa, Mexico City, México

Josep Illa
University of Lleida, Lleida, Spain

INTRODUCTION

Ammonia is an important environmental pollutant emitted to the atmosphere by a great variety of agricultural and industrial processes.[1] Besides short range odor nuisance, ammonia emissions have been related to the eutrophication and acidification of natural ecosystems. The treatment of gas streams containing low ammonia concentration can be achieved through gas biofiltration, which represent an attractive technology due to the relatively low cost of installation and operation. However, the performance and stability of these systems is seriously compromised by the deterioration of the organic packing materials commonly used so far.

Biofiltration of ammonia involves several complex phenomena interacting simultaneously, such as dispersion, absorption, adsorption, and biodegradation of volatile compounds. In order to understand the relative importance of each process a mathematical modeling approach is required.[2] Some assumptions have been taken in earlier models in relation to the packing material, such as plug flow dispersion pattern of the gas phase, no influence of the packing on the pH, and Henry's law gas-liquid interface equilibrium. In the present study, the validity of these assumptions was tested on expanded clay particles (EC) as model inert biofilter packing material. EC is increasingly being used as carrier material for biofiltration of odorous gases arising from processes related to the decomposition of organic matter.[1] This information is fundamental for the definition of accurate biofiltration numerical models.

MATERIALS AND METHODS

TRACER PULSE STUDIES

Free air space (FAS) of EC was determined by a self constructed picnometer.[3] The gas mixing regime was studied in a 2 L EC packed column, operated at an empty bed residence time (EBRT) of 20 to 60 seconds, by measuring the residence time distribution (RTD) of a tracer pulse. Ethene (1 mL) was used as tracer gas and the outflow concentration was recorded continuously with a PID detector (RAE Systems Inc., USA). RTD curves were modeled as series of CSTR[4] and the mean pulse residence time was used to estimate the effective gas volume, value that was compared with the FAS for the quantification of "dead zones" within the EC packing.

SORPTION EXPERIMENTS

Ammonia adsorption and absorption was assayed in batch tests. Approximately 15 g of EC were placed into a Boston flask (250 mL) sealed with Mininert Teflon valves (VICI Precision Sampling, USA). In a second set of flasks, 3 mL of a solution of hydrochloric acid at pH 3.8 was also added. Three flasks were prepared for each of the two treatments, dry and humidified EC. Increasing amounts of ammonia gas were successively added into the headspace, and the concentration of ammonia at the equilibrium (25°C) was measured by injecting 1 mL of headspace into a portable PID detector (RAE Systems Inc., USA). Numerical models on ad/absorption of ammonia were fitted against the experimental data. Calculations on the absorbed ammonia were based on mass and charge balance equations. Four ionic dissociation processes were considered: water, carbonic acid (to bicarbonate and carbonate species), and ammonia. The effect of EC on the pH was determined upon additions of hydrochloric acid (0.1 M) on 150 mL of stirred demineralized water that also contained 7.64 g of expanded clay, and by continuously monitoring the pH (meter GLP 22, probe 52 03, CRISON Instruments, Spain). The ionic behavior of EC was modeled considering three different hypothesis: (*i*) EC was equivalent to a monoprotic weak acid, in which the dissociation constant (pk_a) and total acid equivalent concentration ($S_{a,0}$) were the model fit parameters; (*ii*) EC behaved as a diprotic weak acid, with two dissociation constants (pk_{a1}, pk_{a2}) and one total base equivalent concentration ($S_{a,0}$) as model fit parameters; and (*iii*) EC was equivalent to two monoprotic weak acids, being the model fitted with four parameters (pk_{a1}, $S_{a1,0}$, and pk_{a2}, $S_{a2,0}$).

RESULTS AND DISCUSSION

The used EC particles had an average diameter and standard deviation of 14.8 ± 1.2 mm. The FAS of the biofilter bed was 77 ± 2% of the packed volume, value that was distributed between the particle interstitial space (34 ± 5%)

and the porosity of the material (38%). RTD of tracer pulses demonstrated that the effective gas volume was of about 36 to 42% of the FAS, depending on the gas flow. Hence, the real gas residence time was less than the half of the EBRT value. This indicates that diffusion into the fine porosity of EC might be the limiting step for the adsorption of ammonia under operational EBRT values that are common for biofiltration. RTD studies also demonstrated that gas dispersion, modeled as a series of N CSTR (not shown), was fairly close to the ideal plug flow assumption ($N = 7–11$).

The adsorption of ammonia onto EC was accurately modeled according to the Langmuir isotherm (Fig. 1). The estimated EC maximum ammonia adsorption capacity of 0.12 mg g^{-1}dw was significantly smaller than the values determined previously for other organic materials.[5]

The observed alkalinizing effect of EC (Fig. 2) was modeled most precisely ($r^2 = 0.998$) as being equivalent to two monoprotic weak acids (p$k_{a1} = 3.168$, $S_{a1,0} = 7.748 \cdot 10^{-1}$ mM, and p$k_{a2,0} = 6.083$, $S_{a2,0} = 3.818 \cdot 10^{-2}$ mM), in relation to

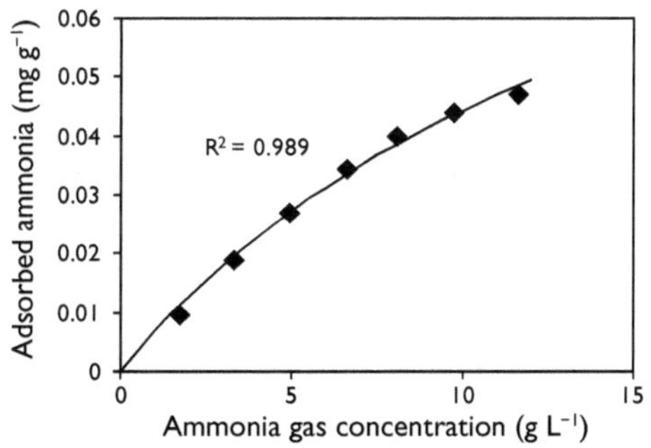

Figure 1 Ammonia adsorption onto EC; the Langmuir isotherm equation (line) was adjusted to measured results (dots).

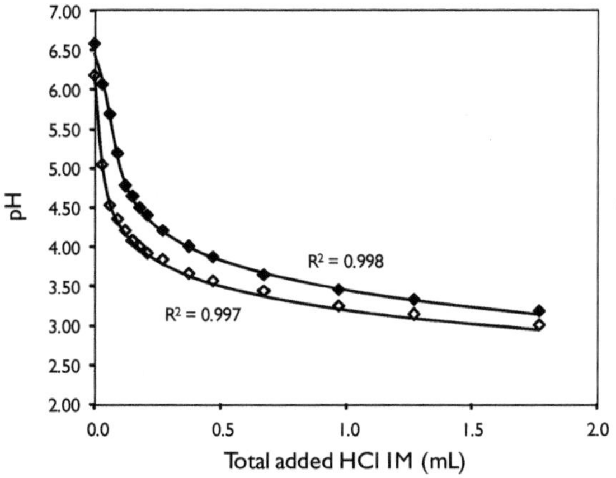

Figure 2 Evolution of the pH upon additions of HCl into 150 mL of demineralized water alone (empty dots), and amended with 7.6 g dw of EC (solid dots). Lines refer to simulated results.

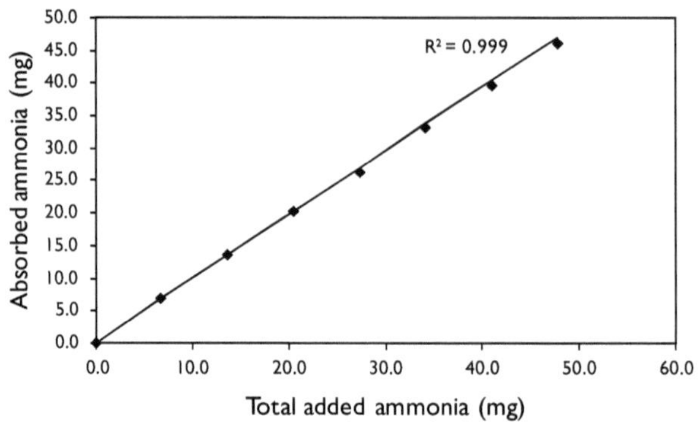

Figure 3 Experimental measurements (solid dots) versus modeled data (line) on the absorption of ammonia into the water phase (3 mL, initial pH = 3.8) in batches containing approx. 15 g dw of EC.

the model fittings that considered EC as a monoprotic or a diprotic weak acid ($r^2 = 0.975$ and $r^2 = 0.983$, respectively). Ammonia absorption was linearly correlated to the total mass of added ammonia within the tested experimental range, and there was a good agreement between experimental and modeled results (Fig. 3).

EC has been reported as an interesting substitute of organic packing materials in the biofiltration of odorous gases. However, the applicability of EC for the removal of ammonia might be limited due to its poor adsorption capacity and ionic behavior, which might buffer any induced strong acidification of the packing in order to improve the absorption process. However, this buffering capacity might also be positive for the microbial populations that are active in the bed, and that are responsible for the biodegradation of volatile compounds, including the oxidation of ammonia into nitrate.

REFERENCES

1 Kennes C and Veiga MC. Bioreactors for Waste Gas Treatment. In *Environmental Pollution*, pp. 320. Edited by Alloway BJ and Trevors JT. Dordrecht: Kluwer Acdemic Publishers (2001).

2 Baquerizo G, Gamisans X, Gabriel D and Lafuente J. A dynamic model for ammonia abatement by gas-phase biofiltration including pH and leachate modelling. *Biosystems Engineering* 97:431 (2007).

3 Solé-Mauri F. Mathematical Modelling of the Composting Process. University of Lleida, 148 pp. (2006).

4 Prenafeta-Boldú FX, Illa J, van Groenestijn JW and Flotats X. Influence of synthetic packing materials on the gas dispersion and biodegradation kinetics in fungal air biofilters. *Applied Microbiology and Biotechnology* 79:319–327 (2008).

5 Pagans E, Font X and Sanchez A. Adsorption, absorption, and biological degradation of ammonia in different biofilter organic media. *Biotechnology and Bioengineering* 97:515–525 (2007).

Chapter 56

REMOVAL OF SILOXANES FROM BIOGAS USING ADSORPTION

Alice Prochazkova, Veronika Vrbova & Karel Ciahotny
Department of Gas, Coke and Air Protection, Institute of Chemical Technology Prague, Prague, Czech Republic

INTRODUCTION

Biogas produced by anaerobic decomposition of organic materials at wastewater treatment plants or waste disposal sites is generally burned in internal combustion engines of cogeneration units. Biogas contains besides main components like methane (typical concentration 50–65% vol.) and carbon dioxide (concentration of 30% vol.) also components which often complicate its following using. Such components are first of all low molecular siloxanes. Siloxanes are family of man-made organic compounds that contain silicon, oxygen and methyl groups. Siloxanes are used in the households in make-ups for personal hygiene, health care and in industrial products. As a consequence of their widespread use, siloxanes are found in wastewater and in solid waste deposited in landfill. At wastewater treatment plants and landfills, low molecular weight siloxanes volatilize into digester gas and landfill gas. When this gas is combusted to generate power in internal combustion engines of cogeneration units, siloxanes from biogas are oxidated into silicium dioxide, whose deposits on the walls of cylinders have a negative effect on the service life of internal combustion engines.

TEORETICAL PART

BIOGAS

Biogas is produced in different environments, e.g., in landfills, sewage sludge and biowaste digesters during anaerobic degradation of organic material.

Methane which is the main component of biogas is available renewable energy source, but also a harmful greenhouse gas if emitted into the atmosphere. Methane, upgraded from biogas, can be used for heat and electricity production or as biofuel for vehicles to reduce environmental emissions. Biogas from sewage digesters usually contains from 55% to 65% methane, to 45% carbon dioxide and <1% nitrogen. Biogas from organic waste digester usually contains from 60% to 70% methane, from 30% to 40% carbon dioxide and <1% nitrogen while in landfills methane content is usually from 45% to 55%, carbon dioxide from 30% to 40% and nitrogen from 2% to 15%. Typical biogas also contains hydrogensulphide and other sulphur compounds, compounds such as siloxanes and aromatic and halogenated compounds [1].

Biogas containing organic silicon compounds is harmful to engines because the silicon compounds are oxidated into silicon oxides during combustion, causing scuffing to engine parts.

SILOXANES

Siloxanes are widely used in industrial processes and costumers products. Some of them reach the wastewater. Siloxanes are not decomposed in the activated sludge process and partly concentrate in the sludge [2].

The term siloxane consists of parts of the words: **sil**(icon) + **ox**(ygen) + (meth)**ane** (Fig. 1.). Siloxanes are a subgroup of silicons containing Si-O bonds with methyl, ethyl or other organic groups bound to the silicon atom. The structure of a siloxane can be linear or cyclic.

But the central structure element is always the same. Each siloxane is composed of several [-(-MeSiO-)n-]- groups. The linear compounds are shortened with L and completed with number of silicon atoms. Similarly the cyclic ones are abbreviated with D and the related silicon number.

Figure 2 shows the structure of OMCTS and DMCPS which are the most important siloxanes for our measuring.

Common volatile siloxanes are listed in Table 1. Due to the length of the names of the various siloxanes, abbreviations are commonly used to identify the compounds. Siloxanes that are cyclic in structure have a single abbreviation

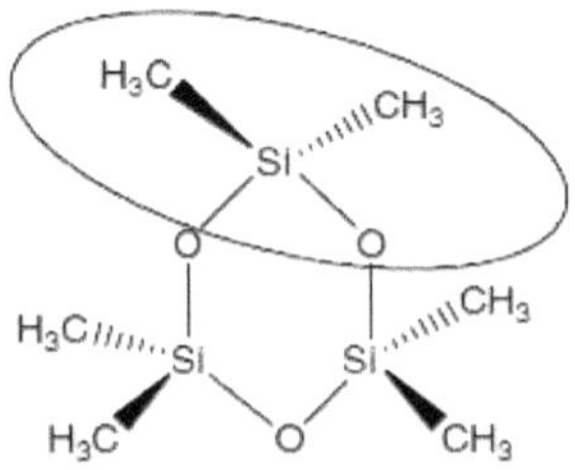

Figure 1 Structure of siloxanes [2].

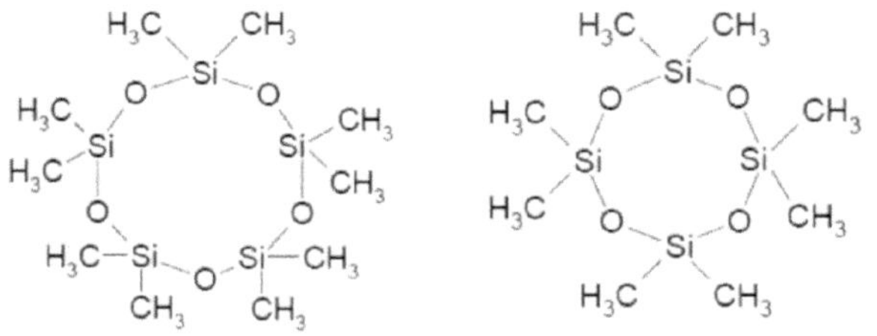

$[Si(CH_3)_2O]_5$ $[Si(CH_3)_2O]_4$

OctaMethylCycloTetraSiloxane DecaMethylCycloPentaSiloxane

Figure 2 Structure of OMCTS and DMCPS [2].

Table 1 List of siloxanes [3].

Name	*Formula*	*MW*	*Vapor pressure (mmHg/ 77°F)*	*Abbreviations*	*Boiling point °F*	*Water Solubility (mg/l), 25°C*
Hexamethylcyclotrisiloxane	$C_6H_{18}O_3Si_3$	222	10	D_3	275	1,56
Octamethylcyclotetrasiloxane	$C_8H_{24}O_4Si_4$	297	1,3	D_4	348	0,056
Decamethylcyclopentasiloxane	$C_{10}H_{30}O_5Si_5$	371	0,4	D_5	412	0,017
Dodecamethylcyclohexasiloxane	$C_{12}H_{36}O_6Si_6$	445	0,02	D_6	473	0,005
Hexamethyldisiloxane	$C_6H_{18}Si_2O$	162	31	L_2, MM	224	0,93
Octamethyltrisiloxane	$C_8H_{24}Si_3O_2$	236	3,9	L_3, MDM	–	0,035
Decamethyltetrasiloxane	$C_{10}H_{30}Si_4O_3$	310	0,55	L_4, MD_2M	–	–
Dodecamethylpentasiloxane	$C_{12}H_{36}Si_5O_4$	384	0,07	L_5, MD_3M	–	–

of D. Siloxanes that have a linear structure have two abbreviations using either an L or M nomenclature. Table 1 also identifies the molecular weight, vapor pressure, boiling point, chemical formula, and water solubility of these compounds. Siloxane concentrations are generally higher in digester gas than in landfill gas. As a result, it is somewhat easier to reliably quantify siloxanes in digester gas. Landfill gas may contain significant quantities of other siloxane compounds such as D_3 and D_6, plus L_2 through L_5. D_4 and D_5 may represent only slightly more than a majority of the siloxanes in some landfill gases to over 90 percent of the total in others [3].

SILOXANE REMOVAL TECHNOLOGIES

Although exist more of possible siloxane removal technologies, carbon adsorption is still the only proven method now in commercial operation. The following provides a discussion of the capabilities of this technology and other potential removal technologies.

Adsorption

The adsorption process is based on the accumulation of molecules on the surface of a solid material. Activated carbon has been widely used for the removal of a variety of substances from air and gas for decades. Adsorption capacity is defined as the point where siloxane breakthrough can be detected [3].

Absorption

For absorption are used high boiling organic solvents. Spray and packed columns, e.g., with Raschig rings, are installed for siloxane absorption. The main problem is that siloxane is easily stripped from the solvent at higher gas flow rates. Therefore the siloxanes should be chemically absorbed [3].

Cryogenic condensation

In order to receive a well-ammable gas the raw biogas needs to be treated in a drying step. In most cases the moisture is removed by cryogenic condensation. The formed condensate contains a part of the present siloxanes. In the study of Schweigkoer and Niessner [4] the biogas was cooled to 5°C. But 91% of the initial siloxane concentrations remained in the biogas. Hagmann et al. [5] measured a removal rate of 25.9% for cooling to –25°C and of 99.3% for freezing to –70°C. Cooling the biogas to such low temperatures demands a great amount of energy. For this reason the costs increase.

EXPERIMENTAL PART

For an effective elimination of the negative impacts of siloxanes is suitable used adsorption. In the Department of gas, coke and air protection at Institute of Chemical Technology in Prague have been tested a lot of adsorption materials for removing of siloxanes from biogas. Simultaneously with testing in laboratory have been done measurement of siloxanes content in biogas produced at wastewater treatment plants in Prague where is situated adsorption technology for the siloxanes removal from biogas.

TEST CARRIED OUT UNDER LABORATORY CONDITIONS

In laboratories at ICT Prague were tested 10 different adsorbents for removal of siloxanes from model gas mixtures. The adsorption isotherms were determined to find the adsorbent with the best adsorption behavior. In the Table 2 are information about all using adsorbents.

Testing apparatus shows Figure 3.

Table 2 List of adsorption materials.

Adsorbent	*Basis*	*Form*	*Surface area* m^2/g	*Total pore volume (ml/g)*	*Micropore volume (ml/g)*
Trockenperlen	Silica gel	Globular	679	0,470	0
MS 13 X	Zeolite	Globular	485	0,342	n.a.
Alumina + 5% Mn	Impregnated Alumina oxide with Mn	Globular	278	0,373	0
AP 460	Activated carbon	Pressed cylinders	872	0,410	0,273
SC 40 + 10% Mn	Impregnated Alumina oxide with Mn	Pressed cylinders	958	0,545	0,052
USU 20	made from PET	Irregular	556	0,569	0,121
HS 3	Activated carbon impregnated with Cu, Cr, Ag	Granular	800	0,485	0,256
Chezacarb BG-6	Carbon black	Pressed cylinders	731	1,141	n.a.
Chezacarb SH-6	Carbon black	Pressed cylinders	634	1,010	n.a.
Envisorb B+	Silica gel and activated carbon	Globular	723	0,852	0,03

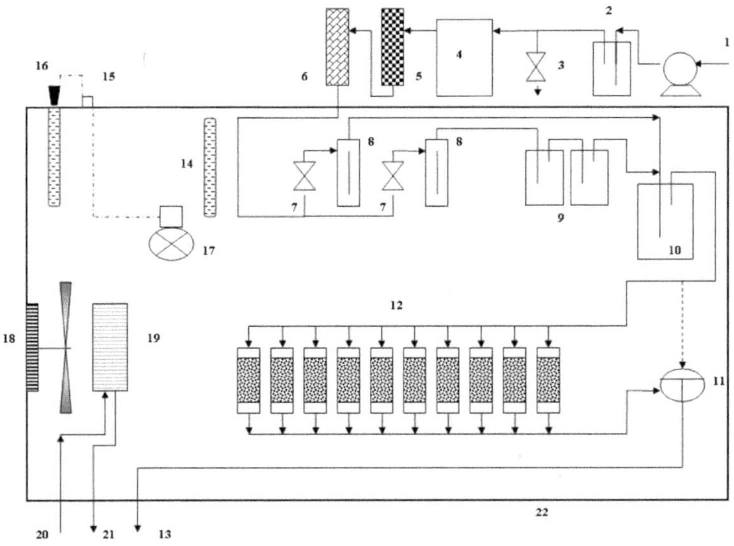

1 pump;
2 bottle for pressure impulse compensation;
3 regulating valve;
4 condensation drier;
5 drying with silica gel;
6 drying with molecular sieves;
7 needle valve;
8 capillary flowmeter;
9 saturated siloxane in washing bottle;
10 mixing bottle;
11 three-way cock;
12 adsorbent samples;
13 fume hood;
14 control mercury thermometer;
15 relay;
16 Vertex thermometer;
17 infrared lamp;
18 fan;
19 cooler;
20 inflow cooling water;
21 outflow cooling water;
22 temperature-controlled box.

Figure 3 Laboratory apparatus for measurement of adsorption isotherms.

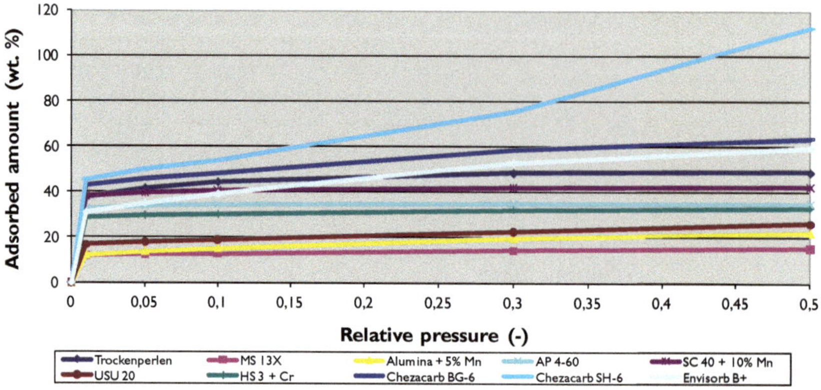

Figure 4 Adsorption isotherms for OMCTS by 20°C.

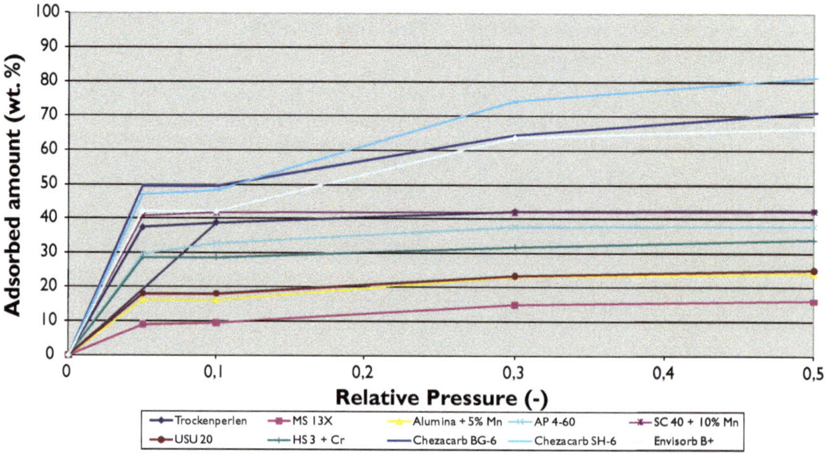

Figure 5 Adsorption isotherms for DMCPS by 20°C.

MEASURING CARRIED OUT AT WASTEWATER TREATMENT PLANT IN PRAGUE-BUBENEC

The wastewater treatment plant in Prague Bubenec is the biggest treatment plant in the Czech Republic. In this sewage treatment plant is treated 95% of the total amount of the municipal wastewater produced in the city Prague. Up to 10,4 m^3/s wastewater flow into the plant. Therefore a huge area for the different treatment steps is used. Mechanical, chemical and biological cleaning devices are spreaded at island over the river Vltava. In each step of the aerobic treatment a sedimentation of particulate material takes place. The formed sewage sludge is removed from the bottom of the different vessels and is stored in several digesters. Due to the anaerobic digestion in these bioreactors biogas is produced. By the digestion process up to 4000 m^3/h of the biogas are recovered from the sludge. This volume has a great energetic potential and a huge economic interest. Therefore

Table 3 Comparison of siloxane concentrations in different solvents after its saturation using the same amount of biogas.

Solvent	*OMCTS (mg/m³)*	*DMCPS (mg/m³)*
n-hexane	0,6	9,1
Acetone	0,7	8,9
Methanol	0,8	8,7
Toluene	0,6	9,2

the biogas is used for the generation of electricity and head by gas engines. The power generation unit of the wastewater treatment plant consists of 5 Deutz gas engines and 5 downstream generators. The power output of each engine is 1,1 MW_{el}. The generated electricity is supplied to the local electrical grid. It should be mentioned that a little amount of the digester gas is sometimes used for heat generation. But the power generation is the most profitable usage of the biogas. Therefore is necessary to remove harmful organosilicon compounds from biogas before its using. The siloxane removal takes place in the adsorption unit. Like adsorption material is used activated carbon AP 4-60 (Chemviron Carbon).

Most important sampling technique for siloxane in biogas is the absorption in adequate solvents. The used organic liquids are methanol, hexane and toluene. The sample gas velocity is limited to 1l/min and it is necessary to cool the absorption unit to 0°C to reduce the vaporization of the solvent by the sampling of biogas. Thereby an increasing absorption capacity is reached. The measured concentrations of some used solvents are shown in Table 2. Absorption of siloxanes removed from biogas in different solvents was measured in Prague-Bubenec (Temperature: 0°C; pressure: 101.325 kPa) [6]. The results shows tab. 3.

Obviously, hexane has to be cooled to deeper temperatures. Toluene offers the better absorption results at 0°C. A vacuum pump draws the biogas through the absorption unit consisting of several gas washing bottles. When using the optimal solvent volume of 20 to 25 ml a perfusion of at least 100 dm^3 biogas is necessary. The siloxane concentration is calculated from the volume of solvent and streamed biogas. The disadvantages of this measuring method are essential cooling and the long-term sampling. Out of the solution of solvent and biogas components little samples for GC/MS are prepared. All solute organic compounds are determined with this measuring device. The results are calibrated with the standard values and the displayed curve is used for the calculation of the siloxane content. The absorption technique is an easy method to measure the siloxane content.

Figures 6 and 7 illustrate quantity of siloxane in raw biogas and behind adsorption units.

Raw biogas pass through two condensers for removal of humidity from gas. First condenser is tube-type cooler using as cooling agent supply water, in the second one is for cooling used glycol. Nearly 60% of siloxanes is

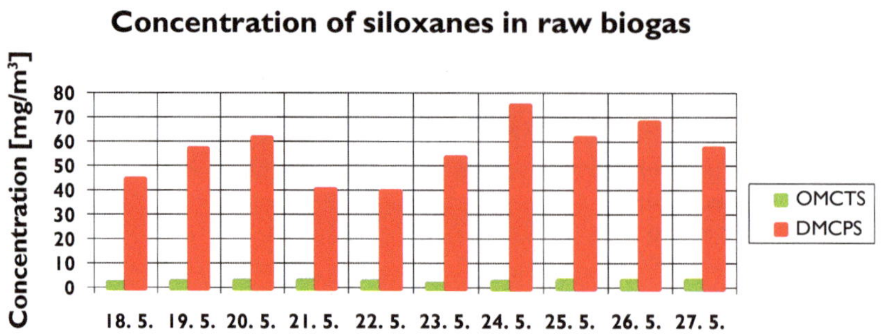

Figure 6 Siloxanes in raw biogas.

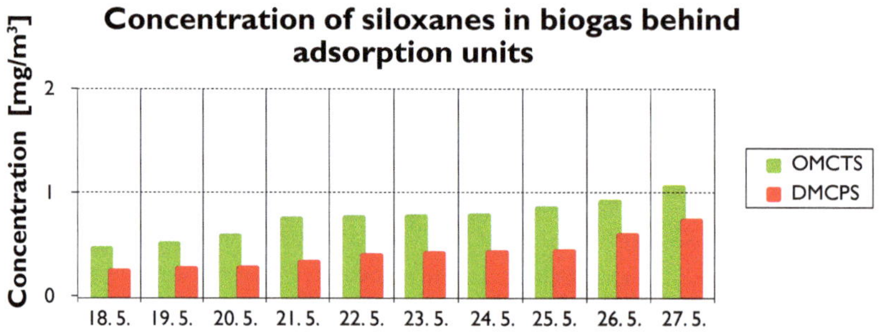

Figure 7 Siloxanes in biogas behind adsorption units.

Table 4 Quantity of siloxanes in organic condensate (analysed by GC/MS).

Important compounds	*Part of total*	*Concentration*
	%	ug/ml
Decane	0.068	
OMCTS	0.013	26,50
D-Limonene	0.108	
Undecane	1.109	
DMCPS	3.166	18425,3
Dodecane	2.275	
Dodecamethylpentasiloxane	0.788	
Tridecane	3.232	
Pentaethylencyclopentasiloxane	1.187	
Tetradecane	4.315	
Pentadecane	3.400	
Hexadecane	1.158	
Heptadecane	1.632	
Octamethyltrisiloxane	0.658	
Octadecane	0.818	

removed by condensator devices. Organic condensate produced in this cooler was analyzed by GC/MS and were found almost 166 compounds. The most important compounds are in table 4.

CONCLUSION

For an effective elimination of the negative impacts of siloxanes in biogas is suitable used adsorption. In the Department of gas, coke and air protection at Institute of Chemical Technology Prague have been tested a lot of adsorption materials for removing of siloxanes from biogas. Simultaneously with testing in laboratory have been done measurement of siloxanes content in biogas produced at wastewater treatment plants in Prague where is situated adsorption unit for siloxanes removal from biogas.

The obtained results shows that siloxanes are and will be big problem for lots of waste treatment plants and waste disposal sites and it is very important to find the best technology for their elimination.

ACKNOWLEDGEMENTS

Resolution of this problems were realised with financial support from research programme MSM 6046137304 which is solved at Faculty of Environmental Technology at ICT Prague. Authors of this article thanks for financial support to Ministry of Education in the Czech Republic.

REFERENCES

1 S. Rasi, A. Veijanen, and J. Rintala. Trace compounds of biogas from different biogas production plants. *Energy*, 32:1375–1380, 2007.
2 R. Dewil, L. Appels, and J. Baeyens. Energy use of biogas hampered by the presence of siloxanes. *Energy Conversion and Management*, 47: 1711–1722, 2006.
3 E. Wheless, J. Pierce. Siloxanes in Landfill and Digester Gas Update. SWANA Landfill Gas Symposium, March 2002.
4 M. Schweigkofer and R. Niessner. Removal of siloxanes in biogases. *Journal of Hazardous Materials*, 83:183–196, 2001.
5 M. Hagmann, E. Hesse, P. Hentschel, and T. Bauer. Purification of biogas-removal of volatile silicones. In *Proceedings Sardinia 2001, 8th International Waste Management and Landfill Symposium*, 2001.
6 A. Prochazkova, O. Prokes, K. Ciahotny, D. Tenkrat, J. Cermakova. Analyza a moznosti odstranovani siloxanu z biomethanu. *Plyn*, 88:264–267, 2008.

Chapter 57

BIODEGRADATION OF GAS – PHASE MIXTURES OF STYRENE AND ACETONE IN A BIOFILTER INOCULATED WITH THE FUNGUS *SPOROTHRIX VARIECIBATUS*

Eldon R. Rene, Radka Špačková, María C. Veiga & Christian Kennes
Faculty of Sciences, Chemical Engineering Laboratory, University of La Coruña, La Coruña, Spain

ABSTRACT

Steady – state biofilter performance, using the fungus *Sporothrix variecibatus*, for the treatment of a gas – phase mixture of styrene and acetone (SA) was studied at different flow rates (0.05–0.4 $m^3 h^{-1}$) and concentrations (0.01–12.9 $g m^{-3}$). The maximum elimination capacity (EC_{max}) of SA appeared to be a strong function of the inlet loading rates (ILR) of the two pollutants and these values were found to be 270 $g m^{-3} h^{-1}$ and 257 $g m^{-3} h^{-1}$ for styrene and acetone respectively. The highest EC_{total} was 363 $g m^{-3} h^{-1}$ at an inlet loading rate of 449 $g m^{-3} h^{-1}$, where nearly 97.5% of styrene and 75.6% acetone were removed. Later, batch studies were conducted with SA, as single compounds and as mixtures using cultures of *Sporothrix variecibatus*, a thermally dimorphic fungus. An analysis of the pollutant removal profiles suggests that, the removal of acetone was inhibited by the presence of styrene, while styrene removal was not affected by the presence of acetone.

Keywords: Biofiltration, *Sporothrix variecibatus*, styrene – acetone mixture, maximum elimination capacity, inhibition

INTRODUCTION

Styrene ($C_6H_5CH = CH_2$), an important VOC and acetone ($CH_3 - CO - CH_3$), a manufactured chemical, are commonly emitted during the manufacture of fibre – glass and composites. In recent years, constant awareness of health hazards associated with their emissions and increased attention from the regulatory authorities have helped manufacturing industries to adopt continuous modifications/development in the existing pollution control technologies.

Biofilters are fast emerging as the most comprehendible, versatile and effective option to handle large volumes of VOCs at moderately high loading rates.[1] Typical examples are biofilters packed with perlite as inert carrier material that have been used to treat styrene or alkyl benzene vapours.[2, 3] Styrene removal in biofilters has been investigated using individual or mixed species of bacteria. *Pseudomonas sp.* represents the most common group of isolates capable of styrene degradation and has been shown to produce styrene mono – oxygenase, which plays a major role in styrene degradation.[4] Jang et al.[5] used *Pseudomonas sp.* in a biofilter packed with peat and ceramic beads to treat styrene vapours and was able to show a maximum elimination capacity (EC) of 170 g m^{-3} h^{-1}. Also, acetone biofiltration has been studied in different biological waste gas treatment systems, such as biofilters and biotrickling filters. Przybylska et al.[6] investigated acetone removal in a biotrickling filter inoculated with two strains of bacteria, viz., *Burkholderia cepacia* and *Pseudomonas baumannii* and achieved a maximum EC of 95.8 g m^{-3} h^{-1}. This paper presents the performance of a perlite based biofilter, inoculated with *Sporothrix variecibatus*, for the treatment of gas – phase styrene and acetone at different concentrations and flow rates.

MATERIALS AND METHODS

The original inoculum used to inoculate the biofilter consists of a previously isolated *Sporothrix variecibatus,* from a styrene treating biofilter.[7] *Sporothrix sp.* was acclimated under batch conditions using equal headspace concentrations of both styrene and acetone (4 g · m^{-3}) in a well defined mineral salt media under shaking conditions (175 rpm) and at 30°C.

The biofilter was made of glass having a diameter (ID) of 8 cm and 70 cm in height. The packing in the biofilter consisted of sieved perlite beads (4–6 mm), which was packed to give a total filter bed volume of 0.0019 m^3. After inoculating the filter bed with well – acclimated cultures of *Sporothrix sp.*, the bed moisture was maintained constant by periodic addition of fresh mineral salt medium[2] from the top (once every 3 d).

Styrene concentration in gas samples were measured by gas chromatography on an HP 5890 gas chromatograph, using a 50 m TRACER column and a FID detector.

RESULTS AND DISCUSSION

The biofilter was acclimated with low loading rates of both styrene and acetone vapours (0.05 m^3 h^{-1}) for 14 d to obtain sufficient biomass in the filter bed. The relative humidity of the air stream was maintained at around 92%

or more. After acclimation, the concentrations of both styrene and acetone in the feed were changed to identify the influence on the performance of the biofilter under different flow rates (0.05–0.4 m^3 h^{-1}) that corresponds to empty bed residence times varying between 17 to 137 s. The elimination capacity reflects the capacity of the biofilter to remove the pollutants. A plot of the EC *vs* the inlet styrene and acetone load showed that, under steady state conditions, a linear relation between the two variables was observed with a maximum EC of 270 g m^{-3} h^{-1} and 257 g m^{-3} h^{-1} for the SA mixture, though only 68.4% and 75.6 of styrene and acetone was removed in the biofilter. Also the critical load to the biofilter was observed to be a strong function of the residence times. For instance, the critical load ($ILR_{critical}$) for 100% removal was 147.8 and 119 g m^{-3} h^{-1} at EBRT of 137 s for styrene and acetone, while this value decreased considerably to 60.3 and 18.9 g m^{-3} h^{-1} at an EBRT of 17 s.

During batch experiments, the concentrations of styrene and acetone were measured through headspace analysis and later calculated for liquid phase concentrations based on Henry's constant at the operating conditions. Sample aliquots were also collected to measure OD_{600} and CO_2 produced. Fig. 1 shows the biodegradation profiles of acetone and styrene, when present as a mixture under batch conditions (175 rpm, 30°C, pH – 5.9). Biomass growth with these two compounds was almost instantaneous followed by steady utilization of the substrate. At low concentrations (235 and 95 mg l^{-1} of SA), SA were utilized completely in less than 78 hours by the *Sporothrix sp.*, while at higher concentrations (399 and 159 mg l^{-1} of SA), it required 331 and 235 h for complete removal.

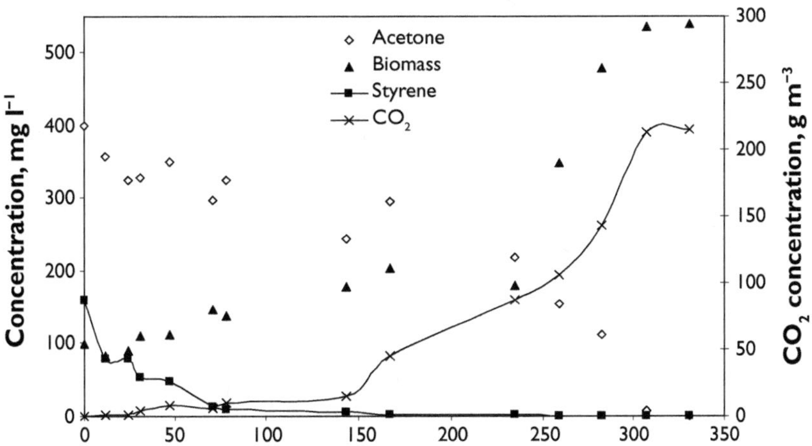

Figure 1 Biodegradation, biomass growth and CO_2 generation profile of high initial concentrations of styrene and acetone as mixture.

It was also observed that, though acetone was removed in the early hours of experimentation, the removal was even more rapid when styrene concentrations were depleted inside the flask. This confirms the fact that, the presence of styrene hinders the removal of acetone in mixture, while styrene removal is faster in the presence of acetone (71 h for 95 mg l^{-1}). Also, this could be due to the fact that the *Sporothrix sp.*, used in this work was isolated from a styrene degrading biofilter and subjected to constant styrene vapours even during its storage on agar plates.

CONCLUSIONS

The results from this study show that the co – treatment of styrene and acetone vapours from off gas emissions can be performed with high efficiencies by means of a biofilter. The total $ILR_{critical}$ was found to be 240 g m^{-3} h^{-1}, while the total EC_{max} was 363 g m^{-3} h^{-1}.

REFERENCES

1 Kennes C, Rene ER and Veiga MC. Bioprocesses for air pollution control. *Journal of Chemical Technology and Biotechnology.* In Press, DOI: 10.1002/jctb.2216, (2009).

2 Kennes C, Cox HHJ, Doddema HJ and Harder W. Design and performance of biofilters for the removal of alkylbenzene vapours. *Journal of Chemical Technology and Biotechnology* **66**: 300–304 (1996).

3 Veiga MC and Kennes C. Parameters affecting performance and modelling of biofilters treating alkylbenzene polluted air. *Applied Microbiology and Biotechnology* 55: 254–258 (2001).

4 O'Leary ND, O'Connor KE and Dobson ADW. Biochemistry, genetics and physiology of microbial styrene degradation. *FEMS Microbiology Reviews* **26**: 403–417 (2002).

5 Jang JH, Hirai M and Shoda M. Styrene degradation by Pseudomonas sp. SR-5 in biofilters with organic and inorganic packing materials. *Applied Microbiology and Biotechnology* **65**: 349–355 (2004).

6 Przybylska KP, Zieminski K and Szopa T St. Acetone biodegradation in a trickle – bed biofilter. *International Biodeterioration and Biodegradation* **57**: 200–206 (2006).

7 Rene ER, Veiga MC and Kennes C. Experimental and neural model analysis of styrene removal from polluted air in a biofilter. *Journal of Chemical Technology and Biotechnology* **84**: 941 – 948 (2009).

Chapter 58

BIOFILTERS OF THE NEXT GENERATION

Franjo Sabo
University of Applied Science, Wiesbaden, Stuttgart, Germany

Stefan Prechel
REINLUFT Environmental Engineering Ltd., Stuttgart, Germany

Intelligent plant design significantly increases the rate of degradation of VOCs, pollutants and odours. New intelligent sensors allow a look inside the "black box" biofilter.

The results of intensive research in conjunction with extensive operational experience have revolutionized biofilter technology in recent years. The increasing knowledge and understanding of the backgrounds of biodegradation, mass transfer and flow processes in the filter system now allow the use of the biofilters even on fields of applications that were previously reserved to chemical or thermal processes.

The major progress is being achieved by the combination of different biological techniques, the optimization of biofiltration and the development of intelligent control techniques. These enhancements contribute to a wider acceptance of the biofilter as best available technology – even under the new, more stringent legal requirements.

With all these improvements, the main advantage remains: It is still the most cost-effective air cleaning technique. In this poster session, these developments will be presented with examples.

Intensive R & D works, especially in the process and control engineering of biological systems, have come up with numerous innovations in the biofilter technology. In many areas, these improvements could already replace expensive chemical or thermal exhaust air purification plants. Besides of the environmental advantages, biofiltration significantly lowers the investment and especially the operating costs.

One of the main reasons for biofilters to become state-of-the-art in many application areas can be found in the encapsulation of the biofilter and therefore in the avoidance of external influences to the filter media. In this controllable reaction chamber, the biochemical processes can significantly be improved and stabilized. This makes it possible to use special filter material with high specific surface areas and high microbiological activities.

One of the main reasons for biofilters to become state-of-the-art in many application areas can be found in the encapsulation of the biofilter and therefore in the avoidance of external influences to the filter media. In this controllable reaction chamber, the biochemical processes can significantly be improved and stabilized. This makes it possible to use special filter material with high specific surface areas and high microbiological activities.

During intensive research in collaboration with the University of Stuttgart and the University of Applied Sciences in Wiesbaden, mass transfer and fluid mechanics optimization measures could be realized. The basis is given by gas-tight, specially designed and encapsulated biofilter modules. The dimensions are similar to those of the ISO-standard containers. Thus, the individual modules are easy to transport and manipulate. Furthermore, these chassis are already self-supporting and can be repeatedly stacked as shown in Fig. 1 and Fig. 2.

Figure 1 High-efficiency biofilter plant at DAIMLER AG for 220.000 m^3/h air treatment.

Figure 2 Examples of high-efficiency biofilter plants.

MULTI-STAGED BIOFILTER SYSTEMS

Increased performance can further be achieved by connecting the biofilter modules in series. The classical biofiltration concept consists of a fan, a humidifier and just one biofilter bed. As industrial applications often come up with high concentrations of a complex mixture of substances, a multistage process can achieve much better reduction values. Especially, if the mixture consists of components of which some are hard to degrade and others are easily available to microorganisms, bacteria tend to metabolize the latter first. In multi-level arrangements of the filter units, this phenomenon of diauxie and the competing degradation processes can be transferred to separate regions of the biological plant.

This can be achieved by combining a bioscrubber and biofilters. The scrubber system will already eliminate some of the easily accessible components, whereas the degradation of less soluble substances will take place in the biofilter. Another method is to connect up to three biofilter modules in series. The type of configuration to be applied depends on local circumstances. In practical applications this means that the exhaust gas passes the filter media at twice to three times the normal speed and also takes a longer way through the material. This increases the turbulent flow through the filter bed and results in a much improved mass transfer of pollutants from the gas phase into the liquid phase. Therefore, more pollutants can be absorbed by the filter media and be degraded by the microorganisms. This optimization is relevant in such cases where the elimination capacity of the filter is limited by reasons of mass transport.

Encapsulated high-efficiency biofilters can be operated with much higher loads than other systems. The threat of breakthrough currents can be eliminated by the connection of filter beds in series. Thus, the filter performance remains at high levels for long periods. In the industrial field in which high legal requirements have to be met, the fault tolerance is of critical importance. This is one of the main reasons why these types of construction have proven very successful in practice.

As illustrated in Fig. 1, the biofilter was installed in 2005 at MERCEDES BENZ and put into operation. The peculiarities of the system are the two-stage design and the space-saving, compact arrangement. By stacking the modules eight times high, more than 1100 m^3 of filter material is used in each tower. Together with its "sister plant", these biological systems purify more than 220.000 m^3/h of foundry air.

INTELLIGENT CONTROL SYSTEMS

Problems in the operation of biological systems can occur if the filter dries out completely or partially gets too wet, or scrubber systems no longer provide adequate performance. Unfortunately, few manufacturers of such biological systems have the necessary know-how to equip their systems with appropriate control and monitoring techniques to maintain the initial high efficiency over the years. Even so, in the future more new intelligent, universally applicable,

Figure 3 Visualisation of humidity distribution.

yet easy to use measurement and control systems are needed to take over the regulation and control of biological waste air purification plants. A system of sensors and control techniques that are based on the methods of fuzzy logic and neural networks, is designed specifically for the control of biological waste air purification.

A moisture sensor was developed in several research projects, to enable a cost-effective insight into the interior of the biofilter modules and to display the humidity and flow distribution graphically in real time in the filter material. Figure 2 shows how a distributed arrangement of sensors is set up. Even the flow distribution of air can be plotted. Thus, for the first time, a gaze deep inside the "black box" biofilter has become true.

With this system, operators or authorities are able to gain a three-dimensional picture of the state of the biological system at any time. Critical operating conditions can be avoided by early detection and targeted action. Existing biofilters can easily be fitted with these low-cost sensors at a later date.

Chapter 59

ODOR MITIGATION WITH TREE BUFFERS:
Swine production case study

Steven Trabue, Thomas Sauer, Richard Pfeiffer & Guillermo Hernandez
USDA Agriculture Research Service, National Laboratory of Agriculture and Environment, Ames, IA, USA

John Tyndall
Department of Natural Resource and Ecology Management, Iowa State University, Ames, IA, USA

ABSTRACT

Tree buffers are a potential low cost sustainable odor mitigation strategy used by animal feeding operations; however, there is little to no data on their effectiveness. This study was designed to evaluate how tree buffers affect air flow patterns at a swine facility and measure their effectiveness in controlling odor from the facility. Wind tunnel experiments were used to determine the effect buffers had on wind flow patterns within the facility. Odor mitigation analysis measure both particulate matter and vapor phase odorants before and after tree buffer. Particle concentrations were measured with optical particle counters (OPC) and odorous compound concentrations were measured using sorbent tubes with thermal desorption (TDS) GC-MS analysis. Both particle filters and plant material from tree buffers were analyzed for sorption of odorous compounds with TDS-GC-MS analysis. The wind tunnel experiments indicated that the effect of both buildings and trees on wind velocity and turbulence would potentially enhance trapping of odorous material by slowing wind velocity and increasing mixing near trees. For particulates, tree buffers reduced concentrations by 44% when winds came from the south (main buffer ran east-west on the north side of the facility). Most particles were classified in the smaller size fraction (0.30–0.49 μm); however, the tree buffers removed the larger size particles more effectively than smaller size fraction. The buffer removed volatile fatty acids (VFAs) in the air stream by over 50%, while aromatic odorants were not as easily removed with the summer sampling showing little to no difference in terms of concentrations of these compounds and Fall sampling showing reduced levels. Odorants sorbed to both particulate filters and plant material also support this finding in that VFA were significantly higher than odorous aromatic compounds. Plant material taken from trees in the buffer showed significantly higher loadings of odorous VFAs, phenolic, and indole compounds for samples taken from 8 ft compared to samples taken from either 2 or 4 ft.

INTRODUCTION

Tree buffers (i.e., shelterbelts) are a relatively new approach to lowering odour from animal agriculture. Shelterbelts are thought to lower odour through the interception of odorous compounds and dilution of odours material though lofting of ground level air into upper air streams.[1] In general, shelterbelts as an odour mitigation technique are both cost effective,[2] and environmentally beneficial;[1] however, their effectiveness have not been thoroughly tested with reliable odour metrics. The purpose of this study is to report the effects tree buffers have on wind flow patterns and odour emissions from a swine facility. In this study, wind flow patterns were monitored with towers (10 and 20 m) with continuous micrometeorological measurements and modeled in wind tunnel experiments, while emissions of odorous materials (both particulate and odorous compound concentrations) were monitored before and after tree buffers.

RESULTS AND DISCUSSION

WIND TUNNEL

The wind tunnel experiments demonstrated the potential impact of both buildings and tree buffers on wind speed and air mixing near swine confinement facility. Both building and tree models were showed to reduce air velocity and increase turbulent intensity with these effects persisting up 33 m height in the field, but with more pronounced effects near the ground level. Higher turbulence intensity values indicate greater air mixing with lesser direct horizontal transport of air and any odor or particulates entrained in the air. The reduced wind speed and increased air mixing further supports the significant contribution of tree buffer to odor mitigation due to hypothetically enhance particulate trapping and odor sorption near swine facilities.

PARTICULATE MATTER (PM)

Odor transport on PM is significant transport mechanism from swine facilities to surrounding community.[3,4] The tree buffer produced clear results in reduction of suspended particulates. An overall 44% reduction in particle counts was measured when comparing OPC sensors placed between the northern building and the tree buffer vs. just North from the tree buffer (Figure 1). In addition, based on particles size distribution buffers were preferentially trapping the larger size fraction. In the case of OPC sensor deployed at 20 m North of swine buildings, it was expected to observe higher particle counts but due to proximity to the buildings may have missed the particulate plumes which could

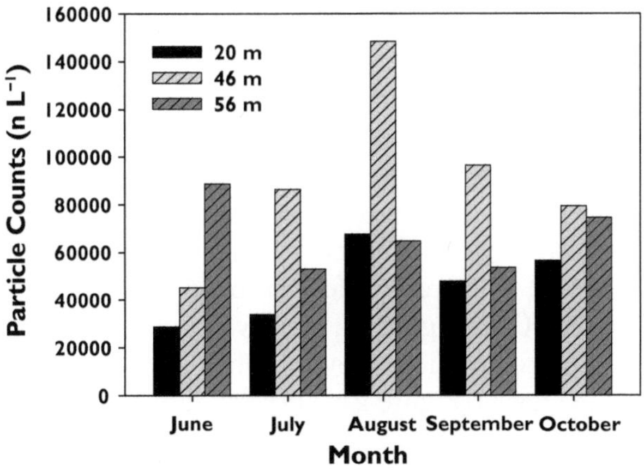

Figure 1 Particle concentration measured at 20, 46, 56 m north of swine facility buildings monitored between June and October 2008.

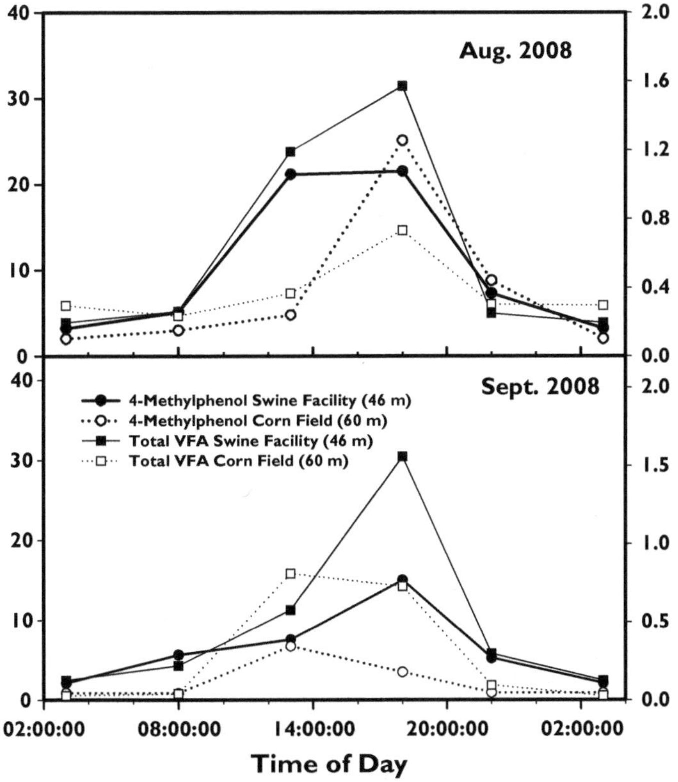

Figure 2 Weekly average concentration of total volatile fatty acids (VFAs) and 4-methylphenol (p-cresol) in air taken before and after tree buffers measured in Summer and Fall 2008.

have been transported above the sensor height (2 m). Wind tunnel predicted air flow patterns showed have placement of buildings and tree buffer facilitated interception of particle movement from buildings into tree buffer.

Key chemical classes associated with odor measured from both the swine facility and building ventilators included: volatile fatty acids (VFA), phenolic compounds, and indole compounds (data not shown). Several research groups have identified volatile fatty acids as key compounds associated with swine odour,[5,6,7] while others have identified both VFAs and 4-methylphenol as a key odorants associated with swine production.[6] The tree buffer removed volatile fatty acids (VFAs) in the air stream by over 50% (Figure 2); however, control of 4-methylphenol was mixed with the Summer sampling showing little to no difference and Fall sampling showing reduced levels (Figure 2). These results demonstrate that tree buffer do remove odours compounds with a preference for VFAs over aromatic compounds, which is anticipated given the leaf surface lipophilic nature.[8]

CONCLUSION

This study presents evidence supporting a beneficial effect of tree buffer planting around confined animal facilities on swine odor abatement; however, the complex, diverse behavior of air-flow circulation and odor compounds near buildings and trees make results far from self-evident and prevent drawing definite conclusions. Additional research can further clarify these unknowns.

REFERENCES

1 Tyndall, J and J. Colletti. Mitigating swine odor with strategically designed shelterbelt systems: a review. *Agroforest Syst.* 69:45–65 (2007).

2 Tyndall, J and J. Colletti. Financial feasibility of using shelterbelts fro swine odor mitigation. *Agroforest Syst.* 76:237–250 (2009).

3 Bottcher, R. 2001. An environmental nuisance: odor concentrated and transported by dust. *Chem. Senses* 26:327–331.

4 Schiffman, S.S., J.L. Bennett and J.H. Raymer. 2001. Quantification of odors and odorants from swine operation in North Carolina. *Agricultural and Forest Meteorology* 108:213–240.

5 Zahn, J., A. DiSpirito, Y. Do, B. Brooks, E. Cooper and J. Hatfield. Correlation of human olfactory responses to airborne concentrations of malodorous volatile organic compounds emitted from swine effluent. *J. Environ. Qual.* 30:624–634 (2001).

6 Wright, D., F. Nielsen, D. Eaton, L. Kuhrt, J. Koziel, J. Spinhirne and D. Parker. Multidimensional GC-MS-olfactometry for identification and prioritization of malodors from confined animal feeding operations. *J. Agric. Food Chem.* 53:8663–8672 (2005).

7 Zhu, J., G. Riskowski and M. Torremorell. Volatile fatty acids as odor indicators in swine manure-a critical review. *Transactions ASAE* 42:175–182 (1999).

8 Reischl, A., M. Reissinger and O. Hutzinger. Occurrence and distribution of atmospheric organic micropollutants in conifer needles. *Chemospheres* 16:2647–2652 (1987).

Chapter 60

DEVELOPMENT OF SUBMERGED AND SOLID STATE BIOREACTORS FOR WASTE GAS TREATMENT

Olga Muter, Alina Mihailova, Dzidra Zarina & Uldis Viesturs
Institute of Microbiology and Biotechnology, University of Latvia, Riga, Latvia

Juris Vanags & Uldis Viesturs
Latvian State Institute of Wood Chemistry, Riga, Latvia

Silvija Strikauska
Latvian University of Agriculture, Jelgava, Latvia

Keywords: submerged and solid state bioreactors, ammonia biotransformation, bacteria immobilisation, enzymatic activity

ABBREVIATIONS

PNNS – *Pseudomonas* spp., *Nitrosomonas* spp., *Nitrobacter* spp., *Sarcina* spp.; SPSS – Solid Phase Submerged System; CLE – Cabbage Leaf Extract; CFU – Colony Forming Units; FDA – Fluoresceine Diacetate.

INTRODUCTION

Over the past two decades, biofiltration had been recognised as a reliable, environmentally friendly and cost effective technology for air treatment. In this study, ammonia was chosen as a model gas. Ammonia is a toxic and strong odour gas, which is produced in cases of ventilating intensive cattle breeding facilities, manure handling and waste composting, and is considered to be a major contributor to acidification and eutrophication of terrestrial and aquatic environments.

The general concept of this work was to compare different fermentation systems and cultures of microorganisms for treatment of contaminated gases. One of the most important tasks in this study was to determine the factors, which are favourable for the microbial growth and immobilisation as well as further ammonia biotransformation in the chosen biofiltration system.

Table 1 Content and methodical sequences of the study.

Task		*Methodical approach*	*Endpoint*
1	Activation of the growth of PNNS association in the combined SPSS	Use of amendments (e.g., CLE, molasses)	mg $N\text{-}NH_4^+$/l; cfu/ml
2	Immobilisation of PNNS association on the ceramsite carrier	Pre-treatment of the carrier with molasses	FDA hydrolysis; N_{total}/gdw; C_{total}/gdw
3	Stimulation of the process of $N\text{-}NH_4^+$ biotransformation by immobilised PNNS association	Addition of C source (e.g., fructose)	mg $N\text{-}NH_4^+$/l;mg $N\text{-}NO_2^-$/l;mg sugars/l

MATERIALS AND METHODS

The developed set of experimental methods was in parallel: 1) 10-L solid state bioreactor with different types of carriers (the subjects of experiments); 2) 8-L and 20-L submerged bioreactors with a magnetic stirrer's drive, stirrers are interchangeable for mixing regime optimisation; 3) Process Controller BIO-2,3 [1, 2]. Ammonia-biodegrading PNNS association, previously isolated from the biological activated sludge of a fish factory wastewater treatment plant, was used in the experiments [3, 4]. To perform a comparative study, the yeast *Saccharomyces cerevisiae* was used. As a contaminant (also for adaption of the system), NH_3 was used, or $(NH_4)_2SO_4$ was directly added to the media as a NH_3 source. The content and methodical sequences of the study are demonstrated in Table 1.

The analytical methods applied in this study are described [3, 4].

RESULTS AND CONCLUSIONS

In our experiments, the combined SPSS has been developed to study the growth, immobilisation and ammonia-biotransformation activity of the PNNS association. The tested organic amendments, i.e., molasses, CLE and fructose had a stimulating effect on the growth of the association and on the 1st phase of the nitrification process. Fructose (0.45%) significantly increased the degradation of ammonium in a concentration range of 0.53 – 1.59 g/l $N\text{-}NH_4^+$.

Among the six materials (chopped straw, pine bark, peat, etc.) tested as a potential carrier for immobilisation of the association, ceramsite granules were chosen as the most appropriate carrier for the future larger scale experiments. Besides, for such simultaneously degradable carriers, we could not solve completely the hindrance phenomena caused by stacking, forming of gas channels and crowding of the mass to be fermented. Probably, composting [5], aerated by contaminated gas, could be also a solution (a variant for solid state fermentation). Ceramsite granules had also significantly higher immobilisation capacity for the PNNS association ($p < 0.01$). The immobilisation process was compared using enzymatic activity (i.e. the activity of FDA hydrolysis per dry material weight) as well as the total nitrogen and carbon of the ceramsite granules with immobilised biomass. Pre-treatment of ceramsite granules with molasses resulted in a significantly higher ($p < 0.05$) ammonium biotransformation activity of the immobilised biomass.

Among the three bioreactors used, the best kinetics of assimilation for such a simple contaminant as NH_3 was shown in the case of the 8-L bioreactor with the double counter-flow turbine using the yeasts. However, the energo-efficiency of the 3 systems mentioned is still the subject of further experiments.

REFERENCES

1 Vanags J., Rychtera M., Ferzik S., Vishkins M., Viesturs U. Oxygen and temperature control during the cultivation of microorganisms using substrate feeding. *Eng. Life Sci.*, 2007, 7 (3), 247–252.

2 Pencheva T., Petrov M., Ilkova M., Roeva O., Vanags J., Viesturs U., Tzonkov S. *Bioprocess Engineering*, Viesturs U., Tzonkov S. (Eds.), Bulgaria, Sofia, 2006, 255.

3 Mihailova A., Muter O., Strikauska S., Limane B., Berzins U., Viesturs U., Zarina D. Comparison of the effect of various amendments on the growth of the targeted bacteria association and ammonia biodegradation. In: The 7th International Conference "*Environmental Engineering*", Ed. D. Cygas, K.D. Froehner, Vilnius 22–23 May, 2008, pp. 218–223.

4 Strikauska S., Zariņa Dz., Mutere O., Viesturs U., Bērziņš A. Effect of various factors to ammonia biodegradation by two stage biofiltration system. In: *Proc. of the II Int. Congress on Biotechniques for Air Pollution Control*, La Coruna, Spain, October 3–5, 2007, 293–302.

5 Viesturs U., Zarina D., Strikauska S., Berzins A. Utilisation of food and woodworking production by-products by composting. *Bioautomation*, 2004, 1, 83–98.

Author index